Estimating for
Building and Civil Engineering Works

692.5   GEDDES SPENCE

To
HMW and ASW

Spence Geddes

Estimating for Building and Civil Engineering Works

Ninth Edition

Edited by
John Williams FRICS
(Chartered Quantity Surveyor)

Butterworth Heinemann Ltd
Linacre House, Jordan Hill, Oxford OX2 8DP
A division of Reed Educational and Professional Publishing Ltd

℟ A member of the Reed Elsevier plc group

OXFORD   BOSTON   JOHANNESBURG
MELBOURNE   NEW DELHI   SINGAPORE

First published 1951
Second edition 1960
Third edition 1963
Fourth edition 1966
Fifth edition 1971
Sixth edition 1976
Seventh edition 1981
Eighth edition 1985
Ninth edition 1996

Whilst every effort is made to ensure the accuracy of the information given in this publication, neither the author nor the publishers in any way accept liability of any kind resulting from the use made by any person of such information.

British Library Cataloguing in Publication Data
A catalogue record for this book is available from the British Library

ISBN 0 7506 2797 2

Library of Congress Cataloguing in Publication Data
A catalogue record for this book is available from the Library of Congress

Composition by Genesis Typesetting, Rochester, Kent
Printed and bound in Great Britain by Hartnolls Limited, Bodmin, Cornwall

Contents

Preface   vi

1   Introduction   1
2   Forms of contract   3
3   Estimating the total cost   10
4   Labour including bonus   16
5   Materials   26
6   Plant   29
7   Preliminaries   36
8   Overheads and establishment charges   46
9   Tendering   48
10 Demolitions and alterations   50
11 Piling   51
12 Groundwork   66
13 Concrete work   118
14 Masonry   152
15 Woodwork   171
16 Structural steelwork   184
17 Roofing   187
18 Surface finishes   200
19 Glazing   206
20 Plumbing   209
21 Quarry work, road, paths and platelaying   216
22 Drainage   245
23 Landscaping   266
24 Weight of materials   278
25 Useful tables   299

Index   309

Preface

This is a new, fully updated edition of a standard reference publication on estimating. The opportunity has been taken to make certain alterations whilst maintaining the essential feature of earlier editions.

The purpose of the book is to deal in a practical and reasonable way with many of the estimating problems which can arise where building and civil engineering works are carried out and to include comprehensive estimating data within the guidelines of good practice.

The early part of the book has been completely rewritten to contain chapters useful to students and practitioners alike for the development of the estimating process resulting in the presentation of a tender for construction works.

The second and major part of the book contains estimating data fully updated for the major elements in building and civil engineering work, including a new chapter on piling, and a wealth of constants for practical use in estimating.

The comprehensive information on basic principles of estimating found in 'Spence Geddes' are still as valid today as the first edition.

In this edition the prevailing rates of labour and costs of materials are taken whenever possible as a round figure. Readers will appreciate in the construction industry that prices are continually changing, rise and fall, and that worked examples should therefore be used as a guide to method of calculation substituting in any specific case the current rates applicable to it.

In the case of plant output dramatic increases have been experienced in productivity over past years and again estimators with their own records should substitute values appropriate to their work.

Of the assistance I have received in the preparation of this new edition, R.D. Wood FRICS and Peter Jolly FFB for their encouragement to begin this task, Alan Wright BSc, FRICS and John Hallett BSc, FRICS my partners at Hugh Knight, Lomas & Associates for their full support. Sherry Haines for her first class work patiently and meticulously typing endless drafts, Charles East FRICS for reading the proofs.

I am indebted to the various bodies and individuals for their contributions and to the publishers and printers who have applied their care and skill to the difficult and intricate task of the preparation of a new edition.

My wife Ann for her continual encouragement and cheerful support from the earliest days to completion of the new edition of 'Spence Geddes'.

John Williams FRICS
January 1996

1 *Introduction*

Estimating the cost of building and civil engineering works of construction takes into account the following elements:

1. Labour, material and plant unit cost of the various items of work as itemised in a Bill of Quantities or Specification of Works.
2. Site charges including special insurances and conditions of contract, usually known as preliminaries.
3. Costs incurred in maintaining and running a head office with its attendant costs in the way of staff salaries, rent, rates, etc. usually known as overheads and establishment charges.
4. Dayworks, provisional sums and prime cost amounts.

Labour rates and plant hire rates per hour are variable, as are the unit costs of materials, but quite apart from such variability other factors influence the cost of work, and have to be taken into consideration in estimating.

Factors affecting rates:

1. The output of labour, both skilled and unskilled, is variable, for one man produces more of a similar kind in a given time than another.
2. The output of mechanical plant is variable, for some plant operators are more efficient than others. This is particularly applicable to plant where the operator is himself chiefly responsible for the output concerned, such as mechanical excavating plant of the skimmer, shovel, tractor and scraper type.
3. Weather conditions have a marked effect upon output. Under wet weather conditions the output of excavation in particular is reduced, for site conditions become soft and heavy, excavating plant tends to become clogged, while internal transport on open sites becomes difficult. Frost also affects output; unless anti-freezers are used, bricklaying, concreting, plastering, etc. cannot be carried out. Excessive heat can also affect output, for a man does not work energetically when he is uncomfortably warm.

   In general, suitable mechanical plant should be used if the volume of work involved merits its use and the site conditions are such that its use is practicable, for in using it, not only is the work carried out at a greater rate of progress, but it is performed at considerably less cost than if carried out by hand.
4. The availability of materials and transport of materials.

Sound estimates are therefore produced from a combination of experience and recorded cost data of similar works carried out, these data having been progressively averaged over various contracts throughout all periods of the year, the unit costs recorded thus representing a fair average for the items of work so costed.

For purposes of estimating cost, data are best recorded as hour constants, i.e. the labour and plant hours taken to produce one unit of the work in question. Such constants are not affected by any fluctuation in the labour rates of pay or hire rate of plant per hour, the data thus always remaining up to date.

The data shown for purposes of estimating throughout this book are tabulated in such a form. The author was introduced to costing and the recording of cost data at an early age, and from many years of progressively averaged costs useful data have been obtained and recorded in the form of labour and plant hour constants. It is these data which form the main basic data shown for estimating purposes throughout the various sections of this book.

The subject of estimating is essentially practical, for, having examined the drawings and the bill of quantities setting out the amount of work involved under the various items and having carefully gone over the site of the proposed work and noted the prevailing conditions, the estimator must not only be capable of producing the estimate, but must also be sufficiently experienced to decide how best to carry out the work from a progress and economy point of view, for it is on such considerations that the estimate is based.

The experienced estimator is in the fortunate position of being able to count upon past experience to guide him in coming to such decisions. He has available his own recorded cost data, and need only resort to a work of this kind in order to confirm his own experience. To be of real value, therefore, a work on estimating must be sound and at the same time comprehensive, so that both the experienced and inexperienced estimator may not only have every confidence in the matter it contains, but also have available data from which to estimate the cost of all those items of work most likely to be met with.

The student of estimating is advised to seize every opportunity to get all the practical experience possible on outside work: to observe and record the various types and sizes of mechanical plant used, and the apportionment of labour to plant and tradesmen carrying out various operations.

In preparing a Bill of Quantities it is usual to do so in accordance with the Standard Method of Measurement of Building Works or/and the Civil Engineering Standard Method of Measurement, the unit measurements being definite units such as cubic metres, square metres, metres, kilograms, etc.

2 Forms of contract

In introducing the subject of estimating it is advisable to consider the various standard forms of contract so that the estimator may be familiar with these forms and then deal with the various stages and operations involved in estimating the cost of the individual elements for the build up of the total cost and finally to the amount of tender itself.

The forms of contract to be used depend on the Employer's requirements.

There are basically three types of contract:

1. Design of Works by the Employer's Architect or other Professional Consultant and carrying out of the works by Contractor.
2. Design of Works by the Contractor and carrying out the works by Contractor.
3. Maintenance and Minor Works including improvements.

1. Design of Works by the Employer's Architect, Engineer or other professional consultant

The choice of Forms for the Employer who requires design of works by the Architect, Engineer or other professional consultants is wide. The various forms are:

(a) JCT Standard Form with Quantities.
(b) JCT Standard Form without Quantities.
(c) JCT Intermediate Form.
(d) JCT Agreement for Minor Building Works.
(e) JCT Standard Form with Approximate Quantities.
(f) JCT Standard Form of Management Contract.
(g) JCT Standard Form of Prime Cost Contract.
(h) GC/Works/1 Standard Form of Contract for Building and Civil Engineering – Lump Sum with Quantities.
(i) GC/Works/1 Standard Form of Contract for Building and Civil Engineering – Lump Sum without Quantities.
(j) GC/Works/1 Standard Form of Contract for Mechanical and Electrical Works and Services, Lump Sum and Term Contracts.
(k) GC/Works/1 Standard Form of Contract for Minor Works, Lump Sum and Prime Cost Contract.
(l) GC/Works/1 Standard Form of Contract for Minor Works, Measured Term and Call Off Contracts.
(m) ACA Form of Building Agreement, British Property Federation Edition.
(n) ICE Conditions of Contract 6th Edition.

If the Employer requires a lump sum price the selection of contract is restricted to (a), (b), (c), (d), (h), (i) and (m).

The Forms of Contract most generally used are:

(a) *JCT Standard Form with Quantities* requires the Employer through his professional consultants to provide at tendering stage a set of drawings and bills of quantities which specify the work in terms of quality and quantity. Following the acceptance of the tender these drawings and the bills of quantities priced by the Contractor become the Contract Drawings and the Contract Bills defining the Works which are the subject of the Contract

Sum. The Contract Bills also provide price data for assistance in the valuation of variations. The form of contract contains detailed conditions regulating the rights and obligations of the parties, the powers and duties of the Architect and the Quantity Surveyor, and procedures appropriate to the variety of situations that may be met with on projects of any size.

(b) *JCT Standard Form without Quantities* requires the Employer through his professional consultants to provide at tendering stage a description of the works in a set of drawings together with either a Specification or Schedules of Work. The Contractor is then required either to price in detail the Specification or the Schedules of Work with the total of that pricing being the Contract Sum for which he is prepared to carry out the work or to state the lump sum he requires for carrying out the work shown on the drawings and described in the specification and to supply either an analysis of that lump sum (called the 'Contract Sum Analysis') or a Schedule of Rates on which that lump sum is based. Whichever priced documents the Contractor is required to provide will be used as a basis for the valuation of variations.

(c) *JCT Intermediate Form of Contract* requires the Employer through his professional consultants to provide at tendering stage a set of drawings together with another document. Where the other document consists of Bills of Quantities or Schedules of Work, the Contractor is required to have priced it; but where the other document consists of a Specification the Contractor is required either to have priced it or to have supplied a Schedule of Rates or a Contract Sum analysis. Following the acceptance of the tender, these drawings and the other document (priced by the Contractor where appropriate) becomes the Contract Documents defining the Works which are the subject of the Contract Sum.

The priced Bills or Specification or Schedules of Work or, as the case may be, the Schedule of Rates or Contract Sum Analysis provide price data for assistance in the valuation of variations. The form of contract contains conditions which are less detailed than those for the With and Without Quantities Forms but are more detailed than those for the Agreement of Minor Building Works. Price range £70 000–£280 000.

(d) *JCT Agreement for Minor Building Works* requires the Employer through his professional consultants to provide at tendering stage either drawings, a specification or schedule or a combination of any of these as required to describe the Works. Following acceptance of a tender, whatever documents are so provided become contract documents defining the works which are the subject of the sum set out in the Agreement to be paid by the Employer. The Agreement contains conditions which are less detailed than those in the Intermediate Form. Price range not exceeding £70 000.

(e) *ICE Conditions of Contract 6th Edition* requires the Employer through his professional consultants to provide at tendering stage a set of drawings, specification, and bills of quantities which specify the work in terms of quality and quantity. Following the acceptance of the tender these drawings, specification and the bills of quantities priced by the Contractor, conditions of contract and the contract agreement become the Contract.

2. Design of Works by the Contractor or part thereof

The choice of Forms for the Employer who requires design of work by the Contractor, or portion thereof, is likely to grow. The various forms are:

(a) JCT Standard Form with Contractor's Design.
(b) JCT Standard Form with or Without Quantities modified by Contractor's Designed Portion Supplement.

(*c*) GC/Works/1 General Conditions of Contract for Building and Civil Engineering, Standard Form of Contract, Single Stage Design and Build Version (1993).
(*d*) ICE Design and Construct Contract.

(*a*) *JCT Standard Form 'With Contractor's Design'* (See Practice Note CD/1 on the SFWCD) where the Employer, having stated his Requirements, requires the Contractor to provide Proposals for the design and construction of the Works and the Contractor's Analysis of the Contract Sum. The contract Form itself does not presuppose any particular mode of tendering procedure but it is assumed that tenderers will be selected on the basis that the Employer is satisfied that those who will supply the Contractor with the necessary professional and technical services in connection with design and detailing will be suitably qualified for preparing the Contractor's proposals and completing the design if the Contractor's tender is accepted.

The two main differences between a contract under this Form and other Forms issued by the Joint Contracts Tribunal where the design of the works is by the Employer's Architect, such as the Standard Form 'Without Quantities' are first, that the Contractor provides the design for the Works; and second, that there is no person who exercises any of the functions ascribed to the Architect or Quantity Surveyor in the Form 'Without Quantities'.

Because of these two main differences, certain provisions are included in the Form With Contractor's Design which do not appear in the 'Without Quantities' Form. The most important of these are concerned with the following:

1. Under the conditions of the Contract the Contractor as designer, has defined liabilities in respect of his design work. The Employer has an option whether to limit the Contractor's liability for the consequential damage arising from any failure by the Contractor to meet his obligations; the exercise of this option by the Employer is a matter for the Employer's own commercial judgement.
2. The absence of any priced Bills of Quantities necessitates the preparation of a new document – The Contract Sum Analysis, the main purposes of which are:

   (*a*) To ensure that if formula adjustment of the Contract Sum to deal with cost changes during the work is adopted, the Analysis provides sufficient information for the operation of formula adjustment.
   (*b*) To assist in the valuation of variations which are called 'Changes in the Employer's Requirements' or 'Changes'.

Provision is also needed to take into account the obtaining of planning permission and any other approvals which may affect the right to develop the site, referred to in the Form as 'Development Control Requirements'.

Matters which under the 'Without Quantities' Form are reserved by both Employer and Contractor to the decisions or opinion of the Architect or Quantity Surveyor (e.g. extension of time, or the valuation of additional loss and expense for disturbance of the regular progress of the Works) are, of necessity, left to one or other party to decide; and if the other party does not accept that decision then the matter must be taken to arbitration.

Because the Employer will almost certainly be a corporate body of one kind or another provisions are made for a person to be named to act as the Employer's Agent for the purposes of the contract. The contract treats the acts of the Agent as those of the Employer.

Unless therefore the Employer specifically otherwise states in his Requirements, the Contractor can and must regard the Employer's Agent as the duly authorised agent of the

Employer for the performance of any of the actions of the Employer under the contract conditions. Whatever qualification the Employer's Agent may possess he acts as the Employer and in no way acts as certifier or valuer between the Employer and the Contractor. The Employer is entitled to remove and appoint his Agent at will but must inform the Contractor of his Agent's identity.

Other provisions in the Form with Contractor's Design are however similar to those in the 'Without Quantities' Form. Thus:

- The Employer has to give a date of possession.
- A Date for Completion has to be inserted in the Appendix and liquidated damages may be provided for payment by the Contractor if he is in default over meeting the Completion Date.
- Provision for Practical Completion is retained, followed by a Defects Liability Period and a defined time when defects have been completed.
- Payment (which under this Form may be by stages or periodically) has to be made during the progress of the Works followed by a final account and a final payment by the Employer.
- The Employer has a right to deduct retention at a percentage agreed with the Contractor.
- The insurance provisions are similar to those in the 'Without Quantities' form including a choice of arrangement for dealing with fire etc. risks to the Works (which risks are collectively referred to as the Clause 22 Risks).
- Partial possession by the Employer before Practical Completion of the whole Works is permitted by agreement.
- Extension of the Contract Completion Date is permitted but certain of the grounds on which extension can be obtained differ because of the nature of the contractual arrangements.
- A choice of fluctuation provisions is given i.e. tax fluctuations; or labour, materials and tax fluctuations; or adjustments by the use of formulae.
- Arbitration is the method provided under the contract for the settlement of disputes.

(b) *JCT Standard Form With or Without Quantities with the modifications to that Form set out in the Contractor's Designed Portion Supplement (CDPS) 1981* (as revised) for the 'With Quantities' Form and in the separate Supplement (1993) for the 'Without Quantities' Form where the Employer and his Architect, for Works designed by the Architect nevertheless require a portion of the Works to be designed by the Contractor and Proposals from the Contractor in respect thereof to be examined prior to the placing of the contract.

(c) *GC/Works/1 General Conditions of Contract for Building and Civil Engineering, Standard Form of Contract, Single Stage Design and Build Version* where the Authority having detailed his requirements requires the Contractor to carry out all or most of the design work as part of the single round of bidding. The resulting design proposed together with a firm lump sum price for both design and construction comprises the tender for the project. This form of procurement is generally reckoned to operate most effectively for relatively simple buildings. Due to the need for the Contractor to commit himself to a lump sum offer, adequate time for design needs to be given in the tendering process, and the number of tenderers should be less than for traditional methods. Single stage Design and Build is most appropriate as a procurement option when the Authority can precisely define its requirements; the Authority does not need ongoing control over the developing design and there is little likelihood of changes to the requirements.

(d) *ICE Design and Construct Contract* where the Employer requires the Contractor to design, construct and complete the Works in conformity in all respects with the provisions of the Contractor. The lump sum price for both design and construction comprises the tender for the project or such other sum as may be ascertained in accordance with the Conditions of Contract.

If the Employer requires a lump sum price the selection of contracts (*a*), (*b*) and (*c*) is suitable.

3. Maintenance and Minor Works including Improvements

The Schedule of Rates form of contract is often used for maintenance and minor works consisting of a detailed schedule setting out the work involved under the various items, the work in each case being carefully described and a rate column affixed.

From the estimator's point of view tendering in this form presents distinct disadvantages and difficulties. First the Schedule of Rates sets out the work in itemised form with no quantities. Second there is no Preliminary Bill incorporated in the contract documents as is the case of a Bill of Quantities.

Therefore, the estimator has no means of computing the value of the work under the various items of work or its total value and can only assess this on an approximate basis from contract drawings if such are available.

It is usual in practice to assess the quantities of the main items of work only, such as excavation, concreting, brickwork, etc. and increase the normal rates tendered for these by an amount sufficient to allow for the cost of those items usually allowed for in a preliminary bill.

From the foregoing it will be seen that this form of contract is to be deprecated and is unsatisfactory from the point of view of all concerned. In normal conditions a Bill of Quantities form of tender is greatly to be desired and tendering in this form is the generally accepted practice.

In adopting the 'Schedule of Rates' form of tender, payment is made to the builder or contractor for the work done by 'measuring up' the work under the various items as set out in the schedule and extending these by the relevant rates, the whole being then brought to a total.

A typical example of a 'Schedule of Rates' dealing with external works is shown below.

Table 2.1 Neesh road contract

	Schedule of rates		*January 1996*
Item No.	*Description*	*Unit*	*Agreed Schedule Rate*
1.	Excavate to reduce levels, maximum depth not exceeding 1.00m.	m^3	£3.06
2.	Extra over excavation for breaking out tarmacadam and concrete base under 175mm overall.	m^2	£1.65
3.	Dispose of excavated material off site.	m^3	£6.73

Where an Employer has a regular flow of maintenance and minor works including improvements to be carried out, he can either contract separately for each piece of work or can engage one contractor to carry out all such work in respect of specified properties for a specific period.

Where the Employer chooses the first alternative, to contract separately for each piece of Maintenance or Minor Works, the Employer may wish to have the work carried out under a contract where his professional consultant is to act throughout. If so the use of the Agreement for Minor Building Works should be considered. If, however, the Employer wishes to deal directly with the Contractor then the 1990 Jobbing Agreement should be considered. The Jobbing Agreement requires the use of a standard Tender and Agreement (JA/T 90) which is sold in pads. It is only suitable for use by Local Authorities and other Employers who place a number of small jobbing contracts with various Contractors and who are experienced in ordering jobbing work and in dealing with Contractor's accounts. It is for work of small value and which should not normally be of a longer duration than one month. The Conditions provide for a single payment following receipt of the Contractor's invoice for checking and approval by the Employer.

Where an Employer does not wish to use the JCT Tender and Agreement (JA/T 90) he can place various orders by means of his own Works Orders with the conditions JA/C 90.

An early start

If the Employer wants an early start and adequately detailed contract documents cannot be prepared no lump sum can be quoted and the selection of contract is based on:

1. JCT Standard Form with Approximate Quantities which requires a Contractor to quote a Tender Sum which is indicative of the likely price. The basis of the contract with the Form is similar to that described in detail above when the JCT (80) Standard Form with Quantities is used except that the work as instructed is completely remeasured. The remeasurement is priced on the basis of the rates set out in the Bills of Approximate Quantities. The Bills of Quantities are prepared in accordance with the Standard Method of Measurement of Building Works as agreed in the industry.
2. JCT Standard Form of Management Contract which requires the Employer to contract on the basis of an estimate of the total cost of his Project (called a Contract Cost Plan Total); in addition he will have to pay a Management Fee which is usually calculated as a percentage of the final cost.
3. JCT Standard Form of Prime Cost Contract which requires the Employer to contract on the basis of an estimate of the total cost of his project; in addition to that cost he will have to pay a Contract Fee as a fixed or percentage fee in respect of the Contractor's on-site overheads and profit.

The principal differences between the Management Contract and the Prime Cost Contract are:

- In the Management Contract the Contractor is only required to manage the carrying out of the work and the contract sets out in detail the duties required of him in so doing; whereas under the Prime Cost Form it is envisaged that the Contractor will carry out at least some of the work with his own labour (which he is not permitted to do in the Management Contract).
- The Management Contract is divided into two periods: the Pre-Construction Period and the Construction Period whereas the Prime Cost Form does not provide for a separate Pre-Construction Period.
- The documentation published by the Tribunal for the Management Contract is comprehensive and includes an obligatory Form of Works Contract whereas the Tribunal does not prescribe any form for use with any Domestic Sub-Contractors who may be employed under the Prime Cost Form. It does however prescribe for Nominated Sub-Contractors in the Standard Form.

4. The Schedule of Rates form of tender requires setting out the work involved under the various items in a detailed schedule. The work in each case being carefully described with a rate inserted for each item.

The Latham Report (1994)

Sir Michael Latham's Report *Constructing the Team* made a proposal for a modern form of contract possibly based on the New Engineering Contract (New Construction Contract) to reduce conflict in the UK Construction Industry.

The new contract takes the form of a suite of interlocking contracts suitable for use on building works of all types. The contract is written in plain English and puts the emphasis on flexibility, co-operation and good project management practice.

The contract is intended to eradicate conflict. Consultants are not specifically referred to in the contract but only the key parties are named. They are the employer, the contractor and sub-contractors, the project manager, the adjudicator and the supervisor.

The main proposals of the Latham Report are:

- Legislation in the form of a Construction Contracts Bill.
- Modern form of contract possibly based on the New Engineering Contract.
- Role of government as 'best practice' client, including improvements to briefing, tendering and payment procedure and development of a quality register of approved firms.
- Adjudication the normal method of dispute resolution.
- Compulsory latent defects insurance and trust fund to safeguard payments.
- Greater use of standard components and co-ordinated project information to encourage co-operation and better communication.

Latham estimated the 'peace dividend' as a 30 per cent fall in construction costs by the year 2000.

The Government proposes in 1996 to introduce legislation to implement parts of the Latham Report.

3 Estimating the total cost

Estimating the cost and tendering for work should not be confused.

Estimating the total cost of the work is derived from the cost of labour; materials; plant; preliminaries; provisional and prime cost sums; dayworks and overheads. The method of estimating the cost of these individual elements and the building up of the total cost is described here.

Tendering includes the element of profit by adding a lump sum or percentage to the total estimated cost of the work and is the subject of a separate chapter.

The estimation of the total cost of the work consists of:

- Estimating the cost of the individual elements of the cost.
- Building up the total cost taking into account each individual element of the cost concerned.

The major elements of the total cost of the work are:

1. The labour cost.
2. The material cost.
3. The plant cost.

The remaining elements which contribute to the total cost of the work are:

4. Dayworks, provisional sums, prime cost sums and contingencies.
5. Preliminaries or site charges.

   (a) Water for the works.
   (b) Small plant and tools.
   (c) The haulage of plant.
   (d) Site offices for the architect or engineer.
   (e) Site offices for the builder or contractor.
   (f) Watching and lighting.
   (g) The control of traffic and the additional cost involved.
   (h) Temporary roads and hoardings.
   (i) The contract guarantee bond.
   (j) Insurances.
   (k) Building fees.
   (l) Hoardings.

6. Establishment, overheads or head office charges.

The plant, labour and material cost

The labour, material and plant cost of a unit of work as itemised in a Bill of Quantities or Schedule of Rates is obtained from the estimating data shown in the various chapters of this book.

The tables shown are in two forms:

1. The main estimating tables.
2. The tables of multipliers.

The function of the main estimating tables is to supply data from which to estimate the labour and plant cost of a unit of work as itemised in a Bill of Quantities or Schedule of Rates. The data refers to the labour and plant hours taken to carry out one unit of the work in question and are given in Hour Constants. They are, therefore, in no way affected by fluctuating labour rates of pay or plant hire rates per hour and are always up to date. Data from which to calculate the material cost per unit are also incorporated in these tables where they can conveniently be done, otherwise they are shown separately.

The function of the tables of multipliers is to make available much useful estimating data in concise form. The multipliers are used in conjunction with the plant and labour hours shown in the main estimating tables. By multiplying the hours shown by the appropriate multiplier the cost of work of a different kind and carried out under different conditions is obtained. For example, stock brickwork is tabulated as brickwork built in cement mortar. By using the multiplier shown, 0.80, with the bricklayer and labour hours shown, the bricklayer and labour hours taken to carry out the work in lime mortar are obtained.

The multipliers thus form a simple and direct way of keeping the data concise and make it possible to give much information in little space.

The labour cost

In estimating the labour cost of a unit of work the cost of labour supervision must be considered.

The direct supervision of labour is carried out by gangers, trade foremen and leading hands, these being in turn supervised by a general foreman who co-ordinates the labour supervision and labour as a whole under the agent and his staff.

In costing work with a view to producing estimating data a decision has to be made as to what labour supervision is a direct labour charge and what is included in the Preliminaries Bill.

This is generally resolved as follows:

(a) Gangers, trade foremen and leading hands are a direct labour charge.
(b) The agent and the site administrative staff and general foreman are included in Preliminaries.

The labour hour constants shown for estimating purposes in the various chapters of this book include for the direct labour charge as shown in (a) and allow for the costs involved in connection with such supervision.

Supervision by the agent and the site administrative staff including that of general foreman being a general site charge is allowed for in the Preliminaries Bill.

The labour cost per unit of work is obtained by multiplying the labour hour constants shown in the estimating tables by the prevailing labour rates of pay per hour (see chapter 4), these rates referring to those of a craft operative or labourer, as the case may be.

The material cost

In estimating the cost of the material involved in a unit of work, apart from its quality, which should be in accordance with the specification, consideration should be given to:

1. The availability of a sufficient supply delivered at regulated intervals and in the correct amount, ensuring both plant and labour are kept actively employed.
2. The cost of the material delivered on the site.

It will be appreciated that if the rate of delivery is insufficient, plant and labour are not worked to capacity, and if delivered in overwhelming amounts the material will have to be double handled involving additional cost.

The data shown in the estimating tables are based on the materials being delivered on the site adjacent to where they are placed in the work in well regulated amounts, and allow for off-loading, handling and placing them in the work complete.

Under certain site conditions material may have to be off-loaded and dumped or stacked in convenience dumps, necessitating their being reloaded and distributed in suitable vehicles throughout the site at a later date. The cost of this reloading, hauling and distributing should be taken into account in assessing the actual cost of the materials delivered on the site and is in addition.

In the event of the materials having to be loaded and hauled to the site from an outside source such as a railway station, wharf, ballast pit, etc. at the expense of the contractor, the cost of this loading and hauling must also be taken into account in arriving at the actual cost of the material on the site of the works.

When the above factors have been taken into account and the cost of the material delivered on the site computed, the material cost per unit of work is obtained by multiplying the amount of the material required in the unit of work in question by the cost of the material per unit amount.

The plant cost

In considering the plant cost involved in carrying out a specific unit of work, two classes of plant have to be considered:

1. Small plant and tools such as barrows, ropes, pails, blocks and tackle, saws, picks, shovels, scaffold poles, scaffold boards, putlogs, steps, trestles, etc. and other small plant of a non-mechanical nature.
2. Mechanical plant such as mechanical excavators, concrete mixers, lorries, dumpers, power hoists, etc.

For small plant and tools the costs are allowed for in estimating by adding a percentage addition to the estimated cost of the work.

For large plant, e.g. mechanical excavators, concrete mixers, dumpers, the mechanical plant cost per unit of work is obtained by multiplying the plant hour constants shown in the estimating tables by the prevailing plant hire rate or working cost of the plant per hour, this plant rate being wholly inclusive of:

1. The hire rate of the plant per hour, allowing for depreciation and repairs and renewals.
2. The wage of the plant driver per hour.
3. The cost of the fuel, lubricating oil and grease consumed per hour.
4. The licence cost of the plant, if any, on a per working hour basis.
5. The plant insurance cost, if any, on a per working hour basis.

Dayworks, provisional sums, prime cost sums and contingencies

Dayworks are incidental to the carrying out of the works and are for unforeseen work.

Dayworks is usually allowed for in the tender as a provisional lump sum or as a provisional number of hours for trade operatives. The contractor binds himself under the contract to carry out all Dayworks at their actual cost plus an agreed percentage addition to allow for establishment charges and profit and/or the actual time to carry out the works at an enhanced rate to include establishment charges and profit.

Provisional sums are for works that cannot be described and quantified in accordance with the rules of the various standard methods of measurement. Provisional sums should be inclusive of profit.

Prime cost sums are for works carried out by nominated sub-contractors or goods supplied by nominated suppliers. Prime cost sums are exclusive of profit and attendance.

Contingencies sum is to allow for possible additional works.

Preliminaries or general site charges

Preliminaries or general site charges refer to costs incurred on the contract sites.

Examples of Items coming under this heading are:

- Salaries paid to the agent and the administrative staff on the site of the works.
- The allowance for Small Plant and Tools.

Establishment, overheads or head office charges

These charges refer to costs incurred in maintaining and running a head office with its attendant costs in the way of staff salaries, rent, rates, etc.

Establishment charges vary considerably with different firms and are influenced by the following factors:

- The annual turnover.
- The class of work carried out.
- The ability and drive of the administrative staff.
- The analysis as to which element of the cost is allowed for as establishment charges.

Profit

See separate chapter – Tendering.

Points of importance to note in using the data shown for estimates

In using the data shown for estimating purposes the following should be noted:

1. From the hour constants and material data shown in the estimating tables are obtained the estimated labour, material and plant cost of the work, exclusive of Preliminaries or General Site Charges, Establishment Charges and the Net Profit desired.
2. In using the tables shown for estimating purposes care must be taken to ensure that no part or parts of the operation or operations involved in carrying out the work is omitted from the calculations, for example:

    - The laying of bricks is dealt with in one table, pointing being the subject of a subsidiary table.
    - Mixing and placing concrete is dealt with under one table, wheeling and transporting it in vehicles being dealt with in subsequent tables.
    - Excavation over areas and in bulk is dealt with in one table, the transport of the excavated material being shown under separate tables.

3. Every element contributing to the cost of a specific item of work must be included in the calculation. For example, in concreting roads the data shown in the tables allow for

mixing, transporting, placing and tamping the concrete; but do not allow for attendant operations such as

- Fixing side forms.
- Fixing expansion joints.
- Fixing reinforcement.
- Curing.
- Bullnosing edges.

These items are usually billed separately, but on occasions they have to be allowed for in the cost of the concrete itself, in which case this is best done by allowing for them on a per square metre of concrete basis.

4. In assessing the cost of the material involved in a unit of work this must be based on the cost of the material delivered on the site adjacent to where it is placed in the work.
5. Preliminaries are allowed for by adding the preliminary costs to the estimated cost of the work.
6. Establishment charges are allowed for by:

- Adding a percentage addition to the estimated labour, material and plant cost of the work, and Preliminaries as billed in a Bill of Quantities or Schedule of Rates.
- Adding a percentage addition to the lump sum costs allowed for in the estimate.

Table 3.1 Alternative ways of allowing for elements of cost

Description	Allowed for in the estimated plant, labour and material cost of the work	Allowed for on a lump sum basis	Allowed for by adding a percentage to the estimated plant, labour and material cost of the work and to the estimated cost further elements which are allowed for on a lump sum basis
Labour cost	x	–	–
Material cost	x	–	–
Plant cost	x	–	–
Provisional Sums	–	x	–
Dayworks and Prime Cost amounts (lump sum basis)	–	x	–
Preliminaries:			
(a) Water for the works	–	x	x
(b) Plant and tools	–	–	x
(c) The haulage of plant	x	x	–
(d) Site offices for the architect or engineer	–	x	–
(e) Site offices for the builder or contractor	–	x	x
(f) Watching and lighting	–	x	–
(g) The control of traffic	–	x	–
(h) Temporary roads	–	x	–
(i) The contract guarantee bond	–	x	–
(j) Insurances	–	x	–
(k) Building fees	–	x	–
(l) Hoardings	–	x	–
Head office charges	–	–	x
Net profit	–	x	x

7. The net profit desired is allowed for by adding a percentage addition to the total estimated cost of the work.
8. In tendering for work the percentage additions for establishment charges and profit are dealt with together.
9. In using the table of multipliers it should be noted that more than one multiplier may be involved.

System of estimating the total cost

In estimating the Total Cost a system should be formulated whereby every element of the cost is allowed for in one way or another, and in doing so the following should be borne in mind:

- It should be foolproof in that no element of the cost is omitted.
- It should be simple to operate.
- It should be quick in operation.

Having evolved a system, it is wise to adhere to it, as by so doing, estimating becomes a fixed routine which makes for both accuracy and speed.

Table 3.1 sets out the alternative ways in which the individual elements of the cost may be allowed for in building up the Total Cost, and this may be used as a guide in formulating a system of estimating, care being taken to ensure that every element of the cost is allowed for in one or other of the ways shown.

The alternative ways in which to allow for the individual elements of the cost are designated by a cross.

4 Labour including bonus

The typical cost to employ craft operative and labourer is shown in Tables 4.1 and 4.2 overleaf.

Labour costs are influenced by several factors, the chief of which are:

1. The amount of similar work carried out by individual operatives in a given time is variable.
2. The quality of labour as a whole varies in different localities.
3. The weather has a marked effect upon labour output.
4. Output is influenced by those who directly control the labour on the work, i.e. foremen, gangers, charge hands, etc.

In estimating the labour cost of an item of work the estimator is only concerned with the 'average labour output' in carrying out the work. The labour outputs shown refer to those taken in carrying out work under normal conditions and are based on any materials involved being delivered on the site adjacent to where they are to be placed in the work.

The data shown may be used as a basis on which to build up a price from first principles, and from it may also be assessed:

1. The amount of work of various kinds carried out by one operative in one hour.
2. From the amount of work of a certain kind which has to be carried out, the number of men required to complete it in a given time.
3. Knowing the amount of a particular material being delivered on the site per hour, the number of men required to place it in the work.

Experience has shown that the payment of bonus acts as an incentive to labour to achieve greater output. This principle of payment by results has been practised in the civil engineering and building industry for many years with satisfactory results. Both industries recognise this system of payment, and although the building industry is of a different nature and involves more component operations than the civil engineering industry, suitable and similar bonus systems of payment are applicable to both.

Bonus systems might be thought to have an adverse effect on the quality of the work, but such is not the case. Unsound work is not a necessary outcome of high output. If a bonus is offered, the labour concerned is naturally prompted to use ingenuity in overcoming delays and to work with more sustained effort. By combining this factor with keen supervision and sound site administration, it is possible to achieve increased output and still maintain a high standard of work. The output above the normal achieved by labour due to the incentive of bonus varies with the class of work carried out, the increase being greater in the case of straightforward work where a comparatively large amount is involved, such as in bricklaying, pipelaying, concreting roads, etc. In actual practice it has been found that the increase in output can be as much as 50 per cent above normal, and in special cases even greater outputs may be achieved.

The bonus systems of payment usually operated on building and civil engineering works of construction are of two main types:

1. Bonus paid on output in excess of a basic output.
2. Work paid for on a piecework basis.

The method of evolving and operating these bonus systems of payment is the subject here. Their operation is closely related to sound site administration; carefully conceived bonus systems

Table 4.1 Typical cost to employ craft operative

		£	£
Standard Wage (39 hour week) based on a total of 1801 hours worked per year	46.18 wks	175.00	8 081.50
Note: Market conditions will determine plus rates and in this example a 25% plus rate is included as a current assessment	Plus rate	25%	2 020.38
			10 101.88
Extra payments under National Working Rules	46.18 wks	2.25	103.91
CITB Levy	1 year	20.00	20.00
Public holiday pay	1.60 wks	175.00	280.00
Employer's contribution to annual holiday pay, accidental injury and death cover scheme	46.18 wks	20.00	923.60
National Insurance	46.18 wks	15.00	692.70
			12 122.09
Severance pay		1.5%	181.83
			12 303.92
Employer's liability and third party insurance		2%	246.08
Total cost per annum			12 550.00
Total cost per hour (1801 hours)			6.97

For the purposes of calculations in this book £7.00 has been used for Craft Operative.

Table 4.2 Typical cost to employ labourer

		£	£
Standard Wage (39 hour week) based on a total of 1801 hours worked per year	46.18 wks	150.00	6 927.00
Note: Market conditions will determine plus rates and in this example a 25% plus rate is included as a current assessment	Plus rate	25%	1 731.75
			8 658.75
Extra payments under National Working Rules	46.18 wks	4.50	207.81
CITB Levy	1 year	18.00	18.00
Public holiday pay	1.60 wks	150.00	240.00
Employer's contribution to annual holiday pay, accidental injury and death cover scheme	46.18 wks	20.00	923.60
National Insurance	46.18 wks	10.00	461.80
			10 509.96
Severance pay		1.5%	157.65
			10 667.61
Employer's liability and third party insurance		2%	213.35
Total cost per annum			10 880.96
Total cost per hour (1801 hours)			6.04

For the purposes of calculations in this book *£6.00* has been used for Labourer.

make for increased output, resulting in a saving in the cost of the work. Their operation is, therefore, to be encouraged as they result in mutual good to both the employee and the employer.

The basic output

The 'basic output', i.e. the datum level of output which has to be achieved before bonus becomes payable, must necessarily be the basic output on which the estimate was based, otherwise the bonus paid would bear no direct relation to the estimated cost of the work.

Example.   Assuming the estimated cost of 225 mm brickwork in walls, exclusive of pointing and scaffolding, is built up in the following manner:

225 mm Brickwork in walls per m^2

Bricklayer 1.90 h at £7.00 per h	=	£13.30 per m^2
Labourer 1.90 h at £6.00 per h	=	£11.40 per m^2
Bricks 116 No. at £150.00 per 1000	=	£17.40 per m^2
Mortar 0.055 m^3 at £60.00 per m^3	=	£ 3.30 per m^2
Estimated cost of brickwork	=	£45.40 per m^2

It will be noted that the time taken by a bricklayer and attendant labourer to carry out $1\,m^2$ of the brickwork in question is 1.90 hours. The basic labour hours per m^2 of brickwork and the basic output of brickwork per bricklayer per hour on which the estimate is based are, therefore, as follows:

Basic labour hours per m^2 = bricklayer 1.90 hours
                            = labourer 1.90 hours

Basic output m^2 per hour $= \dfrac{1.00}{1.90}\,m^2$

$= 0.53\,m^2$ of brickwork per bricklayer per hour.

In operating a bonus system of payment in this case bonus is therefore payable on output achieved in excess of this amount.

Note: This principle of paying bonus on output achieved in excess of a basic output applies to both plant and labour; for example, in the case of a trench excavated by a mechanical trench excavator, the basic output of the plant per hour is considered when computing the actual excavation achieved in excess of the basic output for purposes of calculating the bonus payable.

The computation of the amount of bonus to pay so as to ensure a saving in the cost of the work

In evolving and operating a bonus system of payment, whether bonus is paid on output in excess of a basic output or the work is paid for on a piecework basis, the underlying principle involved is that bonus is payable on the labour hour saving resulting from output achieved in excess of a normal or basic output.

In computing the amount to pay, two factors have to be taken into consideration:

1. The operation of the system should result in an effective saving in the cost of the work.

2. The amount of bonus or the piecework rate paid must be such that it offers a real incentive to labour to achieve output in excess of the normal by ensuring them a just reward for their additional efforts.

Referring to the example shown in connection with brickwork above, it will be noted that the basic output is $0.53\,\text{m}^2$ of brickwork per bricklayer per hour, the basic labour hours per m^2 being: bricklayer 1.90 hours, labourer 1.90 hours. Bonus is therefore payable on output in excess of $0.53\,\text{m}^2$ per hour. This bonus is, however, not credited at the full basic labour hours, i.e. 1.90 hours, but only at a percentage of them, thus ensuring that a labour saving is effected in the cost of the work. The problem is, therefore, to decide what this percentage should be, bearing in mind that it must be sufficient to induce labour to achieve output in excess of the normal.

Experience has shown that it should be in the region of 60 per cent, the remainder contributing to a saving in the cost of the work.

This value of 60 per cent should be looked upon as a minimum, but should not be varied to any great extent, so as to ensure a reasonable degree of uniformity in the scale and amount of bonus paid throughout the industries concerned. In the case of the brickwork mentioned, assuming the percentage is taken as being 60 per cent, the bonus payable on output in excess of the Basic Output would be: 60 per cent of the basic labour hours per m^2, i.e.:

Bricklayer   60 per cent of 1.90 hours = 1.14 hours.
Labourer   60 per cent of 1.90 hours = 1.14 hours.

Note: In paying labour 60 per cent of the labour hour saving resulting from the output achieved in excess of the basic output it might be thought that a labour hour saving of 40 per cent is obtained by the contractor. This is not strictly correct, as any cost incurred in operating the scheme has to be offset against it. This cost, however, is not considered, and in some cases may be very small as the work entailed in operating the scheme is often carried out in conjunction with their other duties by staff already on the site. The contractor, though not benefiting by the whole of the 40 per cent, does therefore obtain an effective saving in the cost of the work which is reflected in any profit obtained.

The distribution of bonus payments or piecework earnings throughout the gang

After computation of the bonus or piecework earnings of a gang the amount involved has to be distributed throughout the gang. This is best done by apportioning it on a share basis in proportion, or in approximate proportion, to the hourly wage rates (basic wage rate and Joint Board Supplement only) of those concerned, and in direct proportion to the hours worked by each, thus:

Classification	Hourly wage rate	Share basis
Foreman	£7.35	1.23
Bricklayer	£7.00	1.17
Ganger	£6.25	1.05
Plant Driver	£6.13	1.03
Labourer	£6.00	1.00

After computation of the amount due to each classification in a lump sum amount, this sum is then apportioned in direct proportion to the hours worked. The method of distributing bonus and piecework earnings on this basis is illustrated by example later in this chapter in connection with bonus systems of payment where bonus is paid on outputs in excess of a basic output and work paid for on a piecework basis.

The measurement of the work done

The payment of bonus on output in excess of a basic output and work paid for as piecework is invariably carried out on a weekly basis. This involves the measurement of the work done under the various operations, and if a weekly system is in operation this would necessarily have to be done every week. The measurement of the work is carried out by the engineer or surveyor.

The measurements are computed as actual quantities, i.e. cubic metres, square metres, etc. as the case may be, the quantities being best recorded on a Weekly Measurement Sheet a typical example of which is shown below. The measurement sheet records:

1. The quantity of work carried out for the current week.
2. The quantity of work previously carried out.
3. The progressive quantity of work carried out to date.

Table 4.3 Typical weekly measurement sheet

WEEKLY MEASUREMENT SHEET

Foreman: Hussey, J.	*Contract*: NEESH ROAD			*Week ending*: 8/9/95
Operation	*Unit*	*Week's Quantity*	*Previous Quantity*	*Quantity to date*
225 mm brickwork in walls etc.	m^2	133	620	753

The tabulation of work in this manner is a simple method of recording the figures, in which current and progressive quantities to date are always available.

Note: If a weekly costing system is in operation on the contract in question, the work would necessarily have to be measured up in this connection, so that no additional expense would be incurred in operating the bonus system of payment so far as measuring up the work is concerned.

The plant and labour allocation

In operating these bonus schemes of payment the plant and labour hours expended on the work have to be allocated and recorded, and assuming that they are operated on a weekly basis this allocation must be carried out each week. The allocation is generally performed by the ganger or foreman concerned, they obviously being in the best position to do so. It will be appreciated that great care must be exercised in carrying out this allocation, and it should be carefully checked by the engineer or surveyor concerned as the work proceeds.

The allocation of the plant and labour hours is best recorded on a Daily Plant and Labour Allocation Sheet, these hours being abstracted at the end of the current week and recorded on a Weekly Plant and Labour Allocation Sheet.

A typical Weekly Plant and Labour Allocation Sheet is shown below, the Daily Plant is payable on output in excess of a basic output, or work is paid for on a piecework basis, each system is dealt with in detail and illustrated by example, the examples being based on the data shown in the Weekly Measurement Sheet and Weekly Plant and Labour Allocation Sheet (Tables 4.3 and 4.4).

Table 4.4 Weekly plant and labour allocation sheet

WEEKLY PLANT AND LABOUR ALLOCATION SHEET											
Ganger: Smith, J.				*Contract*: LENDEN.					*Week ending*: 8th September, 1995		
	Plant hours					*Bricklayers hours*	*Labour hours*	*Labour hours*			*Remarks*
Operation	Trench Exca- vator mach- ine										
Excavate, refill and ram trench for 150 mm dia. water main 1.05 m deep	40					–	560				Piecework
225 mm brickwork in walls etc.	–					160	160				Piecework

Bonus paid on output in excess of a basic output

In operating this system, the bonus payable per unit in excess of the basic output is a percentage of the basic labour hours. As has been shown above, this should approximate 60 per cent and should not deviate from this to any great extent.

Figure 4.1 and Table 4.5 illustrate this system, the labour hours payable as bonus for the range shown being 60 per cent of the basic labour hours.

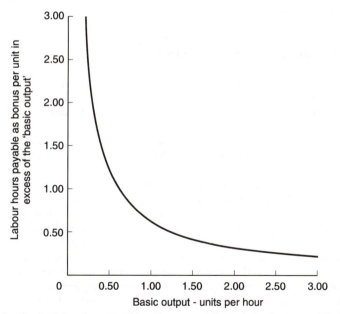

Figure 4.1 Graph showing the labour hours payable as bonus per unit on outputs in excess of basic output of various amounts.

Table 4.5 Labour hours payable as bonus on output in excess of the basic for various basic outputs per hour

Basic output, units per hour	Basic hours per unit	Labour hours payable as bonus on outputs in excess of the basic output, Labour hours per unit
0.25	4.00	2.40
0.50	2.00	1.20
0.75	1.33	0.80
1.00	1.00	0.60
1.25	0.80	0.48
1.50	0.66	0.40
1.75	0.57	0.34
2.00	0.50	0.30
2.25	0.44	0.26
2.50	0.40	0.24
2.75	0.36	0.22
3.00	0.33	0.20

The data shown in those sheets are best abstracted and recorded on a Weekly Bonus Sheet, a separate sheet being kept for each operation.

The manner in which this system is operated is best illustrated by example and for this purpose the data shown in the Weekly Measurement Sheet and Weekly Plant and Labour Allocation Sheet above, relating to the excavation of a trench for a 150 mm diameter water main and 225 mm brickwork in walls, are used.

Work paid for on a piecework basis

In paying for work on a piecework basis the underlying principle involved is similar to that in paying bonus on output in excess of a basic output, that is to say the labour concerned is credited with a percentage of the labour hour saving resulting from output achieved in excess of a basic output.

This should approximate 60 per cent, as explained in the case of a bonus system where bonus is paid on output in excess of a basic output, and should not deviate from this to any great extent.

In paying for work on a piecework basis the work is measured up and paid for at an agreed monetary rate per unit, the labour concerned receiving no other remuneration in the way of ordinary time earnings. The piecework rate is agreed *prior* to the work being carried out. This is fundamentally different from the payment of bonus on output achieved in excess of a basic output, in which case the bonus due is computed *after* the work has been carried out, the labour receiving payment at their ordinary time rates together with any bonus due.

The practice of paying a fixed piecework rate irrespective of the actual output achieved in excess of the basic output is commonly adopted, the rate being based on an assumed output.

In paying piecework rates in this manner the labour is credited with a fixed percentage of the labour hour saving for the actual output achieved, the remainder contributing to an effective saving in the cost of the work.

The computation of the piecework rates payable is best reduced to terms of a percentage of the basic labour hours per unit, the percentage depending upon the assumed output on which it is based.

Figure 4.2 and Table 4.6 illustrate how these percentages are computed. The data shown in the table and graph are based on the labour concerned being credited with 60 per cent of the labour hour saving resulting from output achieved in excess of the normal or basic output.

It will be noted that in crediting the labour with 60 per cent of the labour hour saving, the percentage of the basic labour hours paid as a piecework rate decreases as the output increases.

In order to illustrate the method of computing the scale of piecework rates to pay in connection with a specific operation, the case of a 225 mm wall is considered.

For purposes of this example the following are assumed:

1. Basic output on which the estimate is based is 0.53 m^2 of brickwork per bricklayer per hour.
2. Basic labour hours per m^2 of brickwork on which the estimate is based are:

    Bricklayer   1.90 hours
    Labourer   1.90 hours

3. In computing the piecework rates the labour concerned is credited with 60 per cent of the labour hour saving resulting from output in excess of the basic output.

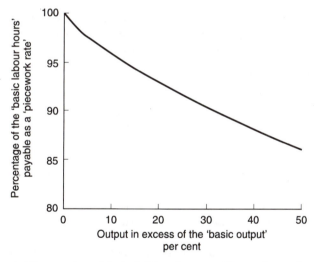

Figure 4.2. Graph showing the percentage of the basic labour hours payable as a piecework rate for various percentage outputs in excess of the basic output.

In Table 4.6 the method of computing the percentage of the basic labour hours payable as a piecework rate is shown. The percentages shown are based on labour being credited with 60 per cent of the labour hour saving achieved, and they are therefore applicable to this example.

The piecework rates payable for the brickwork in question, for outputs ranging from 0 to 50 per cent in excess of the Basic Output, are therefore as shown in Table 4.6

Table 4.6 Scale of piecework rates payable for 225 mm brickwork in walls

Note: 1. The basic labour hours per m^2 of brickwork on which the data are based are: Bricklayer 1.90 hours; labourer 1.90 hours.
2. The labour concerned is credited with 60 per cent of the labour hour saving resulting from output achieved in excess of the basic output, e.g. output in excess of basic output 35 per cent, hours payable as a piecework rate: bricklayer 90 per cent of 1.90 hours = 1.71 hours; labourer 90 per cent of 1.90 hours = 1.71 hours.

Output in excess of the Basic Output %	Percentage of the basic labour hours payable as a piecework rate per unit %	Piecework rate payable per m^2 of brickwork	
		Bricklayer hours	Labourer hours
5	98	1.86	1.86
10	96	1.82	1.82
15	95	1.81	1.81
20	93	1.77	1.77
25	92	1.75	1.75
30	91	1.73	1.73
35	90	1.71	1.71
40	88	1.67	1.67
45	87	1.65	1.65
50	86	1.63	1.63

Referring to this table and assuming that the piecework rate is based on an output of 35 per cent in excess of the basic output, the bricklayer's and labourer's rate of pay per hour being £7.00 and £6.00 respectively, the piecework rate payable per m^2 of brickwork is:

90 per cent of the basic labour hours, i.e. 90 per cent of 1.90 hours = 1.71 hours

The piecework rate payable per m^2 of brickwork is therefore:

Bricklayer 1.71 hours at £7.00 per hour = £11.97 per m^2
Labourer   1.71 hours at £6.00 per hour = £10.26 per m^2
                                              £22.23

Note: The basic labour cost per m^2 of brickwork on which the estimate is produced is: •

Bricklayer 1.90 hours at £7.00 per hour   = £13.30 per m^2
Labourer   1.90 hours at £6.00 per hour   = £11.40 per m^2

Basic labour rate on which estimate is based = £24.70 per m^2

The contribution to the saving in the cost of the work, therefore, is £24.70 − £22.23 = £2.47 per m^2, any cost incurred in operating the system being offset against this.

The work having been measured up and the total piecework earnings of the gang computed, the whole is then distributed throughout the gang on a share basis in the manner previously shown.

Bonusing on small works

The foregoing notes are generally applicable to major contracts lasting for one year or more with 50–100 operatives in various continuous employment during the duration of the contract. Those conditions account for about half of the total number of the construction industry. The other half are engaged in small works, alterations, repairs, improvement grant work, etc. Jobs of this nature involve between two and ten men, in periods varying from six days to six months. Any thoughts of bonusing for teams (or gangs) are impracticable. Each job or part of job must be assessed for bonus target hours separately.

5 *Materials*

With the building and civil engineering industry accounting for £8.7 billion turnover per annum and the materials share being approximately 50 per cent there is obviously a need for efficiency in the expenditure of materials.

In most building and civil engineering contracts the responsibility for sending out enquiries, chasing suppliers who are slow to respond and collating the quantities lies with the buying or purchasing department. Where a buyer is used to obtain quotations the instructions as to what is required are passed from the estimator. The quotations, once received, are passed back to the estimator for inclusion in the estimate. The usual function of the buying department in the production of an estimate is to provide a service to the estimator.

The estimator is required when checking quotations to ensure that materials meet the specification of the contract documents and can be delivered to the site at the times required by the construction programme.

In addition the contractual obligations to be entered into for the supply of the material must be satisfactory. The aspects of a materials quotation to be checked include:

- That the quotation covers the actual material stated on the drawings.
- That the material meets the standards described in the specification.
- That the quantity is appropriate to the total quantity required in the works.
- That the delivery period and programme meet the time required for incorporation in the works.
- That the discount rate (where applicable) is not less than the normal market rate.
- That the trading conditions and terms of payment are acceptable.
- That the time limit which may be applied to the acceptance of the supplier's offer is acceptable.
- Whether the material is offered on a firm price basis or prices current at the date of delivery.

The determination of the material prices for inclusion in the direct cost estimate may be considered to be one of the most precise aspects of estimating. The process of obtaining materials prices as has been described can be seen to consist solely of contacting suppliers who have the material available and negotiating a suitable rate under satisfactory contractual conditions. In addition the estimator has to undertake the more difficult task of determining allowances for material wastage, damage, theft and delivery discrepancies in so far as they affect the costs of the works.

For some materials this may reach particularly high proportions and this aspect must be assessed by the estimator and reflected in the price included in the estimate.

The estimator has also to include an allowance for the off loading and storage of the materials as well as the other allowances. Thus the material prices used in the item build up calculations may not be the quoted price but a higher price to allow for these variables.

An example of including such allowance is: £

Quoted price for bricks per thousand   200

Wastage allowance 5%   10

Allowance for unloading of pallets and distribution by site plant has also
a bearing on the cost of the brickwork. One third of an hour per
thousand is needed for unloading and half an hour for distribution
around the site point.

		£
$\frac{1}{3}$h labour @ £6	=	2
$\frac{1}{2}$h dumper and operator @ £8	=	4

                                                                                                                 6
                                    Total cost of bricks per thousand   £216

The three major resources in construction namely labour, plant machinery and materials have
been subjected to much scrutinising as to how their utilisation may be achieved more
efficiently.

The Just-In-Time (JIT) concept has helped Japanese car manufacturers compete successfully
against American and European car makers and has an application to materials resource in the
construction industry. JIT can be interpreted to mean the production of materials only as
required to meet the demands of the customer. On time delivery therefore implies that the
receiving work station will only receive the material just before it is required.

JIT is already an acknowledged concept for improving productivity in manufacturing but its
direct application to construction must consider the differences which exist between the
construction and manufacturing industry. Some of these differences are:

- The building owner is often deeply involved in the construction process whereas the buyers
  of manufactured products do not generally have access to the manufacturing facilities nor
  deal directly with the production managers concerned. The building owner therefore
  appears to be in a position to influence the construction process.
- Construction often involves large scale and complex assemblies of components on site
  which are both difficult to handle and fasten manually in place. In contrast manufactured
  products are usually assembled readily by mechanical means in factories.
- When conducted outdoors, traditional construction is commonly affected by the vagaries
  of inclement weather which interrupts the smooth flow of building operations.
  Manufacturing on the other hand is normally carried out indoors with ample protection
  from inclement weather.
- Because construction activities are often carried out in sequence, planning for the
  development of different skilled tradesmen on site is critical. Unless schedules are carried
  out carefully, delays and cost overruns are almost inevitable. The conveyor belt mechanical
  systems in manufacturing facilities do not normally pose such difficulties.
- The nature of construction activities may make it difficult to gain access to the building
  under construction. The provision of safety measures on a permanent basis may also be
  impossible during construction. Safety precautions in manufacturing can, on the other
  hand, be taken more readily because of the permanency of production facilities.
- The final product in construction is usually of unique design and differs from work station
  to work station. No fixed arrangement of equipment or aids such as jigs and fixtures is
  therefore possible. The processes and outputs from most manufacturing facilities are on the
  other hand repetitive and standard used.

- The durations of most construction projects or their individual work phases are relatively short. Consequently, the management team and workforce must be assembled quickly and cannot often be restricted before the project or work phase is completed. In addition planning and tooling up for operations are often carried out only once. This contrasts sharply with most long term manufacturing processes which are repetitive in nature.
- Because on site work stations are not permanent, specialised construction crews are required to move from one location to another after a given operation is completed. In manufacturing facilities operations are arranged to be carried out in one place.

Unlike manufacturing, construction appears to be a more complex and fragmented industry involving transient players and one where the building owner is involved in the design and construction process right from the very beginning. Nevertheless the differences between construction and manufacturing do not mean that productivity improvement techniques adopted by the latter are not applicable in the former. Rather efforts should be made to modify productivity practices in the manufacturing industry for use in the construction industry.

JIT is a relatively new concept introduced only in the early 1980s and, while the benefits of improved productivity are obvious in JIT manufacturing, some modifications would be required to the concept for it to be applied to the construction industry. Construction activities are not only complex, the industry is also fragmented with different professionals, suppliers and contractual arrangements. Nevertheless, where construction is concerned, the JIT concept appears gainfully relevant for the building materials industry with process or repetitive manufacturing processes.

Although it may be more difficult for the JIT philosophy to be applied in one-off building projects, this may not be so where industrialisation which makes use of prefabricated components has been adopted for construction. The use of standard components in fabrication, which mirrors the repetitive process found in manufacturing, therefore offers an environment which is conducive to the application of the JIT concept. However, it needs to be acknowledged that not many clients like buildings which are similar both aesthetically and functionally. Hence although there are advantages associated with prefabricated buildings, prefabrication is unlikely to be popular when building owners have preference for unique structures even at the expense of buildability.

6 Plant

Plant as used by the builder or contractor in construction work may be divided into two classes:

1. Small and non-mechanical plant and tools.
2. Power-driven mechanical plant.

Small and non-mechanical plant and tools consists of such plant as scaffolding putlogs, tubes, barrows, tarpaulins, picks, hoardings, shoring, ladders, scaffold boards, hand hauling tackle, endless tackle, hand rollers, etc.

The cost of such plant is allowed for in estimating either by adding a percentage addition to the estimated cost of the work, this percentage being variable, depending upon the type and value of work carried out, or alternatively, may be dealt with in the Preliminaries Section to which the estimator is referred.

Power-driven mechanical plant consists of such plant as lorries, backhoe loaders, concrete mixers, compressors, cranes, excavators, dumpers, tractors, rollers, etc. The number of hours the plant works per annum varies considerably for different types of plant. A roller for instance working an average of 1800 hours per annum may last for twelve years, while a lorry working an average of 2200 hours per annum might be worn out in five years. The capital depreciation of the roller is, therefore, only $8\frac{1}{2}$ per cent while that of the lorry is 20 per cent.

Recording results and estimating data for power-driven mechanical plant working costs per hour

It should be noted that quite apart from the allowance for depreciation, which is a mathematical calculation, such elements of the cost as repairs and renewals and the fuel and oil consumed by various plants per hour can only be obtained from recorded results, and it is from such records that data for estimating the total working cost of plant per hour are obtained.

These recorded results are obtained in the following way:

A log book is kept with each item of plant which is entered up daily by the operator, who records the following:

1. The number of hours the plant works per day.
2. The number of hours the operator works per day.
3. The amount of fuel put in the tank per day.
4. The amount of lubricating oil put in the engine per day.
5. The amount of grease used per day.
6. General remarks such as minor repairs carried out.

In connection with (1) and (2) the hours worked by the operator may be more than those worked by the plant since the operator may spend time in carrying out minor repairs, oiling and greasing, etc.

Table 6.1 shows a typical plant record sheet which is both a time sheet for the plant and operator hours worked and also a record sheet, showing the fuel and lubricants used, together with a record of any minor repairs etc. carried out.

Table 6.1 Typical plant record sheet

Mechanical Excavator No. 7						*Driver:* N.J. Noel
Contract: Gednol By-pass						*Week ending:* 14th April 1995

	Hours worked		Fuel and lubricants put in machine per day			
Date	*By operator*	*By plant*	*Diesel oil litres*	*Lubricant oil litres*	*Grease kg*	*Remarks*
9/4/95	9	8½	36	1	–	Machine began work on contract
10/4/95	9	8½	27	0.75	0.25	–
11/4/95	9	8½	27	1	0.25	–
12/4/95	6	6	36	0.75	0.25	Rain stopped work for 2 hours
13/4/95	9	8½	27	1	0.25	–
14/4/95	4½	4	27	0.75	–	Choked fuel pipe, stoppage ½ hour
Total for weeks	46½	44	180	5.25	1	

From such records the following are derived:

1. The fuel consumed by the plant per working hour.
2. The lubricating oil consumed by the plant per working hour.
3. The grease consumed by the plant per working hour.
4. The cost of minor repairs and renewals.

Note: Regarding (4) the cost of major repairs and overhauling is obtained from records kept by the plant engineer responsible who records the fitters' hours, replacement parts, etc. involved in carrying out the repairs.

Tabulating plant hire rates per hour

It is customary for firms who own plant to tabulate the working cost per hour or in other words, the plant hire rate of other various plants on a record sheet. These hire rates are then available for the following:

1. Estimating purposes, the estimator having before him the plant hire rates for the various items of plant on which to build up the estimate for any work being tendered for.
2. The hire rates for the plant chargeable to a contract on which it is used.

The plant hire rates on which the estimator bases the estimate are thus the actual rates at which this plant is hired out to the contract on which it is being used.

It is usual to record these plant hire rates at so much per hour or per week, the rates tabulated allowing for the following:

- The capital cost of the plant.
- Depreciation.

- The cost of repairs and renewals.
- The cost of insurance and any licence cost in connection with the plant.

The wages of the operator and the cost of the fuel and lubricants consumed per hour are not usually recorded on those sheets as these are variable.

In assessing the total working cost of the plant per hour, the estimator must add the driver's wages and the cost of the fuel and lubricants consumed per hour.

A typical plant hire table is shown below which gives much useful information as well as the working cost or hire rate of the plant per hour, exclusive of the driver's wages, fuel and oil.

In the table, depreciation has been allowed for on the straight line method and the cost of repairs and renewals, insurances, licences, etc. is also allowed for.

Table 6.2 Typical plant cost record sheet

Messrs Nolged Ltd.
Builders and Contractors
Marshal Road
London W8

										Plant Hire Rates	Date: . . . .

Plant		*Cost of the plant £*	*Estimated working hours per annum*	*Estimated life in years*	*Plant cost per annum £*	*Plant cost per hour £*	*Cost of repairs and renewals per hour £*	*Cost of insurance per hour £*	*Cost of licence per hour £*	*Working cost or hire rate of plant exclusive of operator's wages and the cost of fuel or lubrication per hour £*
No.	*Type*									
7	3 tonne dumper	9 900	2000	3	3000	1.50	0.15	0.25	0.100	2.15
3	Backhoe loader	32 000	2000	5	5000	2.50	0.25	0.75	0.025	3.53

Three methods of dealing with plant depreciation are shown:

1. The straight line method.
2. The interest on outlay and reserve fund method.
3. The depreciation on the written down value method.

1. The straight line method

1. The average number of hours the plant will work per annum.
2. The number of years the plant will work in a really efficient manner with little spent on it in the way of maintenance costs.
3. The residual value of plant at the end of the period of years of really efficient working.
4. The total years of life of the plant.

Example.

Capital cost of plant £6000

Period	Amount
First Year	£2000
Second Year	£2000
Third Year	£2000
	£6000
Credit estimated residual value at end of third year	£1800
	£4200

$$\text{Cost per annum} = \frac{£4200}{3 \text{ year}} \qquad = £1400$$

Average hours worked per annum   =   1800

$$\text{Cost } \frac{£1400}{1800} = £0.777 \text{ per hour}$$

2. The interest on outlay and reserve fund method

Period	Interest on capital outlay at 5% annum		Interest on reserve fund at 5% per annum
First year	5% on £100	= £ 5.00	–
Second year	5% on £100	= £ 5.00	5% on £33.33 = £1.667
Third year	5% on £100	= £ 5.00	5% on £66.67 = £3.333
		£15.00	£5.000
Credit interest on reserve fund		£ 5.00	
		£10.00	

$$\text{Average interest costs per annum} = \frac{£10}{3 \text{ years}} = £3.333$$

$$\text{Annual cost per annum} = (£100 \div 3) + £3.333$$

$$= £36.667$$

Note: By referring to the table shown below it will be seen that the divisor to use for a five year period over which the capital is spread with interest at 5 per cent per annum is 4.329.

$$\text{Annual cost of the plant per annum} = \frac{£6000}{4.329}$$

$$= £1387$$

Hours worked per annum   = 1800 h

$$\text{Cost per working hour allowance for depreciation} = \frac{1387}{1800}$$

$$= £0.771 \text{ per hour}$$

DIVISORS FOR TRANSFORMING CAPITAL COST OF PLANT TO ANNUAL VALUES FOR
CAPITAL OUTLAYS SPREAD OVER FROM 1 TO 10 YEARS AT RATES OF INTEREST RANGING
FROM 3 TO 6 PER CENT

DEPRECIATION BY THE INTEREST ON OUTLAY AND RESERVE FUND METHOD

Period of years over which capital outlay is spread	Interest per annum on capital outlay and reserve fund repayments					
	3%	3½%	4%	4½%	5%	6%
1	0.971	0.966	0.962	0.957	0.952	0.943
2	1.913	1.900	1.886	1.873	1.859	1.833
3	2.829	2.802	2.775	2.749	2.723	2.673
4	3.717	3.673	3.630	3.588	3.546	3.465
5	4.580	4.515	4.452	4.390	4.329	4.212
6	5.417	5.329	5.242	5.158	5.076	4.917
7	6.230	6.115	6.002	5.893	5.786	5.582
8	7.020	6.874	6.733	6.596	6.463	6.210
9	7.786	7.608	7.435	7.269	7.108	6.802
10	8.530	8.317	8.111	7.913	7.722	7.360

3. The depreciation on the written down value method

It will be assumed that a lorry has been purchased costing £20 000, its years of life being assessed at five years. Allowing depreciation at the rate of 25 per cent per annum, draw up the depreciation table for the vehicle.

	£	Depreciation £
Capital cost of plant	20 000.00	
Allow 25 per cent depreciation for first year	5 000.00	
		5 000.00
Value of plant at end of first year	15 000.00	
Allow 25 per cent depreciation for second year	3 750.00	
		3 750.00
Value of plant at end of second year	11 250.00	
Allow 25 per cent depreciation for third year	2 812.50	
		2 812.50
Value of plant at end of third year	8 437.50	
Allow 25 per cent depreciation for fourth year	2 109.38	
		2 109.38
Value of plant at end of fourth year	6 328.12	
Allow 25 per cent depreciation for fifth year	1 582.03	1 582.03
Residual value of plant at end of fifth year	£ 4 746.09	£ 15 253.91

Depreciation rates would vary depending on the use and condition of the vehicle.

Steel scaffolding

Work carried out by bricklayers, painters, plasterers, etc. working from ground or floor level can only be carried out to a certain height, after which scaffolding has to be erected in order that they may reach the work.

The extra labour cost involved in carrying out work at height is allowed for by the use of multipliers, but the cost of the scaffolding must be allowed for in addition.

It should be noted that a number of trades may use the same scaffolding to carry out their work, but from an estimating point of view it is usual either to provide for scaffolding costs under one trade or another, this being done by allowing for the external scaffolding under Masonry and internal scaffolding under Internal Finishes or pricing the scaffolding as an item in Preliminaries.

In the event of scaffolding having to be erected for the specific purpose of carrying out a certain item of work, such as pointing or repairs to existing brickwork, painting at height, etc. the scaffolding should then be charged wholly to the work for which it is required.

Data are shown from which to estimate the cost of steel scaffolding for bricklayers and plasterers, and in this connection the cost of hauling the scaffolding should be borne in mind, together with its hire rate if the scaffolding is hired from a plant hire firm who deal in such a commodity.

Steel scaffolding for bricklayers

Putlog scaffolding. The data shown for putlog scaffolding are for scaffolding with 1.13 m working platforms at 1.30 m lifts, the standards being at 2.25 m centres.

Independent scaffolding. The data shown for independent scaffolding are for scaffolding with 1.13 m working platforms at 1.30 m lifts and standards at 2.25 m centres.

Table 6.3 Steel scaffolding for bricklayers

Note: In each case 112 mm plumbing space is left, and the hours shown allow for the necessary ties, braces, access ladders and scaffolding boards and also for fixing hand-rails and toe-boards to the third lift and over.

	Unit	Labour hours per square metre		
		Erect scaffolding	Dismantle scaffolding	Erect and dismantle scaffolding
Putlog scaffolding	m²	0.53	0.22	0.75
Independent scaffolding	m²	0.65	0.26	0.91

Table 6.4 Multipliers for scaffolding carried out at heights in excess of 12 m

Height of work above ground in metres	Labour hour multipliers
0–12	1.00
12–18	1.10
18–24	1.25
24–30	1.45

Steel scaffolding for plasterer, etc.

Table 6.5 Scaffolding for plasterer, etc.

Notes: 1. Putlog scaffolding with 1.35 m working platforms at 2.25 m lifts and standards at 2.25 m centres.
2. Independent scaffolding with 1.35 m working platforms at 2.25 m lifts and standards at 2.25 m centres.
3. Birdcage access scaffolding with uprights at a maximum of 3.0 m centres and 1.95 m lifts. Boards laid all round to each lift.
4. In each case the hours shown allow for all ties, bracing, access ladders and scaffold boards, also for the fixing of hand-rails and toe-boards to the second lift and over.

		Labour hours per unit		
	Unit	*Erect scaffolding*	*Dismantle scaffolding*	*Erect and dismantle scaffolding*
Putlog scaffolding	m²	0.36	0.14	0.50
Independent scaffolding	m²	0.42	0.17	0.59
Birdcage access scaffolding	m³	0.07	0.025	0.095

Table 6.6 Multipliers for scaffolding carried out at heights in excess of 12 m

Height of work above ground in metres	*Labour hour multipliers*
0–12	1.00
12–18	1.10
18–24	1.25
24–30	1.45

7 Preliminaries

Preliminaries is the work involved in administering a project and providing general plant facilities and site based services.

Preliminaries are generally site charges but may include additional head office costs, finance charges, special insurances and conditions of contract. These charges may be included in the preliminaries but some contractors include them in the rates.

Preliminaries may vary by as much as 7 per cent to 15 per cent of the total contract sum dependent on the site conditions and what a contractor actually allows and prices in the preliminaries.

The proportional cost method was sometimes used for interim valuation purposes but contractors soon realised that the increasing high costs in setting up the site organisation and the recovery by monthly proportional costs was far less than the actual investment cost expended at the beginning of a contract. The proportional method resulted in severe drain on cashflow and placed many contractors in financial difficulties. Therefore preliminaries are now placed into two classes:

1. *Fixed charge.* This is for work the cost of which is to be considered as independent of duration.
2. *Time related charge.* This is for work the cost of which is to be considered as dependent on duration.

The Standard Method of Measurement of Building Works Seventh Edition requires preliminaries to be measured as fixed and/or time related charges and to more accurately reflect the costs to the contractor.

Standard check lists, as shown in the publication *Code of Estimating Practice* by the Chartered Institute of Building for pricing the main preliminaries section of the bill of quantities, can be useful in planning site expenditure and in giving vital information as to the finance needed to start a contract. Naturally the more detailed the preliminaries the greater the amount of information available for both the site and senior management. There is a danger in over standardisation and too much detail, as with any other form of documentation, and care must be taken to avoid information becoming too sophisticated thereby causing unnecessary work for the estimator.

From the standard preliminary check lists, the costs can be abstracted for use in the interim valuations and incorporated into the final valuation.

The proper and timely reimbursement of preliminaries may help keep the building industry from that most doubtful of distinctions of being top of the league for companies that go into liquidation. Bad management may be the cause in some cases, by far the largest percentage results from a lack of cash flow control.

There are other areas which affect cash flow, mostly concerned with efficiency, and these can be mitigated by the use of sensible procedures:

● To keep increased costs under fluctuations contracts up to date with the monthly valuations.
● For variations to contract to be assessed accurately and included in each valuation. They should not be allowed to accumulate over a period of several valuations.
● To clear Daywork sheets for payments monthly.
● To ensure discounts and retention monies are properly claimed against the contractor's own nominated sub-contractor or supplier.

- For collection of monies when properly due.
- To ensure that all loss and expense items under clauses of the JCT conditions of contract are promptly dealt with and paid for.
- For negotiation with a supplier for special terms.

The Preliminary Bill

In the Preliminary Bill are included those items of works temporary or otherwise that are not included in the Measured Bills.

The Preliminaries and General Conditions that are included and priced for in the Preliminaries Bill are:

1. Project particulars
2. Drawings
3. Site and access
4. Description of the works
5. Firms employed direct or executed by Statutory Undertakers
6. Form of Contract
7. Conditions of Contract, Insurances and Appendix
8. Performance bond
9. Supervision/Staffing/Site management
10. Labour
11. Materials
12. Mechanical plant
13. Safety, health and welfare
14. Access and scaffolding
15. Safeguarding the works
16. Water for the works
17. Lighting and power
18. Temporary road
19. Site accommodation
20. Temporary telephones
21. Temporary screens and temporary roofs
22. Temporary fences, hoardings, etc.
23. Name board
24. Setting out, surveying and testing of equipment
25. Protection, drying and cleaning the works
26. Notices and fees to Local Authorities and Public Undertakers
27. Deleterious materials
28. Employer's requirements or limitations
29. CDM Regulations 1994

1. *Project particulars*

Name, nature and location.
Names and addresses of Employer and Consultants are provided.

2. *Drawings*

List of drawings that form part of the Contract.
Procedure for shop drawings.

3. *Site and access*

Location of site and access are described with limitation on working times and access.

4. *Scope and description of the works*

General description of the works to be undertaken and the total gross floor area provided.

A general description is given of the work, access to the site, etc. and from it the estimator can form some idea of the nature of the contract and attendant difficulties, if any. The site of the works should be visited and careful note made of the prevailing conditions, insofar as they affect the tender. Allowance should be made for any difficulties peculiar to the contract which make it other than normal and straightforward. The question of access should not be forgotten, as the builder or contractor may have to go to the expense of laying temporary roads. Plans and specification should be carefully perused and the nature of the ground including contamination carefully assessed, as the latter greatly affects the cost of any excavation involved.

5. *Firms employed direct or executed by Statutory Undertakers*

Details of crossovers by local authority and consideration of statutory services on the site.

6. *Form of Contract*

Details of the form of contract e.g. Intermediate Form of Building Contract with Quantities 1984 Edition with amendments 1:1986, 2:1987, 3:1988, 4:1988, 5:1989, 6:1991, 7:1994, 8:1995 and 9:1995 issued by the Joint Contracts Tribunal.

Supplementary Conditions C (Tax Fluctuations) will apply.

Details of the contract executed under seal or not.

7. *Conditions of Contract, Insurances and Appendix*

The contract documents to provide a complete list of relevant conditions of contract.

Under this item allowance can be made by the estimator for any item which merits a financial allowance which is not allowed for elsewhere in the Preliminary Bill.

Provide details of Joint Names Policy.

Complete required information in Appendix.

8. *Performance bond*

Details of the terms of the bond or guarantee shall be set out. The obtaining of such bond or guarantee and the cost is to be at the expense in all respects of the Contractor. The approved bond or guarantee shall be presented prior to the first payment under this Contract.

9. *Supervision/Staffing/Site management*

The contractor is to allow in the Preliminaries for on and off site management including the Contractor's person-in-charge.

G. Solomon, *CQS* October 1993, states this can include head office, site based and visiting personnel. Three main site-based groups can be identified as:

A. *Works Management.* The works management set up depends on the scale, complexity and type of construction.

These factors have to be reviewed against the proportion of the contract which is to be sublet and so directly supervised by the sub-contractors management.

The contractor allocates staff ranging from a visiting general foreman to a team of agents, sub-agents and engineers dependent on the scale and complexity of the project.

In pricing, allowance will be made for salary costs, payroll burdens, overhead costs and other charges such as company cars. The duration of the supervision requirement will be used to extend the total cost of supervision for inclusion in the tender.

B. *Project control.* The control of cost and time is of increasing importance, particularly where margins are low. In recent times there has been a marked increase in the contribution of both planning and cost control staff.

The charge for these staff is calculated in the same way as for work management using the duration analysed by the planning staff as the time base.

C. *Administrative staff.* The use of administrative staff has given us a response to the need to use modern business systems and new legislation, and on major projects they result in an appreciable on-cost.

The charge for staffing a contract falls within the range of 20 per cent to 30 per cent of the preliminaries for the majority of contracts analysed.

It can be seen that the cost of staffing is an important and large element of a contractor's tender which will require scrutiny during the tender adjudication.

10. *Labour*

The Contractor is to allow in his prices for:

- Importation of labour including travelling time, subsistence, etc.
- Incentive schemes and overtime working
- Guaranteed time
- National insurance
- Annual and Public Holidays
- Sick pay
- Any other disbursement arising from employment of labour

11. *Materials*

Preambles and descriptions of materials, goods and workmanship given in any one section or trade shall apply throughout these Bills of Quantities unless otherwise described.

Details of any samples required.

12. *Mechanical plant*

The Contractor is to provide all necessary plant, tools and vehicles for the proper execution of the works in the agreed time for completion of this Contract.

Plant consists of:

- Large items of mechanical plant, such as mechanical excavators, concrete mixers, lorries, etc.
- Small plant and tools such as barrows, ropes, steps, trestles, picks, mattocks, blocks and tackle, scaffold poles, scaffold boards, putlogs, hoses, saws, pails, lead pots, ladles, etc.

Large items of mechanical plant may be allowed for in the estimated cost of the unit of work, the plant hours per Unit being shown in the tables for use in estimating.

Small plant and tools are allowed for by making a lump sum or percentage addition to the cost of the work, the percentage addition depending on the class and value of the work carried out.

G. Solomon states, *CQS* October 1993, that the use of mechanical plant is continually developing at the expense of the employment of labour.

In future this will involve the use of some of the very sophisticated robotic equipment which is currently being developed. The type of plant used for a contract is dependent upon the

- Nature of the construction project.
- Site conditions.
- Equipment that the contractor has available.

The plant priced within the preliminaries is predominantly general site plant, such as cranes, batch mixers, dumpers and such like.

For the majority of contracts analysed the cost of mechanical plant was in the range of 17 per cent to 26 per cent of the preliminaries. The cost of mechanical plant is calculated by assessing the needs of the contract and producing a histogram for its use.

Decisions will be made regarding the availability of equipment and in particular whether owned or hired plant is employed.

13. *Safety, health and welfare*

The Contractor is to provide suitable safety, health and welfare measures and amenities to comply with all the current Statutory Regulations and the Code of Welfare conditions of the National Joint Council for the Building Industry.

14. *Access and scaffolding*

Access to the works is heavily influenced by the design of the finished project, but the decision on which method to employ rests almost entirely with the contractor. Scaffolding is not the only means of access available.

Use is also made of hydraulic suspended or climbing platform and towers which may significantly reduce costs from traditional methods.

The choice and cost of access systems are influenced by:

- Site layout.
- Shape of the structure to be constructed.
- Nature of the construction frame and elevation.

This is an important area where a contractor's access decision may have a bearing on his success in tendering for work. In the majority of analyses considered, access costs accounted for 15 per cent to 22 per cent of the preliminaries cost.

It is noticeable that in comparison with international construction costs, scaffolding is a far more significant cost burden in the UK than elsewhere. This is an important factor affecting the competitiveness of UK construction.

Construction can, by employing an engineer to assess the access requirements during the tender period, produce economic solutions by taking into account the nature of the access required and the project parameters.

15. *Safeguarding the works*

The Contractor is to safeguard the works, materials and plant against damage and theft and all watching and lighting for the security of the works and the protection of the public and the Employer.

This may be a costly item on certain work and, if required, must be allowed for under this item in the 'Preliminary Bill'. It may mean one or a combination of:

1. Watching and lighting including CCTV system.
2. Artificial light supply, such as in tunnels, night work, etc.
3. Control of traffic by electronically controlled signals.

The cost should be estimated by assessing the time for which any one or more of the above items are necessary, and it should be noted that not only are wages involved, but also fuel. Having assessed the cost, a percentage addition is then added to allow for establishment charges and profit, the total amount being inserted in the Preliminary Bill under the appropriate item.

16. *Water for the works*

The Contractor is to provide water for the works and pay all charges for temporary arrangements.

In estimating the cost of water for the works, two sources of supply have to be taken into consideration:

1. Water available from existing mains adjacent to the work.
2. Water which has to be obtained from an outside source.

Water available from existing mains. Water obtained direct from the mains is charged for by the authority concerned, in one of two ways:

1. By charging for it as a percentage of the cost of the work, this charge approximating 0.35%.
2. By charging for it on a measurement basis, the basis being per 1000 gallons, this charge should properly be ascertained from the Local Water Authority.

The usual method of charging for water by the authority concerned is on a 'percentage of the cost basis', and from an estimating point of view this offers a very simple and direct way of allowing for its cost.

In charging for the water on a measurement basis the cost of installing and removing a meter, together with meter rental, has usually to be borne by the builder or contractor. The meter rental is of a very nominal amount. On whatever basis mains water is charged for, its cost is best allowed for as a percentage of the cost of the work.

Any cost incurred in connection with temporary plumbing, the laying and removing of temporary water services, water storage, etc., should not be overlooked. For data from which to estimate such cost the estimator is referred to the chapter on plumbing.

The method of assessing the percentage addition to the estimated cost of the work to allow for the cost of water is as follows:

Assuming the charge made by the local authority is on a percentage basis, this being 35p per £100, the percentage addition to the estimated cost of the work then is:

$$\frac{35p}{£100} \times 100 = 0.35\%$$

This 0.35 per cent allowance for water may then be included under Preliminaries.

Note: In those cases where water is charged for on a measurement basis the cost of water is also best allowed for in this manner, and for purposes of estimating, a round percentage addition added to the estimated cost of the work of 0.40 per cent is commonly adopted. This is more than sufficient to allow for the cost of water normally charged for on a measurement basis through a meter.

Water not available from the mains. In constructing work in open country or outlying districts it often happens that no water is available from existing mains, and in such cases a suitable source of supply has to be found and arrangements made to convey the water to the site. Such sources of supply may consist of rivers, wells, springs, distant mains, etc. and the water is usually conveyed in suitable containers mounted on lorries, water carts, etc.

The cost of water conveyed to the site in this manner is considerably more than that taken direct from existing mains adjacent to the work, though the water itself may be free of charge.

The cost of water so hauled is best allowed for on a Lump Sum basis, this being computed by assessing the following:

● The total gallons of water required for the contract in question.
● The cost of conveying the water to the site.

The assessment of the total amount of water required may be obtained from the quantities of the individual items of work for which water is required, due allowance being made for waste, water used for other purposes, such as washing down plant, water for site offices, etc. In computing the amount of water required it is wise to make ample provision.

For estimating purposes, the amount of water required for those items of work which are chiefly responsible for its use in quantity may be taken to be as follows:

Concrete	=	80 gallons per cu. metre
Brickwork (one brick thick)	=	15 gallons per sq. metre
Rendering in cement mortar 12.7 mm thick	=	3 gallons per sq. metre
Rendering in cement mortar 19 mm thick	=	4 gallons per sq. metre
Rendering in cement mortar 25 mm thick	=	5 gallons per sq. metre
Plastering 5 mm thick	=	4 gallons per sq. metre
Plastering 12.7 mm thick	=	6 gallons per sq. metre
Plastering 19 mm thick	=	8 gallons per sq. metre

Note: The above quantities allow for normal waste, washing down plant, etc.

Example. Estimate the cost of the water required on a contract, the water having to be pumped into a tank mounted on a lorry and hauled to the site. The source of supply is 2 km distant from the site and the water itself is free of charge. The water is pumped by a small petrol-driven pump which is serviced by the lorry driver. For purposes of this example the following assumptions are made:

1. The quantity of water hauled per trip is 400 gallons.
2. The time taken by the pump to pump 400 gallons is assessed at 6 minutes, i.e. 0.10 hours.

Note: This is also the time the lorry stands while the tank is filled.

3. The hire rate or working cost of the pump, inclusive of fuel and oil, is £2 per hour.
4. The hire rate or working cost of the lorry, including driver, fuel and oil, is £20 per hour.
5. The total quantity of the items of work in connection with which water is required is as follows:

   Brickwork, 100 cu. metres
   Concrete, 50 cu. metres
   12.7 mm plastering, 150 sq. metres
   19 mm plastering, 1000 sq. metres

The estimated cost of the water required is built up in the following manner:

The total quantity of water required. The total quantity of water required, allowing a reasonable and safe margin, is:

1B Brickwork – 400 sq.m at 15 gallons per sq. m	=	6 000 gallons
Concrete 50 cu.m at 80 gallons per cu. m	=	4 000 gallons
12.7 mm plastering 150 sq.m at 6 gallons per sq. m	=	900 gallons
19 mm plastering 1000 sq.m at 8 gallons per sq. m	=	8 000 gallons
Allow for other works, say	=	6 100 gallons
Total water required		25 000 gallons

Cost of pumping and hauling the water per 1000 gallons. The amount of water hauled per load is 400 gallons. Lorry hours per load = time standing while being loaded plus travelling time. Referring to Table 12.11, the travelling time taken by a lorry to haul to a distance of 2 km is 0.27 hours per round trip.

Total lorry hours per load is:

Lorry standing while water is pumped	= 0.10 hours
Lorry travelling	= 0.27 hours
Lorry hours per load	0.37 hours

The cost of pumping and hauling 400 gallons, therefore, is:

Pump hours – 0.37 hours at £2 per hour	=	0.74 per load
Lorry hours – 0.37 hours at £20 per hour	=	7.40 per load
		£ 8.14 per load

$$\text{Cost of water on the site per 1000 gallons} = 1000 \times \frac{£8.14}{400} = £20.35$$

Say £20 per 1000 gallons

Note: In this case the pump hours are taken as being the same as the lorry hours, it being assumed that one lorry only is used on the haulage of water.

Total cost of water required for the works = 25 000 gallons at £20 per 1000 gallons = £500.

Note: The estimator will note that the cost per 1000 gallons, £25, is very much more than in the case of water obtained direct from adjacent mains from the local water authority.

17. *Lighting and power*

The Contractor is to provide lighting and power for the works and for lighting to hoardings, etc. Provide lighting for sub-contractors office and power for their small tools.

18. *Temporary road*

The Contractor is to allow for providing and maintaining all necessary temporary roads, and remove on completion of the works.

19. *Site accommodation*

The Contractor is to allow for providing and maintaining temporary site accommodation for Architect and Contractor including removal on completion.
Site accommodation can consist of:

- Offices
- Canteen and catering facilities
- Toilet and washing facilities
- Locker rooms
- Storage facilities
- Garaging, workshop and hardstandings

The need for site accommodation and other facilities will depend on:

- Number of staff
- Size of labour force
- Size of mechanical plant fleet
- Shape and nature of site
- Client's requirements
- Sub-contractors need where provided by the contractor.
- Whether there are any of the permanent works, existing or redundant facilities that can be used for the duration of the contract. According to G. Solomon, *CQS* September 1993, the cost of site accommodation accounted for between 8 per cent and 14 per cent of the cost of preliminaries.

The pricing of this item needs to consider the cost of accommodation systems available, current legislation and the space needs of the project. As a guideline for calculating office space an average of 10 square metres, which includes circulation space, should be allowed per member of staff

20. *Temporary telephones*

The Contractor is to allow for providing and maintaining temporary telephone facilities.

21. *Temporary screens*

The Contractor is to allow for providing all temporary screens to prevent dust, debris, etc.

22. *Temporary fencing, hoardings, etc.*

The Contractor is to allow for providing all temporary fencing and hoardings.
Cost incurred in connection with fencing and hoardings is best allowed for by assessing the cost involved, inclusive of any fees payable. To this estimated cost is then added a percentage addition to allow for Establishment Charges and profit, the total being entered as a Lump Sum under this item in the Preliminary Bill.

23. *Name board*

The Contractor is to allow for providing, maintaining and clearing away a name board to display the title of this contract and the names of the Employer, the Consultants and the Contractor.
No other advertisements or trade signs will be permitted on the site.

24. *Setting out*

The Contractor is to allow for setting out, surveying and testing of equipment.

25. *Protecting, drying and cleaning the works*

The Contractor is to allow for protecting the works from inclement weather, providing all temporary screening to openings, providing for all costs in connection with drying and controlling the humidity of the works, protecting the whole of the works and removing all rubbish.

The Contractor is to allow for cleaning the Works and is best carried out periodically and refers particularly to building works where the final clearing may be a considerable item. In such cases the cost of this may be estimated and a Lump Sum inserted in this item in the Preliminary Bill to cover such cost, after having added to it a percentage to allow for Establishment Charges and profit. In the case of civil engineering works it is not usual to allow anything under this item.

26. *Notices and fees*

The Contractor is to allow here for the notices and fees.

Building fees are best allowed for as a Lump Sum, this being done by assessing their cost and adding thereto a percentage addition to allow for Establishment Charges and profit. This amount is then included in the Preliminary Bill under this item.

27. *Deleterious materials*

The Contractor is not to use or permit to be used any of the described deleterious materials in the execution of the works.

28. *Employer's requirements*

E.g. Adjoining property, prevention of pollution, etc.

29. *CDM Regulations (1994)*

These regulations popularly known as Condam or CDM were made by the Secretary of State for Employment and came into force on 31 March 1995 to give effect to EU Council Directive 92/57/EEC on the implementation of minimum safety and health requirements of temporary or mobile construction sites.

The Regulations place duties and responsibilities on all those involved in building projects to reduce accidents and increase safety. The regulations apply to almost all construction projects excepting those where:

- Construction phase is no longer than 30 days, or
- Construction phase will not involve more than 500 person days of construction work, and
- Number of people carrying out construction work at any one time is less than 5.

All works of demolition or dismantling of a structure are subject to the Regulations.

8 *Overheads and establishment charges*

Overheads and Establishment charges may be considered under two main headings:

- Head Office charges.
- General Site charges.

and will vary considerably with different firms. They are influenced by:

- The annual turnover.
- The class of work carried out.
- The ability and drive of the administrative staff.
- The analysis as to which elements of the cost are allowed for as establishment charges.

These charges may be allowed for in one of the following ways:

- Allowing for them in the estimated labour, material and plant cost.
- Estimating their cost and allowing for them on a Lump Sum basis.
- Allowing for them by adding a percentage addition to the estimated cost of the work.

Head Office charges are usually expressed as a percentage of the total work or turnover carried out each year and include:

1. Salaries paid to the head office staff, director's fees, attendants, cleaners, etc.
2. The rental of the offices, including rates and taxes and depreciation of the office furniture and effects.
3. Incidental insurances, fire insurances, etc.

   Note: Special care should be taken regarding insurance requirements, particularly if the Employers require coverage. In such circumstances your insurers should be approached for their guidance.

4. Heat, light, power and fuel.
5. Stationery, books, postage and telephone charges.
6. Auditor's fees if the books are audited by an outside firm of accountants.
7. Staff cars.
8. Interest on working capital, loans, bank charges and retention monies. This may vary between 5 per cent and 15 per cent depending on current economic conditions.

To illustrate the method of assessing the percentage addition to allow for Head Office Charges assume that the total amount of Head Office Charges per annum is £75 000, and the Annual Turnover £1500 000

Then the percentage addition to add to the estimated cost of the work to allow for Head Office Charges is

$$\frac{75\,000 \times 100}{1500\,000} = 5\%$$

To illustrate the method of assessing the percentage addition to allow for the loss of interest on working capital and retention monies assume that a business man has an annual turnover of £1600 000, the working capital being £200 000 and the maximum retention monies assessed at £80 000.

Then: £

Working Capital   200 000
Retention Money   80 000
                                          ─────────
                                          280 000

Allow 10 per cent interest on £280 000   = £28 000

Percentage to add to the $$= \frac{£28\,000 \times 100}{£1600\,000} = \frac{28}{16} = 1.75\%$$
estimated cost of the work

This 1.75% is then included.

General Site charges are usually allowed for on a fixed and time related basis and included in the Preliminaries section of the Bills of Quantities, Schedule of Rates and Specification.

9 Tendering

The adjudication of an estimate and allowance for profit to produce a tender is the responsibility of management.

During adjudication, management should give consideration to the estimator's report. This report prepared by the estimator for management should include the following items:

1. A brief description of the project.
2. A description of the method of construction.
3. Note of any unusual risks which are inherent in the project and which are not adequately covered by the Conditions of Contract or Bills of Quantities.
4. Any unresolved technical or contractual problems.
5. An assessment of the state of the design process and the possible financial consequences thereof.
6. Note of any major assumptions made in the preparation of the estimate.
7. Assessment of the profitability of the project.
8. Any pertinent information concerning market and industrial conditions.
9. Any need for qualification of the tender or for an explanatory letter.
10. The terms of quotations from own sub-contractors which have been included in the estimate.
11. The time for which the tender is to remain open for acceptance.
12. Proposed programme.

The tender documents should be examined for the following information:

1. Is sufficient time being allowed for the proper preparation of an estimate?
2. What are the Conditions of Contract? Is it the Standard JCT Form of Contract (current editions); have any alterations been stated and reasons given?
3. Are the general arrangement drawings included?
4. Are the operating conditions defined clearly? Will further information be needed on such matters as

   - stages in which work is to be executed
   - timing
   - access to the site
   - restrictions on hours during which work may be done
   - dangerous or unpleasant conditions

5. What is the value and extent of the project and what is the main contractor's own contribution likely to be?
6. Is the design well developed or have the documents been prepared hastily?
7. Have the Bills of Quantities been prepared in accordance with the Standard Method of Measurement? If not, what method has been used?
8. If the Bills of Quantities incorporate clauses in accordance with the National Building Specification, are they given in full as recommended by NBS Ltd.

If it is found that more information is needed the architect should be asked whether it can be made available and if so at what time and in what form it will be given.

An enquiry Record Form should be completed to summarise for management all that is so far known concerning:

- the client and consultants
- the value of the project
- the Conditions of Contract

Following careful examination of the tenders the contractual risks should be considered including allowance for a Firm Price Tender. For when a firm price tender is to be given, an assessment must be made of likely variations in cost during the proposed contract period.

Records of rates of wages, other labour costs and emoluments and the cost of basic materials will indicate trends. Note should be taken of industrial negotiations and statutory measures which may affect levels of wages, the financial obligations of employers or the costs of basic commodities. Consideration must also be given to indirect charges such as tariffs, fuel and freight.

Regard must be paid to economic and political situations at home and abroad and the state of home and world markets.

The quotations of the contractor's own sub-contractors should be re-examined to ensure that they are firm price offers and that there are no risks of claims for increased costs.

Another area to be addressed at tender stage are the terms of quotations from own sub-contractors. The following conditions should be satisfied:

1. Does the work described in the quotation comply with the Bills of Quantities and the Specification?
2. Are the unit rates consistent throughout the quotation?
3. Are the rates realistic and comparable with those of competitors?
4. Is the sub-contracting organisation financially capable of undertaking the work at the sum quoted?
5. In the case of labour only sub-contractors, is the organisation *bona fide* with regard to the payment of PAYE tax, insurances, training levy, etc.?
6. Will the sub-contractor be able to meet the requirements of the construction programme?
7. Will the sub-contractor be acceptable to the construction team
8. What services are required from the main contractor?
9. Do the conditions of the quotation comply with the terms of the main contract?
10. Are there any onerous conditions applying to the quotation?
11. What are the discounts offered?
12. Is a schedule of rates included with the quotation?

All the above factors should be taken into consideration to ensure a *bona fide* tender is submitted to the client.

10 Demolitions and alterations

Demolitions

Demolition works vary considerably from one scheme to another, depending on access, type of construction, method of demolition, whether any asbestos and contaminated materials, etc. Therefore it is advisable to obtain specific quotations for each scheme under consideration.

Shoring etc.

Shoring and strutting for large openings

The requirements for shoring and strutting for the formation of large openings are dependent on a number of factors; for example, the weight of the superimposed structure to be supported, number (if any) of windows above; number of floors and the roof to be strutted, whether raking shores are required, and the depth to a load bearing surface.

Items would best be built up by assessing the use and waste of materials and the labour involved, including getting timber from and returning to yard, cutting away and making good, overheads and profit.

11 Piling

Piling is carried out in connection with both permanent and temporary work. The driving of piles which are to remain permanently in the work is usually associated with civil engineering works such as harbours, wharves, jetties, quays, sea and river defence works, etc. In such cases the piles are part and parcel of the works of construction and are left in place.

Temporary piling work is carried out as an aid to the construction of permanent work, e.g. the piling of cofferdams in which are built such works as bridge abutments, or sheet piling driven in water-bearing ground prior to excavating it, the piling being shored with walings and struts as the actual excavation is carried out. After the permanent work has been completed the temporary piling is withdrawn, it being usual to charge to the work a proportion of the material cost of the piles to allow for depreciation, together with the plant and labour cost of driving and extracting the piles and any attendant haulage costs in connection with the plant.

In recent years due to severe shrinkage in foundation clays, e.g. London basin, there has been an increasing use of piles and mini-piles in underpinning systems (see Figure 11.1).

Various factors, which must be taken into consideration when preparing an estimate, influence the cost of pile-driving.

1. The nature of the site as regards access and whether it is rough and uneven, involving surface excavation to level it for track-laying on which to run a piling frame or derrick, or level with little or no excavation required.
2. The nature of the pile-driving work which has to be carried out and whether the piles are driven with the plant set on the ground, on staging, or on barges, and if in water, whether it is tidal or non-tidal.
3. The nature of the ground in which the piles are to be driven.
4. The type of piles which are to be driven, i.e. whether *in situ* or precast concrete, steel sheet piles or timber piles.
5. The type and size of piling plant used to carry out the work.

From the foregoing the estimator will appreciate that in estimating the cost of pile-driving work, correct assessment must be made of the nature of the work to be carried out as a whole and also of the most suitable type of plant to use if the work is to be carried out efficiently and economically.

Pile Types

Piles may be classified by their method of construction, as replacement piles, or displacement piles.

Replacement piles

1. Percussion bored.
    (a) Weighted head with steel cutting edge for boring in cohesive soil.
    (b) Weighted head with steel cutting edge and flap for boring in granular soil.
    (c) Steel lining tubes with internal drop hammer.
    (d) Steel lining tubes with compressed air supply.
2. Flush bored pile.
    Using crane mounted telescopic rotary square kelly bar.

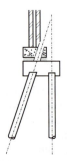

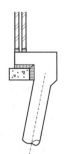
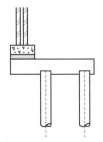

1. Pile and knuckle   2. Angle pile   3. Hammer head   4. Cantilever pile pair

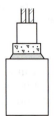
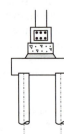
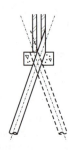

5. Pile pair and beam   6. Jack pile   7. Piles and beam with continuous beam   8. Angle pile pair

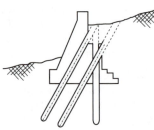

9. Traditional underpinning   9b. Beam and base underpinning   10. Reticulated stabilisation

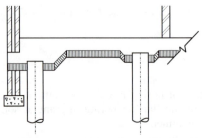
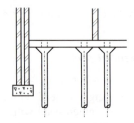

11. Cantilever piled slab with anti-heave   12. Stabilisation of existing slab

Figure 11.1 Typical underpinning systems (Courtesy of Foundation Piling Group)

3. Small diameter rotary bored pile.

Using drilling rig mounted on crane or lorry with auger type or drilling bucket to suit conditions.

4. Large diameter Rotary bored pile.

Using drilling rig mounted on heavy duty crane power unit with auger type or drilling bucket to suit conditions.

Displacement piles

1. Pre-formed.

(a) Timber piles, which should always be treated with a suitable preservative before being driven, with mild steel shoe and fixing straps.

(b) Concrete piles with rig mounted on crane. Piles are percussion driven using a drop or single acting hammer (West's Hardrive Precast Modular Pile).

(c) Steel piles which may be standard steel sheet pile sections percussion driven. Used mainly in connection with marine structures.

or

Steel screw piles. Rotary driven and used for dock and jetty works where support at shallow depths in soft silts and sands is required.

or

Steel tube piles fabricated out of strip with butt welded helix and internal drop hammer acting on concrete plug (BSP Cased Pile).

2. Partially preformed piles.

Precast concrete and *in situ* concrete composite piles. These percussion driven piles are used on medium to large contracts where bored piles would not be suitable owing to running water or very loose soils (West's Shell Pile)

3. Driven *in situ* piles.

Used on medium to large contracts as an alternative to pre-formed piles particularly where final length of pile is a variable to be determined on site (Franki Driven Insitu Pile).

4. Cast *in situ* piles.

An alternative to the driven in situ piles (Vibro Cast Insitu Pile).

Pile driving plant

Pile-driving plant consists of:

1. Piling frames.
2. Hanging leaders.
3. Pile-driving hammers.

The type of plant and hammer selected to carry out the work depends on the nature of the work and the type of pile which has to be driven.

Piling frames are made in two types, heavy or light. They are obtainable in various heights and are made in both timber and steel. The frames are made in standard sizes, the leaders also being standard, so that a change from one type and size of hammer to another is possible.

Steel frames are commonly used, these being mounted on swivelling travelling wheels, fitted with screw jacks, so that the frame leaders to which the pile is attached and which guide the pile may be plumbed to a true vertical position. Adjustable types of frame are also available, and with this type of frame the leaders may be canted to an angle, thus permitting the piles to be driven on the rake.

Hanging leaders are constructed of channels suitably held together so as to form leaders for the pile. They are suspended from the jib of a crane or derrick and are particularly useful for driving stage or sheet piles.

Pile hammers are of two main types:

1. (a) Drop hammers in which the hammer is raised off the head of the pile and allowed to drop on to it, thus driving the pile into the ground.
   (b) Single acting hammers in which the hammer is raised by steam or compressed air sliding up or down a fixed piston on guide legs.
2. (a) Double-acting hydraulic hammers which rest on the head of the pile and keep up a rapid succession of short blows which vibrate the pile into the ground.
   (b) Diesel hammers powered by a measured amount of fuel.

The blows administered by drop hammers are heavier than those of a hammer that rests on the head of the pile, and are comparatively slow and ponderous. When using a drop hammer the heads of the piles should be protected by a pile-helmet to prevent damage to the head of the pile. The maximum height of drop should not be excessive, but this can be controlled so that the pile can be started gently with a light blow before the maximum height of the drop is given. This type of hammer is very suitable for driving concrete or timber piles.

Drop hammers may simply take the form of a weight attached to the leaders by toggle bolts so that they can move freely up and down them, being raised by a hand or a power-winch to the requisite height, whereupon they are released by a hand-controlled trip gear arrangement, allowing them to fall on to the head of the pile.

Power-driven hammers of this type are available. Power hammers may be actuated by steam or compressed air.

Double-acting hydraulic and diesel hammers are suitable for most classes of piling work. The advantages of this type of hammer are:

- It keeps up a rapid succession of blows on the head of the pile so that it may be looked upon as vibrating the pile into the ground and not driving it in, as is the case with drop hammers.
- It requires very little head room in which to work.
- It can be hung from the jib of a derrick or crane with or without hanging leaders.

Pile extractors. Temporary piling work is often used to facilitate the construction of permanent works of construction, and in such cases the piles are withdrawn after the permanent work is completed. This is done in one of three ways:

1. By a direct pull from a winch. This is only successful where the piles are short and are not hard-driven.
2. By hydraulic jacks acting on a fixed surface around the pile. This is a slow method and is only used when the piles are hard set and have been in the ground for a number of years.
3. By double-acting and diesel hammers fitted with extracting gear with special jaws and dies to grip the piles.

Method 3 is that most generally used. The extractor works on the principle that it acts as a hammer which hits upwards and not downwards into the ground so that the pile is, in effect, knocked out of the ground. The effect of the rapid and successive blows is to set up a vibration which helps to break the frictional resistance of the ground around the piles and also, in the case of sheet piling, the friction in the interlocks.

Selecting the most suitable plant for the work

To drive piles efficiently and economically it is essential that the correct types of piling plant and hammers are used. The onus of deciding which is the most suitable plant to use for a specific contract is to a large extent with the estimator, for he must decide on how the work should be carried out and what plant should be used and, having done so, build up the estimate of the cost. In order to guide the estimator, suggested combinations of plant are shown for carrying out piling work under various conditions. It is emphasised that these should be looked upon as a guide as to which type of plant would be suitable. The estimator must use judgement and take into consideration the prevailing site conditions in assessing the type of plant it is considered would best suit the work under consideration.

1. *Piling in level ground* might take the form of driving piles in bad ground for building foundations etc. For carrying out such work a piling frame mounted on a lorry or running on rail track would be suitable, the size of frame and weight and type of hammer being selected to suit the piles.
2. *Piling in rough ground.* On rough, uneven sites a piling frame running on track or rollers would not be economical, as extensive site levelling might be necessary to lay the track on which to run the frame. A more economical plant set up would be a crane fitted with hanging leaders, the crane moving under its own power on caterpillar tracks.
3. *Piling in water where the piles are accessible to a crane working from the ground.* Provided the piles are so sited that a crane set up on the ground can reach the whole of the piles which have to be driven, a suitable and economical method of carrying out the work is to use a crane fitted with hanging leaders. The type of hammer used may be either of the double-acting hammer or diesel hammer. Work for which this plant would be suitable might be driving piles in a river or canal, the crane being set up on the bank. If necessary the crane could be run on track laid along the river bank, should the piles have to be driven over a length necessitating the moving of the plant parallel to the river.
4. *Piling in water where the piles are inaccessible to a crane set up on the ground.* In carrying out pile-driving work in water where the piles are so sited that they cannot be driven by a crane fitted with hanging leaders working from the ground, the piles are driven in one of two ways:

    (*a*) Staging piles may be driven by a crane fitted with hanging leaders, suitable baulks, walings and struts being attached to these piles so as to form a staging. The crane is then run on the staging, and the remaining staging piles driven. On the completion of the staging a piling frame of suitable size is then erected on it, and the permanent piles driven by this means.

    The estimator should note that if this procedure is adopted the staging and the staging piles have to be removed on the completion of the work and both the erection and dismantling cost of the complete staging must be allowed for. For estimating purposes the data shown for pitching, driving and extracting timber piles may be used for the work in connection with the staging piles, while that shown for fixing timber walings etc. may be used for the work entailed in constructing and removing the staging itself.

    (*b*) The piles may be driven from a pontoon or barge, in which case the plant is erected on them. In carrying out pile-driving work by this method hammers of the double acting hammer or diesel hammer type are used. Hammers of the drop type are impracticable as they create too much movement of the floating platform. Either a derrick fitted with hanging leaders or a piling frame mounted on the pontoon or barge is the usual method adopted.

Selecting the type and weight of hammer to use

The size and type of hammer to use depend on two things:

1. The nature of the ground into which the piles have to be driven.
2. The type of piling being used.

Double-acting hydraulic or diesel hammers are the most effective in the lighter grounds, such as sand, gravel, silt, peat, etc. In heavier soil, such as clay, single-acting drop hammers delivering slower but heavier blows are more suitable. It is sound policy to make sure that the hammer used is not too light; it is better to use a hammer a little on the heavy side. The tables shown may be used as a guide to assess the size and type of hammer to use with different classes of piling.

Table 11.1 The type of hammers to use with various classes of piling

Hammer	Type of piles to be driven
Air hammers	Less suitable for new wider and thinner steel sheet piling profiles
Diesel hammers	All types of steel sheet piling
Hydraulic hammers	Very efficient and controllable for all types of piling.

Table 11.2 The weight of single-acting and drop hammers to use with various classes of piling

Note: These hammers are most suitable for driving timber and reinforced concrete piles. They are generally used for driving steel piles only with medium or heavy sections in ground such as clay, rather than in granular soils such as gravel and sand.

Type of piling to be driven	Weight of hammer to use
Timber piles and steel sheet piling	The weight of drop hammer should be at least equal to the weight of the pile and its driving cap. Drop hammers and single acting steam hammers must not be used without a driving cap and dolly to protect the pile head. The drop should be limited to a maximum of 1.33 m.
Concrete piles	The weight of the hammer should be about half the weight of the pile, and not less than one third of its weight.

Estimating the cost of pile-driving work

In estimating the cost of pile-driving several factors contribute to the total cost of the work, all of which must be taken into account in preparing the estimate. These factors are:

1. The cost involved in hauling the piling plant.
2. Any cost involved in the way of site excavation, levelling, etc. in readiness for the erection of the piling plant, should this be necessary.
3. The cost of erecting and of the dismantling of the piling plant on the completion of the work. This also applies to any cost involved in driving staging piles and erecting staging on which to run the piling plant.
4. The cost of preparing the piles, as in the case of timber piles which have to be ringed, shoed, and have the necessary toggle holes bored in them prior to driving them.
5. The cost of pitching and driving the piles to the requisite depths and withdrawing steel tubes where necessary.
6. The cost of extracting the piles in the case of temporary piling works such as cofferdams where the piles have to be withdrawn on completion of the work.

The tables for estimating purposes

The tables shown for use in estimating the cost of pile-driving work consist of the main estimating tables and tables of multipliers for use in conjunction with them.

The data shown for erecting and dismantling the piling plant allow for erecting the plant on reasonably level ground on sleepers or on rail track to a length not in excess of 30 metres. If the nature of the site is such that extensive excavation or site levelling is required the cost of this should be allowed for in addition as should the cost of laying track in excess of 30 metres.

The cost of hauling the piling plant, sleepers, track, etc. must be included, as this is part and parcel of the cost of the work.

The cost of the haulage, erection, and dismantling of the plant is best allowed for on a 'lump sum' basis. The estimator may, however, allow for it by increasing the rates quoted for the piling work by an amount sufficient to cover the cost. This pre-supposes that the number of piles to be pitched and the total length to be driven is as set out in the Bill of Quantities, and as there is always a possibility that it may not be so, and a danger that it may be less, the author considers it wise to cover it by a lump sum, since, by so doing its cost is correctly allowed for.

Regarding off-loading, handling, pitching and driving the piles the estimator will note that separate data are shown for the following:

1. Off-loading, handling and pitching the piles.
2. Driving the piles.

These data have been purposely shown separately as the length of a pile off-loaded, handled and pitched may be considerably more than that portion of it which is driven. Thus a pile may be of an overall length of, say, 12 m while it may only be driven for a length of 5 m.

By showing the two operations separately it is a simple matter to estimate the cost of off-loading, handling, pitching and driving a pile of any overall length driven to any given number of metres.

The total cost of driving a pile complete is the sum of the off-loading, handling and pitching cost per pile, plus the driving cost per pile.

The table of multipliers shown for use with the data shown in the main estimating tables, permits of estimating the cost of carrying out pile-driving work under different prevailing conditions, such as driving in tidal water, driving from pontoons or barges, extracting piles of different types, etc.

Table 11.3 Erecting and dismantling piling plant

Notes: 1. The labour hours shown are for erecting the plant on reasonably level ground and allow for normal levelling prior to laying track etc.
2. If the site conditions are such that extensive excavation and levelling is required, the cost of this should be allowed for in addition. See Chapter 12.
3. The hours shown in the table allow for laying sleepers or rail track on which to erect the piling plant, this track not being in excess of 30 lin.m. If track in excess of this length has to be laid the cost of laying and removing it should be allowed for in addition. See Chapter 21.
4. For erecting and dismantling the plant on pontoons or barges a labour multiplier of 1.25 should be used with the data shown.
5. the cost involved in hauling the piling plant must be assessed and must also be allowed for in the estimate.

		Description		
Type of plant	*Unit*	*Erect Labour hours*	*Dismantle Labour hours*	*Total Labour hours*
Standard type steel piling frame, 6 m high.	Each	187	153	340
Standard type steel piling frame, 10 m high.	Each	325	270	595
Standard type steel piling frame, 15 m high.	Each	440	360	800
Standard type steel piling frame, 20 m high.	Each	500	410	910
Standard type steel piling frame, 25 m high.	Each	570	470	1040
Light type steel piling frame, 6 m high.	Each	170	140	310
Light type steel piling frame, 10 m high.	Each	300	250	550
Light type steel piling frame, 15 m high.	Each	400	340	740
1-ton crane, or derrick	Each	300	230	530
3-ton crane, or derrick	Each	350	280	630
5-ton crane, or derrick	Each	400	320	720

Table 11.4 Concrete piling

Notes: 1. The data shown are for piling in non-tidal water such as canal or river piling. Use the table of multipliers shown for other classes of work.
2. The cost of hauling the piling plant and the cost of erecting and dismantling it are not allowed for in the hours shown.
3. For extracting concrete piling and sundry works in connection with concrete piling, see Table 11.6
4. The piles are assumed to be driven in normal ground, such as clay, gravel or sand.

	Size of piles and piling plant, and labour hours per linear metre					
	Diameter 225 mm		*Diameter 300 mm*		*Diameter 375 mm*	
Description	*Piling plant hours*	*Labour hours*	*Piling plant hours*	*Labour hours*	*Piling plant hours*	*Labour hours*
Off-load, handle and pitch piles – vertical	0.23	2.53	0.26	3.15	0.33	4.26
Drive piles – vertical	0.13	1.06	0.16	1.31	0.20	1.57

Table 11.5 Table of multipliers for use with Table 11.4 for concrete piling

Description	Plant and labour hour multipliers
Driving piles in non-tidal water	1.00
Driving piles in tidal water	1.20
Driving piles from pontoons or barges in non-tidal water	1.30
Driving piles from pontoons or barges in tidal water	1.60
Driving piles in ground as distinct from in water, such as in piling for building foundations	0.75
Driving piles – raking	1.25
Extracting piles, using automatic pile extracting hammer	1.00

Table 11.6 Sundry work in connection with concrete piling

Description	Labour hours Each
Bend and hook bars, 12 mm diameter	0.15
Bend and hook bars, 20 mm diameter	0.20
Bend and hook bars, 25 mm diameter	0.25
Bend and hook bars, 32 mm diameter	0.30
Bend and hook bars, 40 mm diameter	0.35
Break down heads of 225 mm diameter	3.00
Break down heads of 300 mm diameter	4.00
Break down heads of 375 mm diameter	5.00

Table 11.7 Cubic metres of concrete in piles of various sizes and lengths and the weight of the piles in kilograms

Note: The above weights are exclusive of helmets or shoes and also the weight of any reinforcement in the piles.

	Cross-section of pile in mm, content of pile in cubic metres, and weight in kg					
	Diameter 225 mm		Diameter 300 mm		Diameter 375 mm	
Length of pile in metres	m^3	Weight in kg	m^3	Weight in kg	m^3	Weight in kg
3	0.154	0.354	0.270	0.630	0.422	0.970
6	0.308	0.708	0.540	1.260	0.844	1.940
9	0.462	1.062	0.810	1.890	1.266	2.910
12	0.616	1.416	1.080	2.520	1.688	3.880

Steel sheet piling

In the example at the end of this chapter the building up of the estimated cost follows the sequence:

1. Estimate the cost of any site preparation, should this be necessary. This preparation might take the form of excavating or levelling the site prior to erecting the piling plant. (In this example it is assumed that no extensive levelling is necessary).
2. Estimate the cost involved of hauling, erecting and dismantling the piling plant. (In this example the cost of hauling the plant only has been assessed at £500).
3. Estimate the cost of driving any staging piles and constructing staging, should this be necessary. (In this example no staging is required, since the piles are all accessible to a crane working from the river-bank).
4. Estimate the number of straight and corner and junction piles required to carry out the work. From this the total weight of piles required in kgs may be calculated and, knowing their price per kg, the total cost of the piles.
5. Estimate the cost of off-loading, handling, pitching and driving.
6. Estimate the cost of extracting the piles.
7. Estimate the material cost of any timber required in the way of walings, struts, etc. and also the cost of fixing them in the work.
8. Estimate the cost of removing the walings, struts, etc.
   on the completion of the work.
9. Assess the recovering value of the piles and timber on their removal.

In estimating the cost of steel sheet piling work consideration must be given to:

- The length of piles for which the various pile sections are suitable.
- The number of sheet piles required to construct the work.

The length and type of sheet piles to be used in permanent work are invariably stated in the contract documents, as is the depth to which they have to be driven. In constructing temporary works, however, such as cofferdams, the estimator must select the piling to use, giving consideration to the length of pile which is required and the depth to which it has to be driven.

The table shows the length of piles for which 'Larssen' steel sheet piling of various sections is suitable. The data shown may be used as a guide for piling other than 'Larssen' knowing the length, weight and width of the piles which have to be used.

Table 11.8 The length of piles for which the various sections are suitable

Pile section 'Larssen piling'	Width in mm	Thickness in mm	Weight per kg/m	Maximum length of pile in m
LX8	600	8.2	54.6	12
LX12	600	9.7	63.9	17
LX16	600	10.5	74.1	20
LX25	600	15.6	94.0	23
LX32	600	21.5	113.9	26

Table 11.9 Steel sheet piling

Notes: 1. The data shown are for piling in non-tidal water, such as canal or river piling. Use the table of multipliers shown for other classes of work.
2. The cost of hauling the piling plant and the cost of erecting and dismantling it are not allowed for in the hours shown.
3. For extracting sheet piling and sundry works in connection with sheet piling.
4. The data shown are for 'Larssen' steel piling. The weight and dimensions of the various sections of piling are shown. If the piling has to be carried out with steel piling other than the 'Larssen' type, the estimator may use the data shown on which to base the estimate, knowing the cross-section, width, and weight of the piles which are to be used in the work.
5. The piling is assumed to be driven in normal ground, such as clay, gravel or sand.

| | | | | | Piling plant and labour hours per lin. metre | | | |
| | Description | | | | Off-load, handle and pitch piles vertical per lin. metre | | Drive piles vertical per lin. metre | |
Piling section 'Larssen' piling	Weight per lin. m in kg	Width in mm	Weight per m² in kg	Unit	Piling plant hours	Labour hours	Piling plant hours	Labour hours
LX8	54.6	600	91.0	Lin. m	0.152	1.522	0.115	0.918
LX12	63.9	600	106.4	Lin. m	0.164	1.640	0.121	0.971
LX25	94.0	600	156.7	Lin. m	0.198	1.968	0.148	1.181
LX32	113.9	600	189.8	Lin. m	0.216	2.165	0.164	1.312

Table 11.10 Multipliers for use with Table 11.9 for steel sheet piling

Description	Plant and labour hour multipliers
Driving piles in non-tidal water	1.00
Driving piles in tidal water	1.20
Driving piles from pontoons or barges in non-tidal water	1.30
Driving piles from pontoons or barges in tidal water	1.60
Driving piles in ground as distinct from in water, such as in piling for building foundations	0.75
Driving piles – raking	1.25
Extracting piles, using automatic pile extracting hammer	0.70

Table 11.11 Sundry works in connection with steel piling

Description	Unit	Gas consumption		Plant hours	Man on oxy-acetylene flame hours	Carpenter or timberman hours	Labourer hours
		Oxygen m^3	Acetylene m^3				
Burn off tops of:							
6 mm thick piles	Lin. m	1.32	0.40	0.50	0.50	–	0.50
9 mm thick piles	Lin. m	2.97	0.50	0.53	0.53	–	0.53
12 mm thick piles	Lin. m	4.95	0.63	0.56	0.56	–	0.56
16 mm thick piles	Lin. m	6.27	0.73	0.59	0.59	–	0.59
20 mm thick piles	Lin. m	7.59	0.86	0.63	0.63	–	0.63
22 mm thick piles	Lin. m	9.57	1.06	0.66	0.66	–	0.66
'Extra Only' for pitching and driving corner piles	Lin. m	–	–	0.17	–	–	1.32
'Extra Only' for pitching and driving junction piles	Lin. m	–	–	0.20	–	–	1.58
Fix struts and puncheons	m^3	–	–	–	–	17.66	17.66
Fix walings	m^3	–	–	–	–	14.13	14.13

Steel sheet piling example

Example.   Calculate the total cost of constructing a cofferdam for a bridge abutment using 'Larssen' LX8 steel piling. The overall length of the piles to be used is 5 m and they have to be driven in the bed of a non-tidal river to a depth of 2.5 m. The pile driving has to be carried out using a 3 tonne derrick crane set up on the river bank, the crane being fitted with hanging leaders. The whole of the piles are accessible to the crane working from the bank so no staging is required.

The piles when driven are shored with walings and struts.

The length of the cofferdam required is 17 m with two 6 m returns to the river bank. On completion of the works the piling is withdrawn. The river bank is level and firm and no extensive excavation or site levelling is necessary. The estimated cost desired is the total cost of the cofferdam, including haulage, erecting and dismantling the piling plant, together with the driving of the piles and their removal.

For purposes of this example the following plant, labour, and material rates are assumed:

Working cost of the piling plant including hammer, fuel, oil, etc. and crane driver.

Working cost of the piling plant including hammer, fuel, oil, etc. and crane driver	= £20.00 per hour
Timberman's rate	= £6.20 per hour
Labourer rate	= £6.00 per hour
Cost of sheet piling	= £600 per tonne
Cost of timber walings, struts, puncheons, etc.	= £220/m^3

Assessed quantity of timber to be fixed in the cofferdam:

Walings   = 2.25 m^3

Struts, puncheons, etc. = 3.00 m^3

Haulage, erection and dismantling cost of the piling plant

For purposes of this example, the cost of hauling the piling plant has been assessed at £500.

From Table 11.3 the cost of erecting and dismantling a 3 tonne derrick crane is:

630 labour hours at £6 per hour = £3780

The cost allows for normal site preparation and track laying.

The haulage, erection and dismantling cost for estimating purposes, therefore, is:

Cost of haulage	£500
Cost of erecting and dismantling the piling plant	£3780
Total	£4280

Number of piles required and their weight and cost

Since LX Larssen piling is 600 mm effective width and the total length of cofferdam is 17 m + 2 end returns each 6 m long, the total number of piles required is given by:

$$\text{Standard piles} = \frac{17\text{m} + (2 \times 6\text{m})}{0.600} = 49 \text{ piles}$$

2 No. Corner piles

Notes: 1. LX8 piling weighs 54.6 kg per lin. m.
2. The weight of suitable corner piles for use with LX8 piling may be taken as 54.6 kg per lin. m.

The total weight of piles required is therefore:

Standard piles	49 no. at 5m long = 245 @ 54.6 kg/m =	13 377 kg
Corner piles	2 no. at 5m long = 2 @ 54.6 kg/m =	109
		13 486 kg

Total weight of piles = 13 486 kg or 13.486 Tonnes

Cost of piles = 13.486 @ £600 per tonne = £8092

Cost of off-Loading, handling, pitching and driving the piles

From Tables 11.9, 11.10 and 11.11 the following costs are obtained:

(a) Off-load, Handle and Pitch LX8 Piling per lin. m

Plant 0.152 @ £20 per hour	= £ 3.040 per lin. m
Labour 1.522 @ £6 per hour	= £ 9.132 per lin. m
Cost of off-loading, Handling and pitching the piles	= £12.172 per lin. m

(b) Drive LX8 Piling per lin. m

Plant 0.115 @ £20	= £2.300 per lin. m
Labour 0.918 @ £6	= £5.508 per lin. m
Cost of driving 1 lin. m of pile	= £7.808 per lin. m

(*c*) 'Extra only' cost of handling, pitching and driving corner piles per lin. m.

Plant 0.17 @ £20 per hour	= £ 3.40 per lin. m
Labour 1.32 @ £6 per hour	= £ 7.92 per lin. m
Extra only cost of corner piles	= £11.32 per lin. m

The total cost of off-loading, handling, pitching and driving the piles is:

Off-load, handle and pitch piles 51 × 5 m long	= 255 m @ £12.17	= £3103.35
Drive piles 51 × 2.5 m long	= 127.5 m @ £ 7.81	= £ 995.78
'Extra Over' for corner piles 2 @ 5 m long	= 10 m @ £11.32	= £ 113.20
Total cost of off-loading, handling, pitching and driving piles		= £4212.33

Cost of extracting the piles

Referring to the table of multipliers, Table 11.10 the cost of extracting the piles is the cost of off-loading, handling, pitching and driving the piles multiplied by 0.70.

The cost of extracting the piles, therefore is:

0.70 × £4212.33 = £2948.63

Material cost of the timber required in walings, struts, puncheons, etc. and the cost of fixing them in the work

For purposes of this example, the timber required for walings, struts, puncheons, etc. has been assessed as being:

Walings:	2.25 m^3 @ £220/m^3	= £ 495
Struts and puncheons, etc.:	3 m^3 @ £220/m^3	= £ 660
Total material cost		= £1155

From Table 11.11 the labour cost of fixing the above timber in the work is:

Fix walings per m^3

Timberman	14.13 @ £6.20 per hour	= £ 87.61
Labourer	14.13 @ £6.00 per hour	= £ 84.78
		£172.39

Fix struts and puncheons per m^3

Timberman	17.66 @ £6.20 per hour	= £109.49
Labourer	17.66 @ £6.00 per hour	= £105.96
		£215.45

The total cost of fixing walings, struts, puncheons, etc. including the material cost is:

The material cost of the timber	£1155.00

Fix walings = 2.25 m^3 @ £172.39/m^3 = £387.88

Fix struts, puncheons, etc.
= 3 m^3 @ £215.45   = £646.35

	£1034.23
Total cost of timber fixed in the work	£2189.23

Cost of removing the timber

From the table shown for sundry works in connection with steel piling, Table 11.11, the cost of removing the timber from temporary works is the cost of fixing the timber in the work multiplied by 0.50.

The cost of removing the timber, therefore, is:

0.50 × £1,034.23 = £517.12

Recovery value of the piles, walings, struts and puncheons

The recovery value of the sheet piles and timber must be assessed by the estimator and credit given for this. Assuming that the recovery value of these has been assessed at 66 2/3% of their original cost, the recovery value then is:

Original cost of piles	£8092
Original cost of timber	£1155
	£9247

Recovery value = 66 ⅔% of £9247 = £6165

Summary of the total cost of constructing the cofferdam

The haulage, erection and dismantling of the piling plant	£ 4 280
The original cost of the piles	£ 8 092
The cost of off-loading, handling, pitching and driving the piles	£ 4 212
The material cost of the walings, struts, puncheons, etc. and the cost of fixing them in the work	£ 2 189
The cost of removing the timbering	£ 517
Total cost	£19 290
Deduct recovery value of piles, walings, struts	£(6 165)
Estimated cost of cofferdam	£13 125

12 Groundwork

Excavation and earthwork at ground level

The bulkage of the material which has to be loaded and transported in vehicles is of considerable significance in the case of excavation, as the estimator is concerned with the cubic metres of *solid* material hauled per load and not the loose material as loaded for transport.

Weather conditions influence the cost of excavation. Under wet conditions the plant tends to bog, and on open sites the transport of the excavated material becomes difficult, with a resultant decrease in output and increase in cost.

Excavation is generally billed by volume, the depth of excavation being stated. In considering the transport cost of excavated material in lorries both the bulkage and the weight factor are important, as vehicle design loads must not be exceeded. In the table shown setting out the cubic metres of various materials hauled per load by various types of vehicles, the weight and bulkage of the materials (Table 12.20) has been taken into account, as has the normal heaping of the loads in the vehicles concerned.

Plant of various kinds is extensively used for carrying out excavation where the volumes are such that its use is merited and where the nature of the work is such that suitable plant can be employed. Throughout this chapter data are shown in connection with the types of mechanical excavating plant most commonly used, from which can be estimated the cost of excavation carried out by them.

Where the volume of work involved is small and the use of mechanical plant, though possible would not be economical, the work is carried out by hand and the estimate based on such.

Table 12.1 Earthworks operations and the various ways in which they may be carried out

Notes: In carrying out operations using mechanical plant the following is assumed:

- The volume of work involved merits the use of the mechanical plant concerned.
- The prevailing site conditions are such that the use of the plant is practicable.

Application	The method and the plant to use in carrying out the operation
Excavation	
1. Excavation in small quantities disposed of on site up to a 25 m maximum haul.	Excavation carried out by hand and wheeled in hand barrows.
2. Excavation in comparatively small quantities hauled and disposed of at a length of haul not in excess of 500 m.	Excavation carried out by hand or mini excavator and hauled in dumpers.
3. Excavation in large areas where the excavated material has to be loaded and transported in vehicles such as lorries, dumpers, etc.	Mechanical excavators with skimmer equipment.
4. Site stripping and levelling operations for large volume of earthworks.	Bulldozers and scrapers.
5. Grading for large earthwork areas.	Tractor and grader.

Table 12.1 continued

Application	The method and the plant to use in carrying out the operation
Excavation – *continued*	
6. Trench foundation and basement excavations for small or medium contracts where the excavated material has to be loaded and transported in vehicles such as lorries, dumpers, etc.	Multi-purpose excavators with excavating front bucket and a rear backacter bucket.
7. Trench, foundations and basement excavations for medium or large contracts where the excavated material has to be loaded and transported in vehicles such as lorries, dumper, etc.	Backacters.
8. Bulk excavation for reduced levels for medium or large contracts.	Draglines.
Backfilling	
1. Backfilling all types of open excavations.	By hand.
2. Backfilling all types of open excavations.	Angledozers or bulldozers, the material being angled or pushed in. This type of plant backfills at a greater rate than by other means.
Spreading	
1. Excavated materials, clinker, gravel, sand, foundation materials, etc.	By hand.
2. Spreading to form dumps.	Angledozer or bulldozer.
3. Spreading materials such as clinker, gravel, sand, hardcore, etc. in large areas.	Bulldozer.
Compacting	
1. Trenches and other backfilled excavations.	By hand rammers or power rammers of the 'Pegson' type.
2. Road foundation materials.	Power-driven rollers, the weight of rollers used depending on the nature of the formation, the class of material used and the finished thickness to which it is laid.
3. Road surfacing materials such as asphalt, tarmacadam, etc.	Tandem rollers.
4. Filling to embankments of large area and consolidating in soil stabilisation work.	Tractor-drawn sheepsfoot roller.
5. Thin layers of soil, sand, gravel, etc. of large area and consolidating in soil stabilisation work.	Tractor-drawn smooth roller.
Note: Compacting filling of gravel, sand, and granular soils.	
6. Compacting concrete.	Vibrating plant and tampers.
Scarifying	
1. Existing road surfaces, waterbound and tarmacadam roads.	Power-driven rollers fitted with scarifier and tynes.
2. Soft, easily penetrated materials.	Tractor-drawn cultivator.
3. Compacted filling.	Tractor-drawn disc plough.

Table 12.1 continued

Application	The method and the plant to use in carrying out the operation
Loading materials	
1. All classes of materials.	By hand.
2. Loose excavated material and materials such as gravel, sand, clinker, hardcore, etc.	Mechanical excavator with skimmer equipment.
3. Loading broken rock, gravel, clinker, hardcore, sand and loose excavated material.	Mechanical excavator with shovel equipment. This is the most suitable plant to use for loading broken rock.
4. Loading gravel, sand, etc. from storage heaps or bins.	Mechanical excavator with grab equipment.
5. Pile of materials ready for loading by other units.	Angledozers or bulldozers.
6. Sand, gravel, tarmacadam, etc.	Loaded direct from elevated storage hoppers.
7. Loading loose materials, sand, gravel, etc.	'Hilift' loading shovels.

Estimating the cost of excavation

In order to estimate the cost of carrying out excavation a knowledge of the various ways in which this can be carried out and which method to adopt in any particular case is essential. The estimator must be conversant with the various types of excavating plant available and be capable of deciding upon which type and size of plant to use in carrying out specific excavation works. It is on such knowledge that the build up of the estimate is produced.

The method adopted to carry out excavation depends on:

1. The type of excavation.
2. The nature of the ground.
3. The volume of excavation.
4. The length of haul to tip.

Type of excavation. Mechanical excavation should be adopted if the volume of bulk or surface excavation warrants it, for by so doing the excavation can be carried out at a fast rate of progress and at low cost. For this type of excavation, plant of various kinds is available. Thus, for excavation over areas, mechanical excavators fitted with skimmer scoop equipment may be used in conjunction with transport vehicles, or, if the volume to be handled is exceedingly large, elevating graders might be used with suitable transport. In the event of the length of haul being comparatively short, the excavation may be carried out by tractors and scrapers, the cost of which excavation is of a very low order.

For bulk excavation, mechanical excavators of the shovel skimmer or dragline type may be used, the plant selected depending upon the nature of the work.

The nature of the ground. In considering the use of mechanical excavating plant to carry out excavation, it should be noted that plant is capable of excavating all normal grounds such as top soil, sand, loamy soil, soft or sandy clay, marl, stiff clay, gravel and soft chalk. It is not capable, however, of excavating rock in solid beds, though, on occasions, it will remove friable rock in layers interspaced with soil, if not too hard and tough a rock is present. The excavation of this class of material is hard on the machines, however, and is to be deprecated unless the rocky ground is really easily removed without unduly jarring the plant and its mechanism.

The volume of excavation. Consideration should be given to the volume of excavation involved, for if this is so small that the use of mechanical excavating plant is not merited the excavation would in such circumstances probably be best carried out by hand in conjunction with suitable transport.

The length of haul to tip. The selection of the correct type of transport vehicle to use to haul excavated material to tip is of considerable importance, for the economical length of haul of different types of transporting plant is variable.

The attendant labour, which is allowed for, is as follows:

1. Excavation carried out by hand in conjunction with lorries or dumpers, a total of 3 men digging and loading per vehicle.
2. Excavation carried out by 0.5 m^3 mechanical excavator with skimmer equipment in conjunction with 5 tonne lorries or 3 m^3 dumpers, 1 banksman, 2 men trimming the formation, and 1 man on the tip, a total of 4 men.
3. Excavation carried out by 0.5 m^3 mechanical excavator with skimmer equipment in conjunction with locomotive and U-shaped skip, running on 600 mm gauge rail track, 1 man in attendance on the locomotive, 2 men on track maintenance, 2 men trimming formations, and 2 men on the tip, slewing track, etc., a total of 7 men.
4. Bulk excavation carried out by a 1 m^3 mechanical excavator with skimmer equipment in conjunction with tractor-drawn 10 m^3 dump truck: 2 men trimming formations and 2 men on the tip, a total of 4 men.
5. Excavation carried out by D6H tractor and scraper: 1 man in attendance.

Build up of excavation rate

It should be noted where mechanical excavation is predominantly used, that a small amount of hand excavation may be necessary.

The build up of a rate for excavation should include an allowance for hand excavation.

Excavation carried out by mechanical excavating plant

1. Excavation carried out by mechanical excavators in conjunction with lorries fitted with mechanical tipping gear is suitable for all lengths of haul and is cheaper than where dumpers are used. Dumpers can, however, run on rough ground or on open wet sites unsuitable for lorries and under such conditions are extensively used.
2. Excavation carried out by mechanical excavators, using locomotives and wagons running on rail track to haul the excavated material, is a cheap form of excavation provided the quantity of excavation is sufficient to justify the use of this form of transport.
3. Excavation carried out by large mechanical excavators in conjunction with crawler-tractor-drawn tip wagons of large capacity is a cheap form of excavation on short lengths of haul, being cheaper than lorries or dumpers. The length of haul should, if possible, not exceed 450 m, as with hauls in excess of this cost rises rapidly. Bottom dump wagons drawn by wheeled tractors of the Caterpillar 824C type are also commonly used when large volumes of excavation are involved, their rate of travel being approximately 24 km per hour. Under suitable travelling conditions the economical one-way length of haul for plant of this type is up to 2.5 km. On the longer hauls, lorries should be used if the site conditions make this possible.

4. Excavation by tractor and scraper on work for which it is suitable is an exceedingly cheap form of excavation on lengths of haul up to 450 m. If the length of haul is in excess of this, mechanical excavators and lorries should be used, or, on very large contracts, elevating graders may be used.

TYPE	ELEVATION	TYPE OF EXCAVATION SUITABLE FOR
FACE SHOVEL	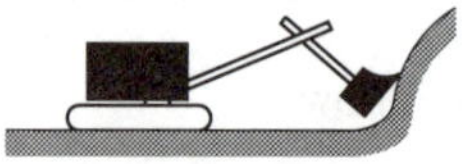	FOR CUTTINGS OR BANKS GREATER THAN 1.5 m HIGH. CAN DIG HARD GROUND AND SOFT ROCK
DRAGLINE	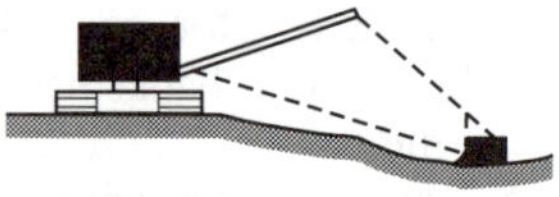	FOR EXCAVATION BELOW THE LEVEL OF ITS TRACKS. CLEARING DITCHES STREAMS, ETC. HAS WIDE DUMPING RANGE AND LONG REACH
SKIMMER	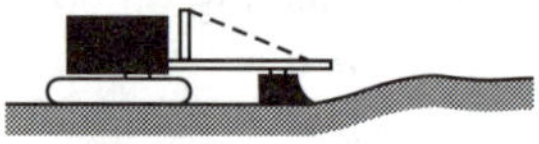	FOR EXCAVATION DOWN TO LEVEL OF ITS TRACKS NOT OVER 1.5 m DEEP SUCH AS AREAS OF EXCAVATION OVER NEW ROADS ETC.
BACKACTER		FOR EXCAVATION OF TRENCHES AND SMALL AREAS OF EXCAVATION BELOW GROUND
GRAB		FOR LOADING OR UNLOADING. DIGGING LOOSE MATERIALS AND PLACING IN HEAPS.
BULLDOZER	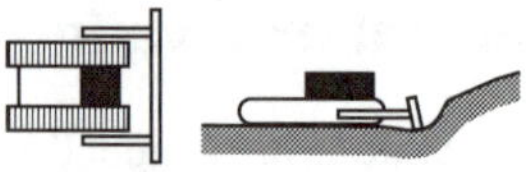	FOR FILLING AND GRADING PUSHES UP TO 100 METRES
ANGLE DOZER	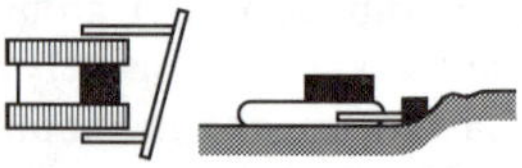	FOR FILLING AND GRADING PUSHES UP TO 100 METRES. BLADE CAN BE SQUARED FOR USE AS A BULLDOZER
TRACTOR AND SCRAPER		FOR EXCAVATING LARGE AREAS LEVELLING AND GRADING. USED FOR EARTH EXCAVATING AND MOVING AND TIPPING FROM 50 TO 500 METRES
ENDLESS BUCKET TYPE TRENCH EXCAVATOR		FOR EXCAVATING TRENCHES. ENDLESS CHAIN OF BUCKETS ON BOOM. CUTS GROUND AND LEAVES A CLEAN TRENCH EXCAVATION

Figure 12.1 Mechanical plant for use in moving earth, showing type of work for which it is suitable.

5. Excavation carried out by mechanical plant is considerably cheaper than hand excavation. If the volume to be excavated merits it and the nature of the work is such that it can be carried out mechanically, excavating plant should be used.

Figure 12.1 shows the kind of excavation for which various types of plant are suitable.

In deciding upon the type of plant on which to build up the estimate, suitable plant owned by the builder or contractor should first be considered. In the event of no such plant being available the estimate may be built up on suitable hired plant, the rate per working hour for this being wholly inclusive of all charges in connection with it, i.e. the driver's wages, fuel, oil, grease and any haulage costs incurred.

The Tables for use in estimating the cost of excavation

Since excavation can be carried out by hand and/or by mechanical plant of different types and outputs and the excavated material can be transported to tip in various types and sizes of vehicles, the length of haul to tip itself being variable, many combinations of excavating and transporting plant are possible, and the correct combination must be used if the excavation is to be carried out economically. The estimator must have available clear and concise data from which to built up an estimate for any likely combination. The various operations involved in carrying out excavation are therefore tabulated under separate tables, from which the cost of excavation carried out by any combination can readily be estimated. In using these tables the estimator should note the following:

1. In the case of excavation carried out by hand, one table (Table 12.2) is shown for excavating and loading the material into barrows, lorries, etc., while Tables 12.10 and 12.11 show the transporting of it by hand-wheeled barrows, lorries, dumpers or wagons. From these tables is calculated the total cost of excavating, loading, hauling and tipping the various soils to the different lengths of haul shown per cubic metre.
2. For excavation carried out by mechanical excavating plant of the type which has transporting plant along with it to haul the excavated material to tip, i.e. skimmer scoops, shovels, draglines, the following Tables apply:

   (a) Tables 12.4, 12.5 and 12.6 show the time taken by the plant to excavate and load 1 m^3 of material into the vehicle which is used to remove it to tip.
   Note: This time is the time the vehicle stands while it has 1 m^3 loaded into it.
   (b) Table 12.7 showing the cubic metres of material hauled per load by vehicles such as lorries, dumpers, etc.
   (c) Table 12.11 gives the time taken by various types of transport vehicles to haul one load of material to tip, off-load and return for the next load, this being shown for various lengths of haul.

Throughout this section detailed examples are shown illustrating the build up of excavation costs which clearly demonstrate the method of using the data shown.

For excavating rock on a large scale and for scarifying existing roads, see chapter 21.

For calculating the working cost or hire rates of plant of various kinds per hour see chapter 6.

Excavation Carried Out by Hand

Excavation by hand is generally associated with barrows as the means of transporting the excavated material, such forms of transport being suitable for comparatively short lengths of haul.

Excavation by hand, however, may also be carried out, using mechanical transport such as dumpers or lorries, but it should be noted that in using such plant the transporting cost is likely to be high, since the vehicle stands for a long period of time while it is being loaded.

Handbarrows are suitable for lengths of wheel up to 20 m; to dumpers up to 300 m and lorries to any range.

In order to illustrate the method of estimating the cost of hand-excavated material, an example is shown for the build up of the excavation cost of loamy soil excavated in bulk digging, barrowed to a 20 m run.

Example 1.   Calculate the cost per cubic metre of reduced level hand excavation not exceeding 1.00 m depth in loamy soil carried out by hand, the excavated material being loaded into barrows and wheeled and tipped at a 20 m run, the prevailing labour rate being £6.00 per hour.

From Table 12.2

Labour hours excavating
bulk loamy soil per cubic metre = 1.60 h

Cost per m^3 = 1.60 h at £6.00 per h = £9.60 per m^3 excavation

From Table 12.8 and 12.10

Labour hours loading and wheeling loamy
soil in barrows to a 20 m run per m^3 = (0.54 h + 0.83 h) = 1.37 h

Cost per m^3 = 1.37 h at £6.00 per h = £8.22 per m^3 load and disposal

Table 12.2 Excavation by hand

Notes: 1. The hours shown are for excavating only, and do not allow for loading, wheeling or transporting, for which see Tables 12.8 and 12.10
2. The shallow trenches referred to are excavations such as those taken out for house foundations, kerb and channel foundations, etc.
3. The rock referred to is rock with many fissures.

		Nature of ground and labour hours excavating only, per cubic metre					
Description *Excavate only* (See notes above)	*Firm* *sand*	*Loamy soil,* *soft or* *sandy clay,* *or marl*	*Stiff* *clay*	*Gravel and* *compact soil* *requiring* *picking*	*Chalk*	*Soft* *stratified* *fissured* *rock*	*Hard* *stratified* *fissured* *rock*
Excavate over areas to reduce levels	1.33	1.60	1.87	2.53	4.00	5.33	8.53
Excavate over areas and load to vehicles	1.60	1.87	2.13	2.80	4.40	5.60	8.80
Excavate shallow trenches	2.13	2.66	3.06	4.00	6.27	8.00	12.27
Excavate small pier-holes	2.66	2.93	4.00	4.80	7.47	8.80	13.43
Excavation in underpinning	2.40	2.66	2.93	4.27	7.73	9.07	15.47
Spread and level in 300 mm layers	0.40	0.53	0.80	0.47	0.53	0.53	0.53
Fill and ram in layers to form banks	0.67	0.80	0.93	0.73	0.80	0.80	0.80

Table 12.3 Breaking out brickwork, concrete, rock and hard road surfaces by hand and pneumatic tools

Notes: 1. The hours shown are for breaking out only. They do not allow for loading the material into vehicles or transporting them.
2. For loading materials, see Tables 12.8 and 12.9
3. For wheeling in barrows, see Table 12.10
4. For transporting in lorries, dumpers, etc., see Table 12.11
5. If the concrete is reinforced use a multiplier of 1.40 with the plant and labour hours shown.
6. For breaking out asphalt or tarmacadam superimposed on concrete, treat the combined thickness as concrete.
7. For scarifying roads with a roller fitted with scarifier, see Table 21.23

| | | Break out, using compressors and pneumatic tools | | | | Break out by hand using points and sledges. Labour hours |
| | | 2-tool compressor | | 3-tool compressor | | |
Description (Break out only, not loaded or transported)	Unit	Compressor hours	Labour hours	Compressor hours	Labour hours	
Remove existing sett paving	m^2	0.10	0.20	0.06	0.18	1.20
Remove existing wood-block paving	m^2	0.04	0.08	0.03	0.09	0.48
Break out brickwork in lime mortar	m^3	0.24	0.48	0.16	0.48	3.33
Break out brickwork in cement mortar	m^3	0.32	0.64	0.21	0.63	5.00
Break out chalk – hard	m^3	0.44	0.88	0.30	0.90	6.66
Break out concrete in areas:						
100 mm thick	m^2	0.11	0.22	0.07	0.21	1.08
150 mm thick	m^2	0.14	0.28	0.09	0.27	1.92
225 mm thick	m^2	0.25	0.50	0.17	0.51	3.60
300 mm thick	m^2	0.40	0.80	0.26	0.78	5.52
Break out mass concrete	m^3	2.66	5.32	1.70	5.10	43.00
Break out sandstone	m^3	1.87	3.74	1.24	3.72	26.60
Break out tarmacadam roads on hardcore:						
150 mm thick	m^2	0.11	0.22	0.07	0.21	1.44
225 mm thick	m^2	0.14	0.28	0.10	0.30	1.80
300 mm thick	m^2	0.18	0.36	0.12	0.36	2.16
Break out asphalt roads on hardcore:						
150 mm thick	m^2	0.13	0.26	0.09	0.27	1.56
225 mm thick	m^2	0.18	0.36	0.12	0.36	1.90
300 mm thick	m^2	0.23	0.46	0.15	0.45	2.40
Break out tarmacadam paths on clinker or hardcore:						
75 mm thick	m^2	0.04	0.08	0.02	0.06	0.72
100 mm thick	m^2	0.05	0.10	0.03	0.09	0.84
Remove existing 50 mm artificial stone paving	m^2	–	–	–	–	0.48
Demolish brick-built property per cubic metre of building	m^3	–	–	–	–	1.70

Mechanical excavators fitted with shovel, skimmer, dragline or grab equipment

In estimating the cost of excavation carried out by the type of mechanical excavating plant shown previously, five items have to be noted in order that the cost may be calculated:

1. The time taken by the plant to dig and load 1 m³ of solid material into the vehicle used to transport it.
2. The time the vehicle stands while it is loaded.
3. The time the vehicle takes to travel to the tip, off-load and return for the next load.
4. The cost of any attendant labour along with the plant, on the tip, trimming, etc.
5. The cost incurred in hauling the plant, unless this is otherwise allowed for.

In connection with transport by lorries, where the weight of load factor has to be considered, the amount of material hauled per load is variable, since the weights of various soils per cubic metre are variable. In the table showing the cubic metres of various soils hauled in lorries per load, this factor is allowed for, as is the bulkage of the respective materials, normal heaping of the loads being allowed for.

Data from which to estimate the cost of excavation, using this type of plant for all combinations of excavating and transporting plant and for various classes of ground hauled to various lengths of haul are shown in the following tables:

1. Tables 12.4, 12.5 and 12.6 showing the time taken by the excavating plant to dig and load 1 m³ of solid material into the vehicle used for transporting it.

   Note. The hours shown also represent the hours the vehicle stands while 1 m³ of solid material is loaded into it.

2. Table 12.7 showing the cubic metres of solid excavation hauled per load, using various types and sizes of haulage plant.
3. The average time taken by the vehicles to haul the load to tip, off-load and return for the next load, Table 12.11

In order to illustrate the method of using these tables an example given, where the haulage is carried out by 15 tonne lorries fitted with mechanical tipping gear.

Example.   Calculate the cost of excavating sandy clay with a 0.5 m³ excavator fitted with skimmer scoop equipment and the excavated material being hauled in 16 tonne tipping lorries fitted with mechanical tipping gear to a tip 0.3 km away. A banksman is in attendance on the excavator while two men are employed on trimming the formation and one man on the tip, i.e. a total of four men in all.

For the purposes of this example the following rates are assumed:

Hire rate or working cost per hour of the 0.5 m³
mechanical excavator, inclusive of driver, fuel and oil   = £22.00

Hire rate or working cost per hour of 16 tonne lorries,
inclusive of driver, fuel and oil   = £10.00

Labour rate   = £ 6.00

From Table 12.4

Hours taken by 0.5 m³ mechanical excavator with skimmer scoop attachment to dig and load 1 m² of sandy clay = 0.071 h per m³.

From Table 12.7

The cubic metres of solid sandy clay hauled per load in a 16 tonne lorry is 9.28 m³ per load.

Table 12.4 Excavators fitted with skimmer scoop equipment

Notes: 1. The hours shown are those taken by the plant to dig and load 1 m³. They also represent the hours the transporting plant stands while 1 m³ of material is loaded into it.
2. For the cubic metres of material hauled per load, employing transporting plant of various types and capacities, see Table 12.7
3. For the time taken by various types of transporting plant to haul to tip at various lengths of haul, see Table 12.11

Bucket capacity of machine, cubic metres	*Nature of ground and plant hours excavating and loading per cubic metre*						
	Firm sand	*Loose excavated material*	*Loamy soil, soft or sandy clay or marl*	*Gravel*	*Broken stone or metal*	*Stiff clay*	*Soft chalk*
0.25	0.110	0.135	0.160	0.173	0.177	0.188	0.221
0.50	0.050	0.063	0.071	0.075	0.080	0.084	0.100
0.75	0.029	0.037	0.040	0.044	0.047	0.051	0.059
1.00	0.023	0.029	0.032	0.035	0.037	0.040	0.045

Table 12.5 Excavators fitted with dragline or shovel equipment

Notes: 1. The hours shown are those taken by the plant to dig and load 1 m³. They also represent the hours the transporting plant stands while 1 m³ of material is loaded into it.
2. For the cubic metres of material hauled per load, employing transporting plant of various types and capacities, see Table 12.7
3. For the time taken by various types of transporting plant to haul to tip at various lengths of haul, see Table 12.11

Bucket capacity of machine, cubic metres	*Nature of ground and plant hours excavating and loading per cubic metre*						
	Firm sand	*Loose excavated material*	*Loamy soil, soft or sandy clay or marl*	*Gravel*	*Broken stone or metal*	*Stiff clay*	*Soft chalk*
0.25	0.084	0.115	0.117	0.127	0.135	0.143	0.168
0.50	0.037	0.047	0.052	0.056	0.060	0.064	0.075
0.75	0.021	0.027	0.031	0.032	0.035	0.036	0.043
1.00	0.016	0.020	0.023	0.025	0.027	0.028	0.032

Table 12.6 Excavators fitted with grabs

Notes: 1. Grabs are only suitable for excavating soft grounds, such as gravel, sand, etc. with hard grounds they are uneconomical. They are chiefly used for loading materials such as ballast, gravel, shingle, broken stone, road metal, sand, etc.
2. For loading loose materials to vehicles mechanically, and from storage hoppers, see Table 12.9

	Nature of ground and plant hours per cubic metre			
	Excavating and loading		*Loading only loose materials*	
Size of grab in cubic metres	*Sand*	*Loam and gravel*	*Sand*	*Gravel, shingle and broken stone*
0.25	0.176	0.400	0.133	0.148
0.50	0.088	0.200	0.067	0.073
0.75	0.059	0.133	0.044	0.050

Table 12.7 Cubic metres of excavated material hauled per load

Notes: 1. In hauling excavated material by lorry the weight of the load hauled should not exceed that which the lorry is designed to carry. The table shows the cubic metres of solid material hauled per load, taking bulkage, normal heaping of the load and the weight factor into account
2. Vehicles such as dumpers and wagons drawn by tractors or locomotives are rated on a volume basis, their struck measured capacity in cubic metres being stated, thus the struck measured capacity of a 2 m³ dumper is 2 m³. The vehicles have a heaped capacity, the load being heaped in the vehicles, and this heaping of the load in conjunction with the bulkage of the material must be taken into account in assessing the cubic metres of solid material hauled per load. In the table the quantity of solid excavated material hauled per cubic metre of struck measured capacity allows for heaping the load and the bulkage of the material Thus if sandy clay is hauled in a dumper whose struck measured capacity is 2 m³, the amount of solid material per load is 2×0.85 m³ = 1.70 m³.

Nature of material hauled	*Weight of the material in the solid m³ per tonne*	*Haulage by lorries. The m³ of solid material hauled per load*			*The m³ of solid material hauled by dumpers or in wagons per m³ of struck measured capacity*
		16 tonne lorry	*24 tonne lorry*	*30 tonne lorry*	
Chalk	0.44	7.04	10.56	13.20	0.70
Loamy soil and soft or sandy clay	0.58	9.28	13.92	17.40	0.85
Stiff clay	0.52	8.32	12.48	15.60	0.80
Gravel	0.57	9.12	13.68	17.10	1.14
Loam	0.66	10.56	15.84	19.80	0.92
Marl	0.57	9.12	13.68	17.10	0.98
Sand	0.66	10.56	15.84	19.80	1.09
Soil	0.62	9 92	14.88	18.60	0 90
Broken rock or metal	0.37	5.92	8.88	11.10	0.66

Table 12.8 Loading loose materials by hand

Notes: 1. The hours shown are for loading only and do not allow for wheeling or
transporting, for which see Tables 12.10 and 12.11
2. The vehicles referred to are vehicles such as lorries, dumpers, etc.
3. The materials referred to are loose – no excavation is involved.

	Labour hours loading into vehicles	
Nature of material	*Per tonne*	*Per cubic metre*
Ashes	0.70	0.35
Asphalt	1.00	1.18
Bricks, loose tipped	0.60	0.46
Bricks, stacked	1.20	1.20
Broken concrete	0.80	0.56
Chalk	0.65	0.49
Clay, loamy	0.75	0.54
Clay, stiff	0.80	0.59
Clinker	0.70	0.27
Concrete, mixed	0.70	0.84
Coke	0.70	0.26
Girders, R.S. joists, etc.	1.00	–
Gravel	0.65	0.55
Gully grates and frames	1.00	–
Hardcore, brick	0.82	0.60
Hardcore, concrete	0.80	0.59
Kerb	0.90	–
Manhole covers	1.00	–
Marl	0.75	0.73
Paving slabs	1.10	–
Pipes, asbestos cement	1.20	–
Pipes, cast iron	1.00	–
Pipes, concrete	1.10	–
Pipes, clayware	1.10	–
Pipes, steel	1.10	–
Precast concrete units	1.10	–
Rails	1.00	–
Road metal	0.60	0.44
Sand	0.60	0.51
Sett paving, granite	0.75	–
Shale	0.70	0.54
Shingle	0.70	0.60
Soil	0.70	0.51
Steel reinforcement	1.20	–
Stone broken, 12 mm down	0.60	0.49
Stone broken, 12 to 63 mm	0.65	0.56
Stone broken, 63 to 100 mm	0.70	0.64
Stone broken, 100 to 150 mm	0.75	0.71
Stone broken, 150 to 225 mm	0.80	0.75
Tarmacadam	1.10	0.68
Timber, carcassing – soft wood	1.20	0.36
Trench timber	1.15	0.34

Table 12.9 Loading loose material mechanically

Notes: 1. The hours shown are those taken by the plant to load 1 m³. They also represent the hours the transporting plant stands while 1 m³ is loaded into it.
2. For the time taken by various types of transporting plant to haul to tip at various lengths of haul, see Table 12.11
3. Knowing the weight of the materials in tonnes per m³, the plant hours loading per tonne may be calculated.

Type of plant used to load	Size of bucket in m³	Nature of material loaded and plant hours loading per m³			
		Ashes, clinker, coke, media, sand	Concrete aggregates, chips, gravel road metal and shingle	Brick, concrete and stone hard-core	Asphalt and tarmacadam
Mechanical dragline or shovel	0.18	0.127	0.153	0.153	0.148
Mechanical dragline or shovel	0.28	0.073	0.077	0.091	0.085
Mechanical dragline or shovel	0.38	0.053	0.056	0.065	0.061
Mechanical skimmer	0.18	0.160	0.169	0.196	0.187
Mechanical skimmer	0.28	0.095	0.100	0.116	0.111
Mechanical skimmer	0.38	0.071	0.075	0.087	0.083
Mechanical grab	0.18	0.316	0.333	0.387	0.373
Mechanical grab	0.28	0.212	0.225	0.260	0.248
Mechanical grab	0.38	0.150	0.157	0.168	0.160
Direct loading from elevated storage hoppers	–	0.040	0.046	–	0.053
'Hilift' loading shovel	0.18	0.105	0.111	0.127	0.120
'Hilift' loading shovel	0.28	0.069	0.073	0.076	0.080
'Hilift' loading shovel	0.38	0.053	0.056	0.059	0.061

Table 12.10 Wheeling in 0.1 m³ handbarrows to a 20 m run

Note. The hours shown refer to barrow runs up to 20 m.

Nature of ground	Labour hours wheeling and tipping per cubic metre
Chalk	1.00
Loamy soil, soft or sandy clay	0.83
Stiff clay	0.91
Gravel	0.69
Marl	0.81
Sand	0.67
Soil	0.79
Hardcore, broken rock or concrete	0.87
Metalling, scarifyings, etc.	0.80

Table 12.11 Time taken per load to haul to tip, off-load and return

Notes: 1. The hours shown are the average hours taken per round trip, including the time taken to off-load. They do not allow for the time the vehicle stands while it is being loaded.
2. The hours shown for haulage by lorries are based on lorries fitted with mechanical tipping gear and assume that the vehicles run on roads or reasonably level firm surfaces.
3. The hours shown for haulage by tractor-drawn wagons or by locomotive and wagons are the hours taken 'per set of wagons', allowing for the off-loading time, it being assumed that sufficient labour is available to off-load without undue loss of time. In the case of locomotive and wagons, shunting time is also allowed for in the hours shown.
4. For the cubic metres of material of various kinds hauled per load, see Table 12.7

Length of haul from site where loaded to point of off-loading in metres and kilometres	*Type of haulage plant used and plant hours per load*					
	Lorries fitted with mechanical tipping gear	*Dumpers*	*Caterpillar tractor-drawn wagons*	*Wheeled tractor-drawn wagons*	*Locomotive and wagons on 600 mm gauge Decauville track*	*Locomotive and wagons on 1400 mm gauge track*
100 m	0.07	0.09	0.14	0.09	0.14	0.17
200 m	0.09	0.11	0.18	0.11	0.18	0.21
300 m	0.11	0.13	0.22	0.13	0.22	0.26
500 m	0.13	0.15	0.27	0.15	0.27	0.31
1 km	0.18	0.23	0.35	0.20	0.41	0.45
2 km	0.27	–	–	0.32	0.58	0.68
3 km	0.34	–	–	–	–	–
4 km	0.42	–	–	–	–	–
5 km	0.49	–	–	–	–	–
6 km	0.56	–	–	–	–	–
7 km	0.62	–	–	–	–	–
8 km	0.68	–	–	–	–	–
9 km	0.75	–	–	–	–	–
10 km	0.80	–	–	–	–	–

From Table 12.11

The time taken by the lorry to haul one load to tip 0.3 km distant off-load and return for the next load = 0.11 h per load.

Time taken to load lorry $= 0.071 \times 9.28 = 0.65$ h

Time of haul $= 0.11$ h

$$\text{Number of lorries required} = \frac{(\text{Time to load}) + (\text{Time of haul})}{\text{Time to load}}$$

$$= \frac{0.65 + 0.11}{0.21} = 3.62 \text{ lorries}$$

Actual number of lorries used = 4

The cost of excavation per cubic metre is therefore built up as follows:

Plant and Labour cost per hour

Mechanical excavator	= 1 h at £22.00 per h	= £22.00
4 lorries	= 4 h at £10.00 per h	= £40.00
3 labourers	= 3 h at £ 6.00 per h	= £18.00
Plant and labour cost per h		= £80.00

Output of mechanical excavator = 0.071 h per m^3

Estimated cost per m^3 = 0.071 × £80.00 = £5.68 per m^3 excavation.

The cost of disposal per cubic metre is therefore built up as follows:

Plant and labour cost per hour.

4 lorries = 4 h at £10.00 per h	=	£40.00
1 labourer 1 h at £ 6.00 per h	=	£ 6.00
Plant and labour cost per h	=	46.00

Output = 0.071 per m^3

Estimated cost per m^3 = 0.071 × £46 = £3.27 per m^3 disposal.

Excavation, using tractors of the caterpillar track type

Excavation by tractors and scrapers on work which is suited to this type of plant is a very cheap form of excavation, and these machines are capable of large outputs per hour on lengths of haul for which they are suited.

In carrying out excavation with this type of plant the following conditions must be fulfilled:

1. The nature both of the work to be carried out and of the ground to be excavated must be such that it is suitable for this type of plant.
2. The length of haul to tip should not be excessive.

The plant consists of two units, the tractor and the scraper.

The scraper is in itself the container in which the excavated material is hauled. While being hauled by the tractor it excavates the ground with a knife-cutting edge, the excavated material passing into the body of the scraper. When full, the cutting edge is raised and the tractor hauls the load to tip, where it is off-loaded.

This type of plant is eminently suited for surface excavation in connection with aircraft runways, large road contracts, site levelling, etc. where the excavated material is cut and filled or disposed of on a length of haul not in excess of 450 m. In the case of longer hauls, excavation with this type of plant is not economical, and in such cases mechanical excavating plant in conjunction with vehicular transport should be used.

Tractors and scrapers are capable of excavating all normal ground such as soil, loamy soil, loamy clay, gravel, etc. With hard grounds such as chalk, soft rock, etc., the ground is generally broken up and loosened by a tractor-drawn ripper or rooter prior to using a scraper which greatly reduces wear on the scraping unit.

Tractors of various makes and horse-power are available, these in turn being capable of handling scrapers of various capacities.

In using this type of plant, labour in attendance is not usually necessary but, on occasions, one may be required. The estimator must judge whether attendant labour is necessary or not, depending on the nature of the work and the prevailing site conditions.

For purposes of estimating the cost of excavation using this type of plant, Table 12.13 gives data for excavating and hauling easily dug sand with tractors and scrapers of all sizes, the data covering lengths of haul up to 480m. For classes of ground other than easily dug sand, multipliers are given in Table 12.14

Example.   Calculate the cost of excavating soft clay, using a caterpillar D6H tractor with a scraper, the excavated material being hauled an average length of haul of 360 m.

For purposes of this example it will be assumed that the hire rate or working cost of the tractor and scraper including driver, fuel and oil is £27 per hour and also that one man is in attendance on the plant.

Referring to the Tables 12.13 and 12.14.

The hours taken by a D6H tractor and scraper to excavate and haul easily dug sand to an average length of haul of 360 m = 0.031 h per m^3.

Referring to the table of multipliers it will be seen that the plant-hour multiplier for soft clay is 1.35.

Plant hours excavating and hauling soft clay = 1.35 × 0.031 h per m^3
$$= 0.042 \text{ h per m}^3$$

Cost per m^3 of excavating and hauling soft clay is:

Tractor and scraper 0.042 h at £27.00 per h = £1.13 per m^3
Labour   0.042 h at £ 6.00 per h = £0.25 per m^3

Total cost   = £1.38 per m^3

Note: The haulage cost of the plant is not included. This may be allowed for by assessing the haulage cost or getting a firm quotation for it and dividing this cost by the total cubic metres of excavation to be carried out. The resultant figure may then be added to the excavation cost per cubic metre. The plant haulage cost may also be allowed for on a lump sum basis if desired.

Table 12.12 Characteristics of Caterpillar tractors

	Caterpillar tractor type No.	
Description	D6D	D7G
Flywheel kilowatts	104	149
Approximate litres of fuel tank capacity	295	435
Blade straight width, m	3.2	3.66
Blade angle width, m	3.90	4.27
Engine model	3306	3306
Rated Engine r.p.m.	1900	2000
Weight of machine, tonne	14.65	20.09

Table 12.13 Excavation by tractors and scrapers

Note: The hours shown are for excavating firm sand. For other classes of ground use the table of multipliers shown.

Type of tractor and scraper	Rated capacity		Length of haul (one way) in metres and plant hours per cubic metre				
	Heaped m^3	Struck m^3	90 m	180 m	270 m	360 m	480 m
621F Caterpillar	15	11	0.024	0.029	0.041	0.046	0.052
631E Caterpillar	24	16	0.015	0.020	0.026	0.031	0.035
651E Caterpillar	34	25	0.009	0.011	0.015	0.016	0.020

Table 12.14 Multipliers for excavation by tractors and scrapers

Nature of ground	Plant hour multiplier
Firm sand	1.00
Firm soil	1.25
Gravel	1.10
Loamy soil, soft or sandy clay and marl	1.35
Stiff clay	1.50
Soft chalk	3.00
Soft friable rock	4.00

Earth moving by angledozers and bulldozers

Angledozers and bulldozers consist of a tractor to which is fitted a blade which pushes the ground as the tractor forces its way ahead. In the case of the angledozer this blade can be set at right angles to the direction of motion or at an angle, whereas in the bulldozer it is set permanently at right angles to the direction of motion of the tractor.

The angledozer may, therefore, be used as a bulldozer, but not the bulldozer as an angledozer.

The machines are available in various sizes and kilowatt ratings.

Angledozers are suitable for cutting track out of sloping ground, such as hillsides, and are also used for trench filling or moving earth bodily in a line parallel to the direction taken by the machine.

Bulldozers may be used for digging ramps or short cuttings and also pushing ground or materials. The maximum length of push should not exceed 90 m, as they are uneconomical for greater distances.

In both machines the output is greater when working downhill than on the level or uphill. Data are shown in the tables for various sizes of machines, carrying out the work on the level and on gradients of 1 in 7 up and 1 in 7 down. The data shown are for moving sand.

For ground other than sand, a table of multipliers is given.

Table 12.15 Earth moving by angledozers

Notes: 1. The hours shown are for moving sand. Use the table of multipliers shown for other classes of ground.
2. The data shown are applicable also to refilling trenches by angledozer.

Type of tractor	Gradient and plant hours per cubic metre		
	Level	1 in 7 up	1 in 7 down
D4C	0.023	0.037	0.017
D6D	0.017	0.028	0.013
D7G	0.013	0.021	0.011
D8N	0.011	0.017	0.008

Table 12.16 Multipliers for earth moving by angledozers

Nature of ground	Plant hour multiplier
Sand	1.00
Broken rock	2.50
Chalk	3.00
Gravel	1.25
Loamy soil, soft or sand, clay or marl	1.33
Stiff clay	2.70
Soil	1.25

Table 12.17 Earth moving by bulldozers

Notes: 1. The hours shown are for moving sand. Use the table of multipliers shown for other classes of ground.
2. The data shown are also applicable to refilling excavations by bulldozer.

Type of tractor	Length of push in metres	Gradient and plant hours per cubic metre		
		Level	1 in 7 up	1 in 7 down
D4C	7	0.016	–	–
D4C	18	0.035	0.059	0.028
D4C	36	0.053	0.087	0.040
D4C	54	0.073	0.120	0.056
D4C	72	0.100	–	–
D4C	108	0.136	–	–
D6D	7	0.012	–	–
D6D	18	0.027	0.044	0.020
D6D	36	0.040	0.067	0.031
D6D	54	0.055	0.088	0.041
D6D	72	0.076	–	–
D6D	108	0.107	–	–
D7G	7	0.009	–	–
D7G	18	0.015	0.025	0.012
D7G	36	0.023	0.039	0.017
D7G	54	0.033	0.055	0.025
D7G	72	0.044	–	–
D7G	108	0.061	–	–
D8N	7	0.007	–	–
D8N	18	0.012	0.020	0.009
D8N	36	0.017	0.029	0.013
D8N	54	0.025	0.041	0.020
D8N	72	0.035	–	–
D8N	108	0.048	–	–

Table 12.18 Multipliers for earth moving by bulldozers

Nature of ground	Plant hour multiplier
Sand	1.00
Broken rock	2.50
Chalk	3.00
Gravel	1.25
Loamy soil, soft or sandy clay or marl	1.33
Stiff clay	2.70
Soil	1.25

Table 12.19 Excavator – sundry works

Note. For grass seeding to large areas, see chapter 23.

Description	Unit	Labour hours
Clay puddling in 150 mm layers	m^3	10.00
Clay puddling in 225 mm layers	m^3	8.00
Fill sandbags	Each	0.25
Trim formations	m^2	0.13

Table 12.20 The bulkage of materials after excavation

Nature of material	Bulkage
Sand	1.10
Gravel	1.20
Vegetable soil	1.30
Loamy soil, soft or sandy clay or marl	1.40
Chalk	1.40
Stiff clay	1.50

Excavation and earthwork below ground

Earthwork support

The cost of earthwork support is influenced by

1. The nature of the ground and the depth of the excavation.
2. The type of excavation which has to be timbered.
3. The material cost of the timber and the assessment made as to how many times it can be used before it is unfit for further use.

This assessment influences the allowance for use and waste, and in this connection it should be noted that the quality of the timber purchased and the treatment it receives from those who fix and remove it has much to do with its lasting properties.

The nature of the ground influences the cost, for in firm ground less timber is required than in moderately firm ground, but in loose or water-bearing ground considerably more timber is necessary than in either of these cases.

Where much water has to be contended with, steel sheet piling is often used in lieu of timbering, the piles being driven and the excavation removed afterwards, walings and struts being fixed to hold the piles as the excavation proceeds. This method of shoring excavations in water-bearing ground of large water content has much to recommend it, provided that the amount of work involved merits its adoption. It greatly simplifies the excavation and tends to speed up the work with a saving in pumping and excavation costs.

In estimating the cost of timbering, the estimator will note that the labour cost per square metre of face in similar classes of ground is variable, depending upon the type of excavation and the depth.

Allowance for use and waste of timber

The cost of earthwork support is made up of two elements:

1. The labour cost of fixing and removing the timber.
2. The material cost of the timber itself.

The method of allowing for the material cost of the timber, knowing its initial cost per cubic metre, is as follows:

1. The number of times the timber can be used before it is unfit for further use is assessed.
2. The cost of the timber is allowed for by dividing its initial cost per cubic metre by the assessed number of times it can be used, and writing it off in the rates tendered for the timbering of the excavations.

   For example, assuming the initial cost of the timber is £220 per m^3 and that, for purposes of estimating, it can be used ten times, then:

   The allowance for use and waste $= \dfrac{£220}{10} = £22$ per m^3. This £22 per m^3 multiplied by

   the cubic metre of timber required per metre of trench or per square metre of face in the case of shafts, basements and tunnels, is the allowance to make in the estimate for the use and waste of the timber concerned.

In connection with the timbering of tunnels the estimator should carefully note whether allowance has to be made for leaving it in the tunnel in whole or in part, and if so, allow for its cost accordingly. In many instances no mention is made as to whether the timber has to be left in or not, and the onus of deciding how much timber can be removed, if any, is put upon the estimator, who must assess this for himself, taking into consideration the probable nature of the ground.

In assessing the amount of timber recoverable from a tunnel, it is wise to err on the side of safety by not setting too high a value on the number of uses which may be expected from it.

Use of compressors and pneumatic tools

In hard ground such as rock, compressors and pneumatic tools are used to advantage, as the rock is broken out at considerably less cost by this means than by hand.

Data are shown for breaking out rock and excavating, using such plant in connection with the excavation of trenches, shafts and basements, the excavated material being staged to the surface by hand. Data applicable to both two-and three-tool compressors are shown.

Excavating trenches, shafts, basements and similar types of excavation by hand

The Seventh edition of Standard Method of Measurement dispenses with 1.5 m stage lifts of excavation and in its place uses the following depth classifications. Maximum depths not exceeding 0.25 m, 1.00 m, 2.00 m, and thereafter in 2.00 m stages. This revision is based on the supposition that the vast majority of larger contracts excavation work is invariably carried out by machine and the maximum depths stated will serve to indicate to the estimator the type of excavating machine(s) that will be required. Data for hand digging in 1.5 m lifts are, however, left in the text, because inevitably the estimator will come across some contracts where owing to lack of space work must be carried out by hand in stage lifts.

In order that the estimator may be conversant with the method and operations involved in carrying out excavation below the ground and in excavating and staging it to the surface by hand, this is dealt with in detail.

In excavating below ground a man is considered as being capable of throwing the excavated material to a height of 1.5 m or, as it is termed, in 1.5 m lifts, thus in excavating down to a depth of 1.5 m and not in excess of this the man throws the excavated material directly to the surface. In excavating at a depth greater than 1.5 m, i.e. in the second or 1.5 to 3.0 m lift, the man is incapable of throwing to the surface, so a stage or platform is inserted in the excavation at a level of 1.5 m below the ground, on to which he throws the excavated material. He is thus able to excavate down to a level of 3.0 m below the ground, throwing on to this stage. The excavated material is thrown from the stage to the surface by another man, referred to as a stageman, and is cleared back from the side of the trench by a third man, so as to make room for the excavated material as it is thrown out.

This process of staging at 1.5 m intervals is continued as the depth increases, thus an excavation 3.0 m deep has a stage at a level of 1.5 m below the ground, an excavation of 4.5 m deep, one at a level of 1.5 m and one at 3.0 m below the ground and so on, the excavated material being thrown in turn from stage to stage, and finally to the surface.

One stageman per stage and one man clearing on top handle the material excavated by two men excavating in normal grounds other than rock.

In rock-bearing ground where the rate of excavation carried out by hand is slow it is uneconomical to keep stagemen standing idle waiting for the excavated material to be thrown up to them. In such cases it is usual to break out and loosen the rock in the trench and, when a sufficient amount has been excavated, place stagemen on the stage in order to stage it to the surface. In such grounds it is wise to consider the use of compressors and pneumatic tools to break out the rock, as this greatly increases the rate of output.

Table 12.21 The build up of labour hours per cubic metre to excavate, backfill and compact a trench in soft clay

| | Labour hours per cubic metre | | | |
| | First lift | Second lift | Third lift | Fourth lift |
Labour operations	0 to 1.5 m	1.5 to 3.0 m	3.0 to 4.5 m	4.5 to 6.0 m
Excavate and throw to a 1.5 m height	2.80	2.80	2.80	2.80
Stage in 1.5 m lifts	–	1.40	2.80	4.20
Clear from top of trench	–	1.40	1.40	1.40
Backfill and compact	1.47	1.47	1.47	1.47
Total labour hours per m³	4.27	7.07	8.47	9.87

Table 12.22 Excavate, backfill and compact trenches by hand in dry ground

Notes: 1. The labour hours shown are those taken to excavate, backfill and compact trenches per cubic metre for the various lifts, exclusive of timbering, the excavated material being staged to the surface.
2. The labour hours shown for the various lifts are applicable to the whole volume of excavation contained in the lift to which they refer.
3. For timbering trenches, see Table 12.26
4. For excavating trenches in rock, using compressors and pneumatic tools, see Table 12.23
5. For excavating trenches, using mechanical excavating plant of the backacter or endless bucket type, see Table 12.24
6. The rock referred to is rock with many fissures and of a stratified nature.

	Nature of ground and labour hours excavating, backfilling and compacting trenches in lifts of 1.5 m per cubic metre						
Depth of excavation in metres	*Firm sand*	*Loamy soil, soft or sandy clay or marl*	*Stiff clay*	*Gravel or compact ground requiring picking*	*Chalk*	*Soft stratified fissured rock*	*Hard stratified fissured rock*
First lift, 0 to 1.5 m	3.70	4.27	4.67	4.80	5.57	8.63	11.20
Second lift, 1.5 to 3.0 m	6.27	7.07	7.70	8.27	10.13	12.80	15.73
Third lift, 3.0 to 4.5 m	7.53	8.47	9.27	10.00	12.27	14.93	18.00
Fourth lift, 4.5 to 6.0 m	8.80	9.87	10.80	11.73	14.40	17.00	20.30

Table 12.23 Excavate, backfill and compact trenches in rock, using pneumatic tools

Notes: 1. The plant and labour hours shown are those taken to break out, excavate, backfill and compact the trench per cubic metre, the excavated material being staged to the surface.
2. The plant and labour hours shown for the various lifts are applicable to the whole volume of excavation contained in the lift to which they refer.
3. The data shown are for a 2-tool compressor with two tools in use. If a 3-tool compressor is used the compressor hours per cubic metre are two-thirds of those shown, the labour hours on tools and the labour hours remaining the same.
4. In the case of stratified fissured rocks little or no timbering of the trench is necessary, but on occasions a pair of poling boards held by a strut may be required here and there if there is a tendency for isolated slips or falls to occur.
5. The stratified fissured rock referred to is soft and hard rock with many fissures, capable of being readily broken out by pneumatic tools. The solid rock is for rock in solid beds, which definitely necessitates the use of pneumatic tools or blasting to remove it economically.

	Nature of ground and compressor and labour hours excavating, backfilling and compacting trenches in lifts of 1.5 m per cubic metre								
	Soft stratified fissured rock			*Hard stratified fissured rock*			*Solid rock*		
Depth of excavation in metres	*Compressor hours*	*Labour hours on pneumatic tools*	*Labour hours*	*Compressor hours*	*Labour hours on pneumatic tools*	*Labour hours*	*Compressor hours*	*Labour hours on pneumatic tools*	*Labour hours*
First lift 0 to 1.5 m	0.40	0.80	4.80	0.53	1.06	5.87	2.23	4.46	6.40
Second lift, 1.5 to 3.0 m	0.44	0.88	8.80	0.59	1.18	9.15	2.49	4.98	11.20
Third lift, 3.0 to 4.5 m	0.48	0.96	9.60	0.64	1.28	12.27	2.72	5.44	13.60
Fourth lift, 4.5 to 6.0 m	0.52	1.04	11.20	0.69	1.38	14.40	2.93	5.86	16.00

In order to illustrate the build up of the labour hours taken per cubic metre to excavate below ground in 1.5 m lifts, a build up is shown in excavating a trench in soft clay.

The estimator should note that the labour hours per cubic metre for the various lifts are applicable to the whole volume of excavation in the lift to which they refer.

The excavation of trenches, shafts and basements is generally billed on a cubic metre basis.

The method of estimating the cost of excavating, backfilling and compacting trenches at any specific depth per cubic metre of trench exclusive of timbering is as follows:

1. The cost of excavating, backfilling and compacting the trench per cubic metre for the 0 to 1.5 m, 1.5 to 3.0 m and 3.0 to 4.5 m lifts, etc., is estimated.
2. The volume of trench excavation in cubic metres per metre of trench in each 1.5 m lift is assessed. Thus a trench 3.9 m deep would have 1.5 m of its depth in the 0 to 1.5 m lift, 1.5 m in the 1.5 to 3.0 m lift and 0.9 m in the 3.0 to 4.5 m lift.
3. The cubic metre of excavation per metre of trench in each lift is then multiplied by the estimated cost per cubic metre of excavating, backfilling and compacting the trench in the relevant lifts, and the sum of these is the total cost per metre of trench.

Example.   Estimate the cost of excavating, timbering, backfilling and compacting a trench per metre in sandy clay, the depth of the trench being 3.6 m and its width 0.75 m. For purposes of this example the following prevailing rates are assumed:

Timberman's rate, £7.00 per h
Labourer's rate, £6.00 per h
Initial cost of trench timber, £220.00 per m^3

Excavating, backfilling and compacting the trench per cubic metre.

The total depth of the trench is 3.6 m, the depth of the various lifts being as follows:

Depth of trench in the first lift, i.e. 0 to 1.5 m   = 1.5 m
Depth of trench in the second lift, i.e. 1.5 to 3.0 m = 1.5 m
Depth of trench in the third lift, i.e. 3.0 to 4.5 m   = 0.6 m

    Total depth of trench   = 3.6 m

Referring to Table 12.45 the volume of excavation per metre run in a trench 0.75 m wide and 3.6 m deep under the various lifts is as follows:

First lift, 0 to 1.5 m   = 1.13 m^3 (depth 1.5 m)
Second lift, 1.5 to 3.0 m = 1.13 m^3 (depth 1.5 m)
Third lift, 3.0 to 4.5 m   = 0.45 m^3 (depth 0.6 m)

    Total   = 2.71 m^3

Referring to Table 12.22 the labour hours taken per cubic metre to excavate, backfill and compact trenches in sandy clay at the various lifts are tabulated. These hours extended by the prevailing labour rate of £6.00 per hour give the cost of excavating, backfilling and ramming the trench per cubic metre in the various 1.5m lifts thus:

First lift   0 to 1.5 m, 4.27 labour hours at £6.00 per h = £25.62 per m^3
Second lift   1.5 to 3.0 m, 7.07 labour hours at £6.00 per h = £42.42 per m^3
Third lift   3.0 to 4.5 m, 8.47 labour hours at £6.00 per h = £50.82 per m^3

Therefore the cost of excavating, backfilling and compacting the trench per metre, exclusive of timbering is:

First lift	0 to 1.5 m, 1.13 m^3 at £25.62 per m^3	= £28.96 per m
Second lift	1.5 to 3.0 m, 1.13 m^3 at £42.42 per m^3	= £47.95 per m
Third lift	3.0 to 4.5 m, 0.45 m^3 at £50.82 per m^3	= £22.87 per m

The cost of excavating, backfilling and compacting the trench = £99.78 per m

The cost of excavating, backfilling and compacting per cubic metre = £36.82 per m^3

Timbering the trench per metre.

Sandy clay comes under the category of moderately firm ground.
Referring to Table 12.26 for timbering moderately firm ground the following data for estimating purposes are obtained:

1. Timberman and labour hours timbering trench 3.6 m deep per m = 2.38 h each.
2. Quantity of timber required to timber the trench per m = 0.54 m^3.
3. The number of uses of the timber to allow for use and waste = 15.

Since the initial cost of the timber is £220 per m^3, the number of uses 15 and the amount required 0.54 m^3, the allowance for 'use and waste' per metre of trench is:

$$\frac{£220}{15} \times 0.54 \text{ m}^3 = £7.92 \text{ per metre}$$

The cost of timbering the trench per metre, therefore, is:

Timberman	= 2.38 h at £7.00 per h	= £16.66 per m
Labourer	= 2.38 h at £6.00 per h	= £14.28 per m
Allow for use and waste of timbering		= £ 7.92 per m
Cost of timbering trench		= £38.86

The total cost of excavating, timbering, backfilling and compacting the trench per metre is:

Excavating, backfilling and compacting	= £ 99.78 per m
Timbering	= £ 38.86 per m
Total cost	= £138.64 per m

Note. In the above example the estimated cost per cubic metre of the timbered excavation averaged throughout the whole depth of the trench is:

The cost of excavating, timbering, backfilling and compacting the trench per metre divided by the total cubic metres of excavation in the trench 3.60 m deep.

$$\frac{£138.64}{2.71} = £51.16 \text{ per m}^3$$

Excavate trenches, using plant of the backacter and multi-bucket type

In suitable ground and where the volume and depth of excavation merits it the excavation of trenches is commonly carried out by suitable mechanical plant.
The backacter type of plant consists of a mechanical excavator fitted with backacter equipment. The same plant may be fitted with equipment other than the backacter, such as

shovel, skimmer, dragline, etc. The plant is, therefore, useful in that it is universal and capable of being used on classes of work other than trenching simply by fitting the other attachments.

The trench excavated by a backacter tends to be rough on the face, necessitating a certain amount of trimming prior to carrying out the timbering, this roughness being due to the action of the bucket, which drags the sides as the machine excavates. This type of plant is capable of excavating trenches in suitable ground to a depth of approximately 3.6 m.

The multi-bucket plant, sometimes referred to as trenching plant, is designed solely for use in excavating trenches. The machine works on the principle of a dredger and, in suitable grounds, cuts a clean trench with smooth faces, necessitating little or no trimming before the timbering is fixed in position. These machines are available in two main types:

1. Plant capable of excavating trenches to a depth of approximately 1.5 m.
2. Plant capable of excavating trenches to a depth of approximately 4.2 m.

The former type of plant is extremely useful for trenching for water mains, ducts, cables, etc., where a trench of narrow width is required and the depth of trench is comparatively shallow, while the latter is ideal for excavating trenches for sewers, where a greater depth of excavation has to be carried out.

Mechanical excavation of trenches is carried out in good ground which is dry, and in no way loose, it being the object when using such plant to excavate the trench to the depth required or to the maximum depth possible prior to placing the timber in position. Grounds such as loam, soft or sandy clay, stiff clay, firm sand and gravel, etc., are ideal. Grounds such as soft chalk or ground containing a small quantity of broken stratified friable rock may also be excavated mechanically, always provided that the toughness of the rock present is such that it does not jar the mechanism of the plant. In really hard rocky ground, plant of this type is unsuitable; it may do the work, but at the expense of straining the machine and its mechanism. Such grounds wear down the teeth of the buckets and distort the buckets themselves, making it uneconomical to use them on account of the wear and tear involved.

Earthwork support for trenches excavated by mechanical excavating plant

The timbering of trenches carried out by mechanical excavating plant usually consists of runners, these being placed in position immediately behind the machine as it excavates, where they are held in position by struts. The runners may then be securely held afterwards if required by walings. The runners are generally well spaced (depending on the nature of the ground) and the timbering of a comparatively light nature. For normal conditions and suitable soils the data shown for timbering trenches in firm ground in Table 12.26 should be used for estimating purposes.

The plant hours shown are not based on the maximum outputs of these machines, but on what has been found from experience to be a fair average output. By referring to the table the estimator will note that the plant hours per cubic metre decrease with the depth down to approximately 1.5 m below the ground, at which depth they commence to increase, this being due to the fact that the output of these machines increases with the depth down to 1.5 m, after which it falls away as timbering has to be carried out and care taken to secure the sides of the excavation as the excavation proceeds. At the greater depths the machine cannot be allowed to run too far ahead until the timber is in position, all of which tends to reduce the output.

The tables shown for trenches, using mechanical plant

In connection with the tables shown the estimator should note the following:

1. The tables shown for excavating the trenches, using backacters and multi-bucket type plant, give the plant and labour hours per cubic metre.
2. The table shown for timbering trenches give the timberman and labourer hours taken to fix and remove the timber per square metre of trench, both sides of the trench being allowed for in the hours shown. In order that the use and waste of timber may be allowed for, the number of uses of the timber is stated as is the amount of timber required per square metre of trench.

Example. Estimate the cost of excavating per cubic metre in sandy clay, using a multi-bucket trench excavator, the depth of trench being 2.7 m and its width 0.75 m. The following prevailing rates are assumed for purposes of this example:

Hire rate or working cost per hour of a multi-bucket trench excavator, inclusive of driver, fuel and oil	£ 25.00 per h
Timberman's rate	£ 7.00 per h
Labourer's rate	£ 6.00 per h
Initial cost of trench timber	£220.00 per m³

Excavating the trench per cubic metre.

Referring to Table 12.45 the volume to be excavated per metre in a trench of 0.75 m wide and 2.7 m deep is 2.04 m³.

Referring to Table 12.24 the cost of excavating per cubic metre at 2.7 m deep, using a multi-bucket trench excavator.

Trench excavator	= 0.120 h at £25.00 per h	=	£ 3.00 per m³
Labour hours	= 2.98 h at £6.00 per h	=	£17.88 per m³
			£20.88

The cost of excavating the trench per cubic metre = £20.88

Timbering the trench per square metre.

Referring to Table 12.26 for timbering in firm ground the following data for estimating purposes are found:

1. Timberman and labour hours timbering a trench 2.7 m deep per metre = 0.41 h each.
2. Quantity of timber required to timber the trench per metre = 0.20 m³.
3. The number of uses of the timber to allow for use and waste = 15.

Since the initial cost of the timber is £220 per m³, the number of uses 15, and the amount required 0.2 m³, the allowance for use and waste per metre of trench is:

$$\frac{£220}{15} \times 0.2 \text{ m}^3 = £2.93 \text{ per m}$$

The cost of timbering the trench per square metre, therefore, is:

Timberman	0.41 h at £7.00 per h	= £2.87 per m
Labour	0.41 h at £6.00 per h	= £2.46 per m
Allow for use and waste of timber		= £2.93 per m
Cost of timbering trench		= £8.26 per square metre

Note. In the example shown, the cost of hauling the plant has not been included.

Table 12.24 Excavate, backfill and compact trenches, using backacter plant

Notes: 1. The data show the plant and labour hours taken to excavate, backfill and compact trenches per cubic metre, the backfilling and compacting being carried out by hand.
2. The plant and labour hours shown at the various depths are applicable to the whole volume of excavation from ground level to the depths shown.
3. The hours shown do not include for timbering, for which see Table 12.26 the data shown for timbering in firm ground are applicable.

Nature of ground and plant and labour hours excavating, backfilling and compacting trenches per cubic metre

Depth of trench in metres	Firm sand		Loamy soil, soft or sandy clay or marl		Gravel		Stiff clay		Chalk	
	Plant hours	Labour hours	Plant hours	Labour hours	Plant hours	Labour hours	Plant hours	Labour hours	Plant hours	Labour hours
0.30 m³ backacter										
0.25	0.091	2.17	0.128	3.07	0.136	3.27	0.153	3.68	0.181	4.35
1.00	0.085	2.04	0.120	2.88	0.128	3.07	0.145	3.49	0.171	4.08
2.00	0.085	2.04	0.120	2.88	0.128	3.07	0.145	2.49	0.171	4.08
4.00	0.100	2.37	0.140	3.36	0.148	3.56	0.169	4.07	0.197	4.75
0.40 m³ backacter										
0.25	0.069	2.08	0.100	3.00	0.107	3.21	0.120	3.60	0.139	4.15
1.00	0.067	2.01	0.093	2.79	0.101	3.03	0.115	3.45	0.133	4.00
2.00	0.067	2.01	0.093	2.79	0.101	3.03	0.115	3.45	0.133	4.00
4.00	0.073	2.20	0.107	3.21	0.115	3.45	0.128	3.84	0.147	4.40

Table 12.25 Excavate, backfill and compact trenches, using multi-bucket excavator

Notes: 1. The data show the plant and labour hours taken to excavate, backfill and compact trenches per cubic metre, the backfilling and compacting being carried out by hand.
2. The plant and labour hours shown at the various depths are applicable to the whole volume of excavation from ground level to the depths shown.
3. The hours shown do not include for timbering, for which see Table 12.26 the data shown for timbering in firm ground are applicable.

Nature of ground and plant and labour hours excavating, backfilling and compacting trenches per cubic metre

Depth of trench in metres	Firm sand		Loamy soil, soft or sandy clay or marl		Gravel		Stiff clay		Chalk	
	Plant hours	Labour hours	Plant hours	Labour hours	Plant hours	Labour hours	Plant hours	Labour hours	Plant hours	Labour hours
0.25	0.091	2.17	0.128	3.07	0.136	3.27	0.155	3.71	0.181	4.35
1.00	0.080	1.92	0.112	2.67	0.120	2.88	0.136	3.27	0.160	3.84
2.00	0.063	1.48	0.088	2.11	0.093	2.24	0.107	2.56	0.125	3.00
4.00	0.085	2.04	0.120	2.98	0.128	3.07	0.145	3.50	0.171	4.08

Table 12.26 Earthwork support to trenches in dry ground

Notes: 1. The timberman and labourer hours shown are those taken to timber the trench per metre, both sides of
the trench being allowed for in the hours shown.
2. The data shown allow for both fixing and removing the timber.
3. The cubic metre of timber shown represent the total cubic metres of timber required to timber the trench
per metre, both sides of the trench being allowed for. For purposes of estimating, allow for use and waste
of timber 15 uses.

	Nature of ground, cubic metres of timber required to timber trench, and timberman and labour hours timbering trenches per metre								
	Firm ground			*Moderately firm ground*			*Loose ground*		
Depth of trench in metres	*Timber* m^3	*Timberman* h	*Labourer* h	*Timber* m^3	*Timberman* h	*Labourer* h	*Timber* m^3	*Timberman* h	*Labourer* h
0.9	0.07	0.10	0.10	0.14	0.40	0.40	0.22	0.80	0.80
1.2	0.09	0.14	0.14	0.18	0.55	0.55	0.27	1.10	1.10
1.5	0.11	0.19	0.19	0.22	0.73	0.73	0.33	1.47	1.47
1.8	0.13	0.23	0.23	0.26	0.92	0.92	0.39	1.85	1.85
2.1	0.16	0.29	0.29	0.32	1.12	1.12	0.48	2.16	2.16
2.4	0.18	0.34	0.34	0.36	1.34	1.34	0.54	2.70	2.70
2.7	0.20	0.41	0.41	0.40	1.58	1.58	0.60	3.17	3.17
3.0	0.22	0.49	0.49	0.44	1.72	1.72	0.66	3.66	3.66
3.3	0.24	0.55	0.55	0.48	2.09	2.09	0.72	4.19	4.19
3.6	0.27	0.63	0.63	0.54	2.38	2.38	0.81	4.77	4.77
3.9	0.29	0.70	0.70	0.58	2.64	2.64	0.87	5.23	5.23
4.2	0.31	0.80	0.80	0.62	2.97	2.97	0.92	5.95	5.95
4.5	0.33	0.88	0.88	0.66	3.30	3.30	1.00	6.60	6.60
4.8	0.36	0.98	0.98	0.72	3.60	3.60	1.08	7.27	7.27
5.1	0.40	1.08	1.08	0.80	3.98	3.98	1.20	7.96	7.96
5.4	0.43	1.18	1.18	0.86	4.36	4.36	1.29	8.72	8.72
5.7	0.45	1.29	1.29	0.90	4.76	4.76	1.35	9.52	9.52
6.0	0.48	1.38	1.38	0.96	5.06	5.06	1.44	10.20	10.20

Shafts, basements and excavations of a similar type

Under the above heading may be classed excavations carried out in connection with shafts,
manholes, small chambers, basements, pump chambers, underground tanks, such as humus and
sedimentation tanks, etc.

The method of excavating is governed to a large extent by the area of the excavation, and, in
general, is carried out in one of three ways:

1. Excavating by hand and staging the excavated material to the surface.
2. Where the area of the excavation is not excessive, excavated by hand and hoisting the
excavated material to the surface by jack roll or hand winch. This method is commonly
used in excavating shafts.
3. Excavating by hand and hoisting the excavated material to the surface by power-driven
crane and skips.

In carrying out excavations of the basement type of sufficiently large area the estimator should
note that it may be possible to do so using a combination of plant. Thus, if the area and the volume
of excavation justify it, the whole or part of the excavation may be removed by a mechanical

excavator of the dragline type. In the event of the excavation being of such depth that the dragline cannot wholly remove it, on account of the sides falling in, it may be used to excavate to the maximum possible depth, after which a power-driven crane may be used with skips to complete the excavation. A combination such as this would, in suitable circumstances, carry out the excavation at considerably less cost than if it were carried out wholly by crane and skips.

In using a power-driven crane to hoist the excavated material, a point which should be borne in mind is that the crane hours per cubic metre of excavation are governed by the number of men excavating, for if the area of the excavation is such that four men can excavate, the crane hours per cubic metre are only one-half of what they would be were two men excavating.

Earthwork support to shafts, basements and excavations of a similar type

Depending upon the area of the excavation the cost of earthwork support per square metre of face varies, it being greater in the case of excavations of large area than of small. For purposes of estimating, tables are shown for earthwork support to both shafts and basements; the data shown for shafts may be taken as applying to excavations whose area is not in excess of 20 m^2 and that for basements for excavations of areas in excess of this.

Excavation and timbering to excavations of the shaft and basement type are generally estimated in lifts of 1.5 m, the excavation being in cubic metres and the earthwork support per square metre of face. On occasions an all-in rate has to be quoted for the timbered excavation per cubic metre, in which case the cost of excavating and earthwork support to the excavation as a whole must be assessed, which cost, divided by the total cubic metres of excavation, gives the estimated cost of the timbered excavation per cubic metre.

The tables shown for excavations of the shaft and basement type

The data shown in the table for excavation do not allow for the breaking out of hard surfaces, such as roads, nor for the reinstatement of surfaces. The estimator is referred to the first part of this chapter, where data are given for breaking out hard surfaces, and to Chapter 21 where data are shown for the reinstatement of surfaces. The reinstatement of footways is dealt with in chapter 21.

For excavating large excavations of the pit type by mechanical excavating plant of the dragline and shovel type, reference should be made to the first part of this chapter.

To illustrate the method of using the tables in building up an estimate of the cost of this type of excavation a typical example is shown.

Example.   A manhole has to be excavated in sandy clay, the dimensions of the excavation being 1.8 m × 1.35 m × 2.7 m deep. The excavation is to be carried out by hand and staged to the surface. Estimate the cost of the following:

1. The excavating per cubic metre at the various 1.5 m lifts.
2. The earthwork support per square metre at the various 1.5 m lifts.
3. The cost of the timbered excavation per cubic metre.

For purposes of this example the following rates are assumed:

Timberman's rate = £7.00 per h
Labourer's rate   = £6.00 per h

Material cost of the timber used for timbering the excavation = £220.00 per m^3

Excavating and staging to the surface per cubic metre.

Referring to Table 12.27, the labour hours taken to excavate and stage sandy clay per cubic metre at the various lifts are shown. The labour rate being £6.00 per h, the cost per cubic metre is given by:

First lift   0 to 1.5 m, 3.03 labour h at £6.00 per h = £18.18 per m³
Second lift 1.5 to 3.0 m, 6.40 labour h at £6.00 per h = £38.40 per m³

Earthwork support per square metre.

Sandy clay comes under the category of moderately firm ground. Referring to Table 12 the data for timbering in moderately firm ground is obtained:

1. Timberman and labourer hours timbering shafts per square metre at 1.5 m lifts:

	First lift 0 to 1.5 m	Second lift 1.5 to 3.0 m
Timberman hours	0.36	0.48
Labourer hours	0.36	0.48

2. Timber required per square metre in moderately firm ground = 0.068 m³.
3. No. of uses of timber to allow for use and waste = 12.

In the case of this example the timber costs £220.00 per m³, the amount required per square metre is 0.068 m³ and the number of uses is 12.

The allowance for use and waste of timber per square metre
$$= \frac{£220}{12} \times 0.068 \text{ m}^3$$

$$= £1.25 \text{ per m}^3$$

The cost of earthwork support per square metre, therefore, is:

First lift 0 to 1.5 m
    Timberman, 0.36 h at £7.00 per h   = £2.52 per m³
    Labourer, 0.36 h at £6.00 per h   = £2.16 per m³
    Allow for use and waste of timber   = £1.25 per m³
Cost of earthwork support 0 to 1.5 m   = £5.93 per m³

Second lift 1.5 to 2.7m
    Timberman, 0.48 h at £7.00 per h   = £3.36 per m³
    Labourer, 0.48 h at £6.00 per h   = £2.88 per m³
    Allow for use and waste of timber   = £1.25 per m³
Cost of earthwork support 1.5 to 2.7m   = £7.49 per m³

The cost of the timbered excavation per cubic metre.

The above is found by assessing the cost of the total volume of excavation and the total square metre of timbering, and dividing the total cost of these two by the total cubic metres of excavation.

The excavation in the first or 0 to 1.5 m lift   = 1.8 m × 1.35 m × 1.5 m = 3.65 m^3
The excavation in the second or 1.5 to 2.7 m lift = 1.8 m × 1.35 m × 1.2 m = 2.916 m^3
The timbering in the first or 0 to 1.5 m lift   = 1.5 [2 (1.8 m + 1.35 m)] = 9.45 m^2
The timbering in the second or 1.5 m to 2.7 m lift = 1.2 [2 (1.8 m + 1.35 m)] = 7.560 m^2

The total cost of the timbered excavation is:

Excavate the first lift	(1.5 m deep) = 3.650 m^3 at £18.18 =	£ 66.36
Excavate the second lift	(1.2 m deep) = 2.916 m^3 at £38.40 =	£111.98
Earthwork support to the first lift	(1.5 m deep) = 9.450 m^2 at £ 5.93 =	£ 56.04
Earthwork support to the second lift	(1.2 m deep) = 7.560 m^2 at £ 7.49 =	£ 56.63
Total cost of the timbered excavation		= £291.01

The total cubic metres of excavation = 6.566 m^3
The cost of timbered excavation   = £291.01/6.566
                               = £44.32 per m^3

Note. In the above example no allowance has been made for refilling and ramming, if any, or loading and hauling the excavated material to tip.

For refilling and ramming, the estimator is referred to Table 12.32. Loading and hauling the excavation to tip is dealt with in the first part of this chapter.

Table 12.27 Excavate shafts and basements by hand in dry ground

Notes: 1. The labour hours shown are those taken to excavate shafts and basements and similar types of excavation per m^3 for the various lifts, exclusive of timbering, the excavated material being staged to the surface.
2. The labour hours shown for the various lifts are applicable to the whole volume of excavation contained in the lift to which they refer.
3. For earthwork support to shafts and basements, see Table 12.31
4. For backfilling and compacting, see Table 12.32
5. For excavating shafts and basements, using jack rolls, hand winches, or power-driven cranes to hoist the excavated material, see Tables 12.29 and 12.30
6. The rock referred to is rock with many fissures and of a stratified nature.

	Nature of ground and labour hours excavating shafts and basements in lifts of 1.5 m.						
	Per cubic metre						
Depth of excavation in metres	*Firm sand*	*Loamy soil, soft or sandy clay or marl*	*Stiff clay*	*Gravel or compact ground requiring picking*	*Chalk*	*Soft stratified fissured rock*	*Hard stratified fissured rock*
First lift, 0 to 1.5	2.93	3.03	3.33	3.73	4.80	7.60	10.53
Second lift, 1.5 to 3	6.00	6.40	7.03	8.87	10.00	12.80	16.00
Third lift, 3 to 4.5	7.53	8.07	8.93	10.93	12.53	15.33	18.80
Fourth lift, 4.5 to 6	9.07	9.73	10.80	12.00	15.07	17.87	21.60

Table 12.28 Excavate shafts and basements in rock, using pneumatic tools

Notes: 1. The plant and labour hours shown are those taken to break out, excavate and stage to the surface per cubic metre.
2. The plant and labour hours shown for the various lifts are applicable to the whole volume of excavation contained in the lift to which they refer.
3. The data shown are for a 2-tool compressor with two tools in use. If a 3-tool compressor is used the compressor hours per cubic metre are two-thirds of those shown, the labour hours on tools and the labour hours remaining the same.
4. In the case of stratified fissured rocks little or no timbering is necessary, but on occasions poling boards, well spaced, may be required if there is a tendency for slips or falls to occur. In the case of solid rock no timbering is required.
5. The stratified fissured rock referred to is soft and hard rock with many fissures, capable of being readily broken out by pneumatic tools. The solid rock is for rock in solid beds, which definitely necessitates the use of pneumatic tools or blasting to remove it economically.

Nature of ground and compressor and labour hours excavating, backfilling and compacting trenches in lifts of 1.5 m per cubic metre

	Soft stratified fissured rock			Hard stratified fissured rock			Solid rock		
Depth of excavation in metres	Com-pressor hours	Labour hours on pneumatic tools	Labour hours	Com-pressor hours	Labour hours on pneumatic tools	Labour hours	Com-pressor hours	Labour hours on pneumatic tools	Labour hours
First lift, 0 to 1.5	0.48	0.96	4.00	0.64	1.28	5.33	2.75	5.50	6.00
Second lift, 1.5 to 3.0	0.52	1.04	7.70	0.71	1.42	10.27	3.00	6.00	11.47
Third lift, 3.0 to 4.5	0.57	1.14	9.87	0.77	1.54	12.80	3.29	6.58	14.40
Fourth lift, 4.5 to 6.0	0.63	1.26	12.00	0.84	1.68	15.33	3.56	7.12	17.44

Table 12.29 Excavate shafts and basements by hand in dry ground and hoist by hand

Notes: 1. The labour hours shown are those taken to excavate and hoist to the surface per cubic metre, exclusive of earthwork support.
2. The hours shown at the various depths are applicable to the whole volume of excavation from ground level to the depths shown.
3. For earthwork support to shafts and basements, see Table 12.31
4. For backfilling and compacting, see Table 12.32
5. For loading and transporting the excavated material to tip, the estimator is referred to the first part of this chapter
6. For excavating shafts and basements by hand and staging the excavated material to the surface, see Table 12.27
7. For excavating shafts and basements by hand and hoisting the excavated material by power-driven crane, see Table 12.30.
8. The rock referred to is rock with many fissures and of a stratified nature.

Nature of ground and labour hours excavating and hoisting by jack roll or hand winch. Per cubic metre

Depth of excavation in metres	Firm sand	Loamy soil, soft or sandy clay or marl	Stiff clay	Gravel or compact ground requiring picking	Chalk	Soft stratified fissured rock	Hard stratified fissured rock
0 to 3	5.07	5.60	6.13	6.93	8.80	11.09	15.36
0 to 6	6.23	7.00	7.33	8.33	11.00	13.87	19.20
0 to 9	7.60	8.40	9.20	10.40	13.20	15.64	23.04
0 to 12	8.88	9.80	10.73	12.13	15.40	19.40	26.88
0 to 15	10.13	11.20	12.27	13.87	17.60	22.20	30.70

Table 12.30 Excavate shafts and basements by hand in dry ground and hoist by crane

Notes: 1. The crane and labour hours shown are those taken to excavate and hoist to the surface per cubic metre, exclusive of earthwork support.
2. The data shown are based on four men excavating. If the area of the excavation is such that only two men can excavate, the crane and the banksman hours per cubic metre are twice those shown, the labour hours remaining the same.
3. The crane and labour hours shown at the various depths are applicable to the whole volume of excavation from ground level to the depths shown.
4. For earthwork support to shafts and basements, see Table 12.31
5. For backfilling and compacting, see Table 12.32
6. For loading and transporting the excavated material to tip, the estimator is referred to the first part of this chapter.
7. For excavating shafts and basements by hand and staging to the surface, see Table 12.27
8. For excavating shafts and basements by hand and hoisting by jack roll or hand winch, see Table 12.29
9. The rock referred to is rock with many fissures and of a stratified nature.

Nature of ground and crane and labour hours excavating and hoisting by power-driven crane. Per cubic metre

Depth of excavation in metres	Firm sand			Loamy soil, soft or sandy clay or marl			Stiff clay			Gravel or compact ground requiring picking		
	Crane hours	Banksman hours	Labourer hours	Crane hours	Banksman hours	Labourer hours	Crane hours	Banksman hours	Labourer hours	Crane hours	Banksman hours	Labourer hours
0 to 3	0.60	0.60	2.40	0.71	0.71	2.84	0.77	0.77	3.08	0.87	0.87	3.48
0 to 6	0.69	0.69	2.76	0.81	0.81	3.24	0.88	0.88	3.52	1.00	1.00	4.00
0 to 9	0.79	0.79	3.16	0.92	0.92	3.68	1.00	1.00	4.00	1.13	1.13	4.52
0 to 12	0.87	0.87	3.48	1.03	1.03	4.12	1.12	1.12	4.48	1.27	1.27	5.08
0 to 15	0.96	0.96	3.74	1.13	1.13	4.52	1.24	1.24	4.96	1.40	1.40	5.60

Nature of ground and crane and labour hours and hoisting by power-driven crane. Per cubic metre

Depth of excavation in metres	Chalk			Soft stratified fissured rock			Hard stratified fissured rock		
	Crane hours	Banksman hours	Labourer hours	Crane hours	Banksman hours	Labourer hours	Crane hours	Banksman hours	Labourer hours
0 to 3	1.11	1.11	4.44	1.73	1.73	6.92	2.40	2.40	9.60
0 to 6	1.27	1.27	5.08	2.00	2.00	8.00	2.76	2.76	11.04
0 to 9	1.36	1.36	5.44	2.26	2.26	9.04	3.12	3.12	12.48
0 to 12	1.60	1.60	6.40	2.53	2.53	10.12	3.47	3.47	13.88
0 to 15	1.77	1.77	7.08	2.80	2.80	11.20	3.84	3.84	15.36

Table 12.31 Timbering shafts and basements in dry ground

Notes: 1. The timberman and labourer hours shown are those taken to timber the excavation per square metre of face.
2. The data shown allow for both fixing and removing the timber.
3. The data shown for shafts are applicable to excavations up to 15 m² in area, and that for basements for areas in excess of this.
4. The total m³ of timber required to timber the excavations per square metre of face is as follows:

	Shafts	*Basements*
Firm ground	0.034 m³ per m²	0.046 m³ per m²
Moderately firm ground	0.068 m³ per m²	0.094 m³ per m²
Loose ground	0.12 m³ per m²	0.154 m³ per m³

5. For purposes of estimating, allow for use and waste of timber as follows:

Shafts	12 uses
Basements	15 uses

6. The tables for earthwork support to shafts and basements are shown in two forms:

(*a*) earthwork support in lifts of 1.5 m, the hours shown for the various lifts being applicable to the whole area of timber in the lift to which they refer;
(*b*) earthwork support at depths ranging from 0 to 15 m deep, the hours shown at the various depths being applicable to the whole area of timber from ground level to the depths shown.

Earthwork support to shafts in dry ground in 1.5 m lifts

Nature of ground and timberman and labourer hours in earthwork support to shafts in 1.5 m lifts. Per square metre

Depth of excavation in metres	Firm ground		Moderately firm ground		Loose ground	
	Timberman hours	Labourer hours	Timberman hours	Labourer hours	Timberman hours	Labourer hours
First lift, 0 to 1.5	0.11	0.11	0.36	0.36	0.72	0.72
Second lift, 1.5 to 3.0	0.12	0.12	0.48	0.48	0.96	0.96
Third lift, 3.0 to 4.5	0.16	0.16	0.60	0.60	1.20	1.20
Fourth lift, 4.5 to 6.0	0.18	0.18	0.72	0.72	1.44	1.44

Earthwork support to shafts in dry ground at depths from 0 to 15 m

Nature of ground and timberman and labourer hours in earthwork support to shafts. Per square metre

Depth of excavation in metres	Firm ground		Moderately firm ground		Loose ground	
	Timberman hours	Labourer hours	Timberman hours	Labourer hours	Timberman hours	Labourer hours
0 to 3	0.11	0.11	0.42	0.42	0.84	0.84
0 to 6	0.13	0.13	0.52	0.52	1.04	1.04
0 to 9	0.17	0.17	0.63	0.63	1.26	1.26
0 to 12	0.19	0.19	0.74	0.74	1.48	1.48
0 to 15	0.22	0.22	0.84	0.84	1.68	1.68

Table 12.31 continued

Earthwork support to basements in dry ground in 1.5 m lifts

Depth of excavation in metres	Nature of ground and timberman and labourer hours earthwork support to basements in 1.5 m lifts. Per square metre					
	Firm ground		Moderately firm ground		Loose ground	
	Timberman hours	Labourer hours	Timberman hours	Labourer hours	Timberman hours	Labourer hours
First lift, 0 to 1.5	0.23	0.23	0.90	0.90	1.80	1.80
Second lift, 1.5 to 3.0	0.31	0.31	1.20	1.20	2.40	2.40
Third lift, 3.0 to 4.5	0.37	0.37	1.50	1.50	3.00	3.00
Fourth lift, 4.5 to 6.0	0.46	0.46	1.80	1.80	3.60	3.60

Earthwork support to basements in dry ground at depths from 0 to 15 m

Depth of excavation in metres	Nature of ground and timberman and labourer hours earthwork support to basements. Per square metre					
	Firm ground		Moderately firm ground		Loose ground	
	Timberman hours	Labourer hours	Timberman hours	Labourer hours	Timberman hours	Labourer hours
0 to 3	0.26	0.26	1.04	1.04	2.10	2.10
0 to 6	0.34	0.34	1.31	1.31	2.63	2.63
0 to 9	0.40	0.40	1.44	1.44	3.16	3.16
0 to 12	0.47	0.47	1.75	1.75	3.65	3.65
0 to 15	0.54	0.54	2.10	2.10	4.20	4.20

Table 12.32 Backfilling and surface treatments to excavations

Note. The data shown are also applicable to trenches

Nature of ground	Labour hours backfilling and ramming wholly by hand. Per cubic metre	Backfilling and ramming, using rammers of the Pegson type. Per cubic metre	
		Rammer and man hours. Per m^3	Labour filling hours. Per m^3
Firm sand	1.20	0.21	0.64
Loamy soil, sandy clay, soft clay or marl	1.47	0.27	0.80
Stiff clay	1.60	0.29	0.88
Gravel	1.33	0.24	0.72
Chalk	1.47	0.27	0.80
Broken rock	1.47	0.27	0.80

Timbering tunnels

For purposes of estimating the cost of timbering tunnels, data are shown for carrying this out in various classes of ground. The data show the timberman and labourer hours per square metre in fixing the timber only, the area of the walls and roof being measured. In the floor of a tunnel the timber fixed consists of footblocks and stretchers only, and the data shown take this into consideration.

The amount of timber required per square metre is also stated, as is the number of uses, to allow for use and waste.

In tunnel work in general it is wise to err on the side of safety in assessing the amount of timber recoverable. In the case of most grounds, apart from firm chalk or rock, a proportion of the timber has to be left in, and this loss of timber must be allowed for.

Cost of tunnel work

The tables shown for estimating the cost of tunnel work are based on one shaft serving two lengths of tunnel with one miner in each length.

Note: 1. If the area of the tunnel is such that two miners can excavate in each length of tunnel and a crane is used to hoist the excavated material, the crane and banksman's hours are then one-half of those shown, the miner and labourer hours remaining the same.
2. The data shown for excavating and hoisting do not allow for spreading and levelling if the excavation is disposed of around the shaft, or loading and hauling it to tip. The estimator is referred to the first part of this chapter where suitable data may be found for these operations.

The tables shown for estimating purposes are as follows:

1. Excavate tunnels and hoist to the surface by jack-roll or hand winch.
2. Excavate tunnels and hoist to the surface by power-driven crane.
3. Timber tunnels.
4. Refill and ram tunnels.
5. Table of multipliers for use with the plant and labour hours shown for tunnels whose length from the shaft exceeds 1.5 m.

In order that the estimator may be conversant with the use of the tables a typical example is shown of the build up of the cost of tunnel excavation carried out in soft clay.

Example.   Estimate the cost per metre of excavating and timbering a tunnel in soft clay, the excavated material being hoisted by a power-driven crane. The following are the particulars appertaining to the work:

- The tunnel is 9 m below ground level, and the length driven from the shaft is 30 m.
- The dimensions of the tunnel excavation are 1.8 m high × 0.9 m wide.
- One miner excavates at each tunnel face.

For purposes of this example the following rates are assumed:

Hire rate or working cost of the crane, including driver, fuel and oil = £ 20.00 per h
Miner's rate = £ 9.00 per h
Banksman's rate = £ 7.00 per h
Labourer's rate = £ 6.00 per h
Initial cost of the timber used in timbering the tunnel = £220.00 per m^3

Excavating the tunnel and hoisting to the surface per metre.

Referring to Table 12.34 the data are shown for excavating and hoisting in tunnels in soft clay for tunnels whose length is not in excess of 15 m. The hours shown multiplied by the assumed rates give the estimated cost per cubic metre, thus:

Crane	1.75 h at £20.00 per h	= £35.00 per m^3
Banksman	1.75 h at £ 7.00 per h	= £12.25 per m^3
Miner	3.50 h at £ 9.00 per h	= £31.50 per m^3
Labourer	3.50 h at £ 6.00 per h	= £21.00 per m^3
	Cost of excavating and hoisting	= £90.75 per m^3

This £90.75 per m^3 refers to tunnels 9 m below the ground whose length from the shaft is not in excess of 15 m. Referring to the table of multipliers, Table 12.38, the multiplier to use for tunnels 30 m in length from the shaft is 1.10.

Therefore the cost of the excavation in the case of this example is:

$$1.10 \times £90.75 \qquad\qquad = £99.83 \text{ per m}^3$$

since the volume of excavation in the tunnel per metre is $2 \times 1 \times 1 \text{ m} = 2 \text{ m}^3$, the cost of excavating per metre $= 2 \text{ m}^3 \times £99.83 = £199.65$.

Timbering the tunnel per metre.

Referring to Table 12.36 the following data for the timbering of tunnels 9 m below ground in moderately firm ground are found:

1. Miner and labourer hours timbering tunnel per square metre = 0.6 h each.
2. Timber required per square metre = 0.068 m^3
3. Number of uses of timber to allow for use and waste = 3.

Note. Regarding (3), if it is definitely specified that all timber has to be left in, the number of uses to allow would be 1. In allowing 3 uses it is assessed that two-thirds of the timber can be removed.

Since the initial cost of the timber is £220.00 per m^3 and the amount required per square metre is 0.068 m^2, the number of uses to allow for use and waste being assessed at 3.

The allowance for use and waste per square metre is:

$$\frac{£220}{3} \times 0.068 \text{ m}^3 = £5.00 \text{ per m}^2$$

The cost of fixing and placing the timber per square metre in tunnels whose length does not exceed 15m from the shaft then is:

Miner	0.6 h at £9.00 per h	= £5.40 per m^2
Labourer	0.6 h at £6.00 per h	= £3.60 per m^2
	Labour cost of fixing only	= £9.00

In the case of this example the tunnel is 30 m in length from the shaft and the multiplier 1.10 is used as before.

The labour cost of timbering per square metre is:

$$1.10 \times £5.00 = £5.50$$

The total cost of timbering per square metre, then, is the sum of the labour cost, plus the allowance for the use and waste of the timber.

Total cost of timbering per square metre is:

Labour cost	£ 9.00 per m^2
Allow for use and waste	£ 5.50 per m^2
Total cost	£14.50 per m^2

The area of timbering to be carried out per metre of tunnel is the perimeter of the two side walls and roof in metres multiplied by 1 m. The area to be timbered per metre of tunnel, therefore, is:

$$1\,(2\,m + 2\,m + 1\,m) = 5\ m^2$$

The cost of timbering the tunnel per metre $= 5\ m^2 \times £14.50 = £72.50$ per m

The total cost of excavating and timbering the tunnel per metre then is:

Excavating and hoisting	= £199.65 per m
Timbering	= £ 14.50 per m
Total cost	= £214.15

The cost of the timbered excavation per cubic metre is obtained by dividing the total cost per metre by the cubic metres of excavation in the tunnel per metre:

$$= \frac{£214.15}{2\ m^3} = £107.1 \text{ per } m^3$$

Note: In the example shown no allowance has been made for loading and transporting the excavated material to tip. It is uneconomical to keep transporting plant waiting while the crane loads direct, as the excavation rate is slow. For purposes of estimating, the loading and transporting cost of the excavated material should be treated as a separate operation. The estimator is referred to the first part of this chapter where suitable data are shown for this, as are data for spreading and levelling the excavated material.

Table 12.33 Excavate tunnels and hoist by jack roll or hand winch

Notes: 1. The miner and labourer hours shown are those taken to excavate, haul through to the shaft, and hoist to the surface per cubic metre, exclusive of timbering.
2. The data shown are based on one shaft serving two tunnels, with one miner excavating in each tunnel.
3. The data shown are applicable to tunnels whose length from the shaft is not in excess of 15 m. For tunnels in excess of this length use the table of multipliers, Table 12.38
4. For timbering tunnels, see Table 12.36
5. For refilling and ramming tunnels, see Table 12.37
6. For loading and transporting the excavated material the estimator is referred to the first part of this chapter
7. For excavating tunnels and hoisting by power-driven crane, see Table 12.34
8. For excavating headings, see Table 12.35
9. The rock referred to is rock with many fissures and of a stratified nature.

Nature of ground and miner and labourer hours excavating tunnels, hauling to shaft and hoisting by jack roll or hand winch per cubic metre in tunnels whose length from the shafts is not in excess of 15 m

Depth of tunnel below ground in metres	Firm sand		Loamy soil, soft or sandy clay or marl		Stiff clay		Gravel or compact ground requiring picking	
	Miner hours	Labourer hours	Miner hours	Labourer hours	Miner hours	Labourer hours	Miner hours	Labourer hours
0 to 6 m	3.07	9.21	3.33	9.99	3.73	11.19	4.13	13.52
0 to 9 m	3.37	10.11	3.67	11.01	4.11	12.33	4.56	13.68
0 to 12 m	3.68	11.04	4.00	12.00	4.48	13.44	4.96	14.88
0 to 15 m	4.00	12.00	4.33	12.99	4.85	14.55	5.37	16.11

Nature of ground and miner and labourer hours excavating tunnels, hauling to shaft and hoisting by jack roll or hand winch per cubic metre in tunnels whose length from the shafts is not in excess of 15 m

Depth of tunnel below ground in metres	Chalk		Soft stratified fissured rock		Hard stratified fissured rock	
	Miner hours	Labourer hours	Miner hours	Labourer hours	Miner hours	Labourer hours
0 to 6 m	5.33	13.32	8.27	16.54	11.47	17.87
0 to 9 m	5.87	14.68	9.09	18.18	12.60	19.65
0 to 12 m	6.40	16.00	9.92	19.84	13.76	21.14
0 to 15 m	6.93	17.33	10.75	21.50	14.90	23.29

Table 12.34 Excavate tunnels and hoist by crane

Notes:
1. The crane, banksman, miner and labourer hours shown are those taken to excavate, haul through to the shaft, and hoist to the surface per cubic metre, exclusive of timbering.
2. The data shown are based on one shaft servicing two tunnels, with one miner in each tunnel. If the area of the tunnel is such that two miners can excavate in each tunnel, the crane and banksman hours per cubic metre are one-half of those shown, the miner and labour hours remaining the same.
3. The data shown are applicable to tunnels whose length from the shaft is not in excess of 15 m. For tunnels in excess of this length use the table of multipliers, Table 12.38
4. For timbering tunnels, see Table 12.36
5. For refilling and ramming tunnels, see Table 12.37
6. For loading and transporting the excavated material to tip the estimator is referred to the first part of this chapter
7. For excavating headings, see Table 12.35
8. The rock referred to is rock with many fissures and of a stratified nature.

Nature of ground and crane, banksman, miner and labourer hours excavating tunnels, hauling to the shaft, and hoisting by power-driven crane per cubic metre in tunnels whose length from the shaft is not in excess of 15 m

Depth of tunnel below ground in metres	Firm sand				Loamy soil, soft or sandy clay or marl				Stiff clay				Gravel or compact ground requiring picking			
	Crane hours	Banksman hours	Miner hours	Labourer hours	Crane hours	Banksman hours	Miner hours	Labourer hours	Crane hours	Banksman hours	Miner hours	Labourer hours	Crane hours	Banksman hours	Miner hours	Labourer hours
0 to 6 m	1.50	1.50	3.00	3.00	1.67	1.67	3.33	3.33	1.87	1.87	3.73	3.73	2.07	2.07	4.14	4.14
0 to 9 m	1.60	1.60	3.20	3.20	1.75	1.75	3.50	3.50	2.00	2.00	3.92	3.92	2.17	2.17	4.34	4.34
0 to 12 m	1.73	1.73	3.46	3.46	1.84	1.84	3.68	3.68	2.05	2.05	4.10	4.10	2.28	2.28	4.56	4.56
0 to 15 m	1.87	1.87	3.74	3.74	1.94	1.94	3.88	3.88	2.14	2.14	4.28	4.28	2.40	2.40	4.80	4.80

Nature of ground and crane, banksman, miner and labourer hours excavating tunnels, hauling to the shaft, and hoisting by power-driven crane per cubic metre in tunnels whose length from the shaft is not in excess of 15 m

Depth of tunnel below ground in metres	Chalk				Soft stratified fissured rock				Hard stratified fissured rock			
	Crane hours	Banksman hours	Miner hours	Labourer hours	Crane hours	Banksman hours	Miner hours	Labourer hours	Crane hours	Banksman hours	Miner hours	Labourer hours
0 to 6 m	2.66	2.66	5.32	5.32	4.13	4.13	8.26	8.26	5.73	5.73	11.46	11.46
0 to 9 m	2.80	2.80	5.60	5.60	4.33	4.33	8.66	8.66	6.00	6.00	12.00	12.00
0 to 12 m	2.93	2.93	5.86	5.86	4.53	4.53	9.06	9.06	6.37	6.37	12.74	12.74
0 to 15 m	3.07	3.07	6.14	6.14	4.83	4.83	9.46	9.46	6.67	6.67	13.34	13.34

Excavate Headings

In excavating headings, one heading only is driven from the shaft or face, and in such a case it is not usual or economical to use a power-driven crane to hoist the excavated material to the surface; the hoisting is generally carried out by hand by means of a jack roll or hand winch, and the data shown are based on this method of hoisting.

Table 12.35 Excavate headings and hoist by jack roll or hand winch

Notes: 1. The miner and labourer hours shown are those taken to excavate, haul through the heading, and hoist to the surface, exclusive of timbering. For timbering headings, see Table 12.36
2. For refilling and ramming headings, see Table 12.37
3. The data shown are applicable to headings whose length is not in excess of this, the table of multipliers, Table 12.38 may be used.
4. For loading and transporting the excavated material the estimator is referred to the first part of this chapter
5. The rock referred to is rock with many fissures and of a stratified nature.

Nature of ground and miner and labourer hours excavating headings, hauling to the face, and hoisting by jack roll or hand winch, per cubic metre, in headings whose length is not in excess of 15 m

Depth of heading below ground in metres	Firm sand		Loamy soil, soft or sandy clay or marl		Stiff clay		Gravel or compact ground requiring picking	
	Miner hours	Labourer hours	Miner hours	Labourer hours	Miner hours	Labourer hours	Miner hours	Labourer hours
0 to 6 m	3.07	12.28	3.35	13.40	3.73	14.92	4.13	16.52
0 to 9 m	3.37	13.48	3.67	14.68	4.11	16.44	4.54	18.16
0 to 12 m	3.68	14.72	4.00	16.00	4.48	17.92	4.96	19.84
0 to 15 m	4.00	16.00	4.33	17.32	4.85	19.40	5.37	21.48

Nature of ground and miner and labourer hours excavating tunnels, hauling to the face and hoisting by jack roll or hand winch, per cubic metre, in headings whose length from the shafts is not in excess of 15 m

Depth of heading below ground in metres	Chalk		Soft stratified fissured rock		Hard stratified fissured rock	
	Miner hours	Labourer hours	Miner hours	Labourer hours	Miner hours	Labourer hours
0 to 6 m	5.35	17.75	8.37	22.00	11.45	23.85
0 to 9 m	5.85	19.55	9.10	24.20	12.60	26.25
0 to 12 m	6.40	21.35	9.80	26.40	13.75	28.60
0 to 15 m	6.95	23.10	10.75	28.50	14.68	31.00

Table 12.36 Timbering tunnels and headings

Notes: 1. The timberman and labourer hours shown are those taken to timber tunnels or headings per square metre, the faces measured being the two side walls and the roof.
2. The data shown are applicable to tunnels or headings whose length from the shaft is not in excess of 15 m. For tunnels or headings in excess of this length use the table of multipliers shown below.
3. The amount of timber required to timber the excavation per square metre of face is as follows:

Firm ground	0.034 m^3 per m^2
Moderately firm ground	0.068 m^3 per m^2
Loose ground	0.129 m^3 per m^2

For purposes of estimating allow for use and waste of timber as follows:

Firm ground	5 uses
Moderately firm ground	3 uses
Loose ground	1 use, i.e. all the timber is left in

	Miner and labourer hours timbering tunnels or headings. Per square metre					
Depth of tunnel or heading below ground in metres	Firm ground		Moderately firm ground		Loose ground	
	Miner hours	Labourer hours	Miner hours	Labourer hours	Miner hours	Labourer hours
0 to 6 m	0.18	0.18	0.54	0.54	1.08	1.08
0 to 9 m	0.19	0.19	0.60	0.60	1.20	1.20
0 to 12 m	0.22	0.22	0.66	0.66	1.32	1.32
0 to 15 m	0.24	0.24	0.72	0.72	1.44	1.44

Table 12.37 Refill and ram tunnels and headings

Note: The data shown are for backfilling and ramming tunnels or headings whose length is not in excess of 15 m from the shaft. For tunnels or headings in excess of this length use the table of multipliers shown below.

	Method of backfilling and ramming, and plant and labourer hours. Per cubic metre		
	By hand	By power-driven crane and skip	
Nature of ground	Labour hours	Crane hours	Labour hours
Sand	7.20	0.67	4.69
Loamy soil, soft or sandy clay or marl	7.60	0.73	5.11
Stiff clay	9.33	0.87	6.09
Gravel	7.47	0.71	4.97
Chalk or broken rock	8.67	0.80	5.60

Table 12.38 Multipliers for tunnels and headings where the length from the shaft is in excess of 15 m

Length of tunnel from shaft in metres	Plant, miner and labourer hour multipliers
0 to 15 m	1.00
0 to 30 m	1.10
0 to 45 m	1.30
0 to 60 m	1.60

Excavating in wet ground

In very wet ground it is common practice to sink sumps clear of the line of the main excavation, these being sunk to a level below that of the main work. Pumping plant of suitable capacity is then placed in position in the sumps to deal with the water in the main excavation. This is done by laying sub-drain pipes consisting of clayware pipes surrounded with shingle or other clean medium in the bottom of the main excavation and leading this drain or drains to the sumps. The data shown for shaft sinking in the various water-bearing grounds are applicable to sump work and may be used for estimating the cost of this.

Earthwork support in wet ground

The cost of earthwork support in wet ground varies both with the depth and the water content. In the tables shown for estimating purposes the timberman and labourer hours are shown for timbering trenches per metre of trench, both sides of the trench being allowed for in the hours shown, and per square metre of face for excavations of the shaft and basement type. The amount of timber required is also stated, as is the number of uses to allow for use and waste.

In estimating the cost of excavating and timbering in water-bearing ground, the estimator must assess and come to a decision on the following:

1. The nature of the ground.
2. The average amount of water in litres per minute which will have to be dealt with throughout the depth of the excavation.
3. The most suitable type and size of pump or pumps to use.

Having come to a decision on those points the estimate of the cost is then built up.

Example.   Estimate the cost of excavating, earthwork support, refilling and ramming a trench per metre in running sand, the following being the particulars pertaining to the excavation in question:

1. Depth of trench 2.7 m and width 0.9 m.
2. Assessed litres of water to be pumped per minute = 2250 litres.
3. Type and size of pump selected to deal with the water, one 125 mm diameter centrifugal.

For purposes of this example the following rates are assumed:

1. Working cost or hire rate of pump per hour,
   inclusive of attendant, fuel and oil   = £ 10.00 per h
2. Timberman's rate   = £   7.00 per h
3. Labourer's rate   = £   6.00 per h
4. Initial cost of timber used to timber the trench   = £220.00 per m^3

Excavating, backfilling and ramming the trench per metre.

The total depth of trench is 2.7 m, the depths in the various lifts being:

Depth of trench in the first lift of 0 to 1.5 m	= 1.5 m
Depth of trench in the second lift of 1.5 to 2.7 m	= 1.2 m
Total depth of trench	= 2.7 m

Referring to Table 12.45 the volume of excavation per metre in a trench 2.7 m deep and 0.9 m wide under the various lifts is as follows:

First lift	$= 1.35 \text{ m}^3$ (depth 1.5 m)
Second lift	$= 1.09 \text{ m}^3$ (depth 1.2 m)
Total m^3 of excavation per metre	$= 2.44 \text{ m}^3$

Referring to Table 12.39 the pump and labour hours taken per m^3 to excavate, backfill and ram trenches in water-bearing sand, where the assessed quantity of water to be pumped throughout the total depth of excavation per minute is 2250 litres, is:

	Pump hours per m³	*Labour hours per m³*
First lift, 0 to 1.5 m	0.48	9.00
Second lift, 1.5 to 2.7 m	0.48	15.45

The cost per m^3, therefore is:

First lift 0 to 1.5 m

Pump hours	$= 0.48$ h at £10.00 per h	$= £4.80$ per m^3
Labourer hours	$= 9.0$ h at £ 6.00 per h	$= £54.00$ per m^3
Cost of excavating, backfilling and ramming		
0 to 1.5 m, inclusive of pumping		$= £58.80$

Second lift 1.5 to 2.7 m

Pump hours	$= 1.35$ h at £10.00 per h	$= £13.50$ per m^3
Labourer hours	$= 1.09$ h at £ 6.00 per h	$= £ 6.54$ per m^3
Cost of excavating, backfilling and ramming		
1.5 to 2.7 m, inclusive of pumping		$= £20.04$

The cost of excavating, backfilling and ramming the trench per metre exclusive of timbering, therefore, is:

First lift	0 to 1.5 m =	1.35 m^3 at £58.80 per m^3	$= £ 79.38$ per m^3
Second lift	1.5 to 2.7 m =	1.09 m^3 at £20.04 per m^3	$= £ 21.84$ per m

Cost of excavating, backfilling and ramming the trench = £101.22 per m

Referring to Table 12.40 for timbering in wet ground where the assessed quantity of water to be pumped per minute is 2250 litres, the following data for estimating purposes are obtained:

1. Timberman and labourer hours timbering trench 2.7 m deep per metre = 2.4 h each.
2. Quantity of timber required to timber the trench per metre = 0.5 m^3.
3. The number of uses of the timber to allow for use and waste = 10.

Since the initial cost of the timber is £220.00 per m^3 the number of uses 10 and the amount required per metre of trench = 0.5 m^3.
The allowance for use and waste per metre of trench is:

$$\frac{£220}{10} \times 0.5 \text{ m}^3 = £11.00 \text{ per metre}$$

The cost of earthwork support to the trench, therefore, is:

Timberman   = 2.4 h at £7.00 per h   = £16.80 per m
Labourer   = 2.4 h at £6.00 per h   = £14.40 per m
Allow for use and waste of timber   = £11.00 per m

Cost of earthwork support to trench   = £42.20 per m

The cost of excavating, backfilling and ramming the trench per metre then is:

Excavating, backfilling and ramming = £101.22 per m
Timbering   = £ 42.20 per m

Total cost   = £143.42 per m

Note: In the above example the estimated cost of the timbered excavation averaged throughout the depth of the trench
per cubic metre is:
  The cost of excavating, timbering, backfilling and ramming the trench per metre divided by the total cubic
  metres of excavation in the trench per metre.

$$\frac{£143.42}{2.44 \text{ m}^3} = £58.79 \text{ per m}^3$$

Table 12.39 Excavate, refill and ram trenches by hand in water-bearing ground

Notes: 1. The data show the pump and labour hours taken to excavate, refill and ram trenches per cubic metre,
exclusive of timbering, the excavated material being staged to the surface.
2. The pump and labour hours shown at the various lifts are applicable to the whole volume of excavation
contained in the lift to which they refer.
3. In estimating the cost per cubic metre of excavating, refilling and ramming trenches in water-bearing
ground, the average litres of water to be pumped per minute throughout the depth of the trench should be
assessed and the pump and labour hours shown under the various lifts corresponding to this assessed
amount used in building up the estimate.
4. For timbering trenches in water-bearing ground, see Table 12.40

Excavate, refill and ram trenches in water-bearing gravel

Pump and labour hours excavating, refilling and ramming trenches in lifts of 1.5 m.
Per cubic metre

Estimated litres of water to be pumped per minute	First lift 0 to 1.5 m		Second lift 1.5 to 3 m		Third lift 3 to 4.5 m		Fourth lift 4.5 to 6 m	
	Pump hours	Labour hours	Pump hours	Labour hours	Pump hours	Labour hours	Pump hours	Labour hours
225	0.23	5.93	0.23	9.73	0.23	11.65	0.23	14.00
450	0.24	6.00	0.24	9.84	0.24	12.00	0.24	14.16
675	0.25	6.07	0.25	9.95	0.25	12.13	0.25	14.32
900	0.27	6.13	0.27	10.05	0.27	12.27	0.27	14.48
1350	0.29	6.27	0.29	10.27	0.29	12.53	0.29	14.80
1800	0.32	6.40	0.32	10.49	0.32	12.80	0.32	15.11
2250	0.35	6.63	0.35	10.72	0.35	13.07	0.35	15.41
2700	0.37	6.67	0.37	10.93	0.37	13.33	0.37	15.73
3150	0.40	6.80	0.40	11.15	0.40	13.60	0.40	16.05
3600	0.43	6.93	0.43	11.37	0.43	13.87	0.43	16.36
4050	0.45	7.07	0.45	11.58	0.45	14.13	0.45	16.68
4500	0.48	7.20	0.48	11.81	0.48	14.40	0.48	17.00
5625	0.55	7.53	0.55	12.36	0.55	15.07	0.55	17.77
6750	0.61	7.87	0.61	12.91	0.61	15.73	0.61	18.56
7875	0.68	8.20	0.68	13.44	0.68	16.40	0.68	19.36
9000	0.77	8.53	0.77	14.00	0.77	17.07	0.77	20.13
10125	0.81	8.87	0.81	14.54	0.81	17.74	0.81	20.92
11250	0.88	9.20	0.88	15.00	0.88	18.40	0.88	21.71

Table 12.39 continued

Excavate, refill and ram trenches in water-bearing sand

Pump and labour hours excavating, refilling and ramming trenches in lifts of 1.5 m.
Per cubic metre

Estimated litres of water to be pumped per minute	First lift 0 to 1.5 m		Second lift 1.5 to 3 m		Third lift 3 to 4.5 m		Fourth lift 4.5 to 6 m	
	Pump hours	Labour hours	Pump hours	Labour hours	Pump hours	Labour hours	Pump hours	Labour hours
225	0.32	7.84	0.32	13.60	0.32	16.36	0.32	19.00
450	0.33	8.03	0.33	13.80	0.33	16.60	0.33	19.40
675	0.35	8.15	0.35	14.00	0.35	16.85	0.35	19.71
900	0.36	8.27	0.36	14.21	0.36	17.11	0.36	20.00
1350	0.40	8.51	0.40	14.64	0.40	17.60	0.40	20.57
1800	0.44	8.75	0.44	15.04	0.44	18.09	0.44	21.15
2250	0.48	9.00	0.48	15.45	0.48	18.59	0.48	21.82
2700	0.52	9.23	0.52	15.87	0.52	19.08	0.52	22.29
3150	0.56	9.47	0.56	16.28	0.56	19.57	0.56	22.87
3600	0.60	9.71	0.60	16.70	0.60	20.70	0.60	23.44
4050	0.64	9.95	0.64	17.11	0.64	20.58	0.64	24.04
4500	0.68	10.19	0.68	17.52	0.68	21.07	0.68	24.67

Excavate, refill and ram trenches in water-bearing chalk

Pump and labour hours excavating, refilling and ramming trenches in lifts of 1.5 m.
Per cubic metre

Estimated litres of water to be pumped per minute	First lift 0 to 1.5 m		Second lift 1.5 to 3 m		Third lift 3 to 4.5 m		Fourth lift 4.5 to 6 m	
	Pump hours	Labour hours	Pump hours	Labour hours	Pump hours	Labour hours	Pump hours	Labour hours
225	0.27	6.67	0.27	11.07	0.27	13.33	0.27	15.60
450	0.28	6.80	0.28	11.28	0.28	13.60	0.28	15.92
675	0.29	6.93	0.29	11.51	0.29	13.87	0.29	16.23
900	0.31	7.07	0.31	11.73	0.31	14.13	0.31	16.79
1350	0.35	7.34	0.35	12.20	0.35	19.69	0.35	17.19
1800	0.39	7.63	0.39	12.67	0.39	15.25	0.39	17.84
2250	0.43	7.92	0.43	13.13	0.43	15.84	0.43	18.54
2700	0.47	8.20	0.47	13.60	0.47	16.40	0.47	19.20
3150	0.51	8.48	0.51	14.07	0.51	16.96	0.51	19.88
3600	0.55	8.76	0.55	14.53	0.55	17.42	0.55	20.51
4050	0.59	9.04	0.59	14.97	0.59	18.08	0.59	21.16
4500	0.63	9.33	0.63	15.47	0.63	18.67	0.63	21.87

Table 12.40 Earthwork support to trenches in water-bearing ground

Notes: 1. The timberman and labourer hours shown are those taken to timber the trench per metre of trench, both sides of the trench being allowed for in the hours shown.
2. The data shown allow for both fixing and removing the timber.
3. The amount of timber shown is the total cubic metres of timber required to timber the trench per metre, both sides of the trench being allowed for. For purposes of estimating allow for use and waste 10 uses of the timber.
4. In estimating the cost of timbering per metre of trench in water-bearing ground, the average litres of water to be pumped per minute throughout the depth of the trench should be assessed and the timberman and labour hours shown in the various lifts corresponding to this assessed amount used in building up the estimate.

Estimated litres of water to be pumped per minute, cubic metres of timber required, and timberman and labourer hours timbering trenches. Per metre of trench

Depth of trench in m	0 to 1100 litres per minute			1100 to 3300 litres per minute			3300 to 6600 litres per minute			6600 to 11000 litres per minute		
	Timber m³	Timberman hours	Labourer hours	Timber m³	Timberman hours	Labourer hours	Timber m³	Timberman hours	Labourer hours	Timber m³	Timberman hours	Labourer hours
0.9	0.13	0.48	0.48	0.16	0.63	0.63	0.20	0.76	0.76	0.23	0.96	0.96
1.2	0.18	0.67	0.67	0.22	0.83	0.83	0.27	1.06	1.06	0.31	1.33	1.33
1.5	0.22	0.89	0.89	0.28	1.11	1.11	0.33	1.39	1.39	0.39	1.76	1.76
1.8	0.27	1.11	1.11	0.33	1.40	1.40	0.40	1.78	1.78	0.47	2.24	2.24
2.1	0.31	1.36	1.36	0.39	1.71	1.71	0.47	2.16	2.16	0.54	2.72	2.72
2.4	0.35	1.62	1.62	0.44	2.04	2.04	0.53	2.58	2.58	0.62	3.26	3.26
2.7	0.40	1.91	1.91	0.50	2.40	2.40	0.60	3.03	3.03	0.70	3.83	3.83
3.0	0.44	2.22	2.22	0.55	2.78	2.78	0.66	3.60	3.60	0.78	4.42	4.42
3.3	0.48	2.53	2.53	0.61	3.18	3.18	0.73	4.01	4.01	0.85	5.07	5.07
3.6	0.53	2.87	2.87	0.67	3.60	3.60	0.80	4.56	4.56	0.93	5.76	5.76
3.9	0.58	3.22	3.22	0.72	4.05	4.05	0.87	5.11	5.11	1.01	6.46	6.46
4.2	0.62	3.60	3.60	0.78	4.51	4.51	0.93	5.70	5.70	1.09	7.20	7.20
4.5	0.66	4.00	4.00	0.83	5.00	5.00	1.00	6.33	6.33	1.17	8.00	8.00
4.8	0.71	4.41	4.41	0.89	5.51	5.51	1.07	6.97	6.97	1.24	8.70	8.70
5.1	0.76	4.83	4.83	0.94	6.05	6.05	1.13	7.64	7.64	1.32	9.75	9.75
5.4	0.80	5.29	5.29	1.00	6.60	6.60	1.20	8.36	8.36	1.40	10.55	10.55
5.7	0.84	5.74	5.74	1.05	7.18	7.18	1.27	9.00	9.00	1.48	11.47	11.47
6.0	0.89	6.22	6.22	1.11	7.77	7.77	1.33	9.73	9.73	1.53	12.41	12.41

Table 12.41 Excavate shafts and basements by hand in water-bearing ground

Notes: 1. The data show the pump and labour hours taken per cubic metre to excavate shafts and basements by hand, the excavated material being staged to the surface, and are exclusive of timbering.

2. The pump and labour hours shown at the various lifts are applicable to the volume of excavation contained in the lift to which they refer.

3. In estimating the cost per cubic metre of excavating shafts and basements in water-bearing ground, the average litres of water to be pumped per minute throughout the depth of the excavation should be assessed and the pump and labour hours shown under the various lifts corresponding to this assessed amount used to build up the estimate.

4. For timbering shafts and basements in water-bearing ground, see Table 12.42

5. The pump hours shown are based on four men excavating, there being one stage man per stage and one man clearing on top for every two men excavating. The pump hours per cubic metre are inversely proportional to the number of men excavating, so that if the area is such that only two men can excavate, the pump hours shown must be multiplied by 4/2 = 2, the labour hours remaining as shown.

Excavate shafts and basements in water-bearing gravel

Pump and labour hours excavating shafts and basements in lifts of 1.5 m per cubic metre

Estimated litres of water to be pumped per minute	First lift 0 to 1.5 m		Second lift 1.5 to 3 m		Third lift 3 to 4.5 m		Fourth lift 4.5 to 6 m	
	Pump hours	*Labour hours*	*Pump hours*	*Labour hours*	*Pump hours*	*Labour hours*	*Pump hours*	*Labour hours*
225	0.93	5.58	0.93	7.44	0.93	9.30	0.93	11.16
450	0.95	5.70	0.95	7.60	0.95	9.50	0.95	11.40
675	0.96	5.76	0.96	7.68	0.96	9.60	0.96	11.52
900	0.97	5.82	0.97	7.76	0.97	9.70	0.97	11.64
1350	1.00	6.00	1.00	8.00	1.00	10.00	1.00	12.00
1800	1.03	6.18	1.03	8.24	1.03	10.30	1.03	12.36
2250	1.05	6.30	1.05	8.40	1.05	10.50	1.05	12.60
2700	1.08	6.48	1.08	8.64	1.08	10.80	1.08	12.96
3150	1.11	6.66	1.11	8.88	1.11	11.10	1.11	13.32
3600	1.13	6.78	1.13	9.04	1.13	11.30	1.13	13.56
4050	1.16	6.96	1.16	9.28	1.16	11.60	1.16	13.92
4500	1.20	7.20	1.20	9.60	1.20	12.00	1.20	14.40
5625	1.27	7.62	1.27	10.16	1.27	12.70	1.27	15.24
6750	1.33	7.98	1.33	10.64	1.33	13.30	1.33	15.96
7875	1.40	8.40	1.40	11.20	1.40	14.00	1.40	16.80
9000	1.47	8.82	1.47	11.76	1.47	14.70	1.47	17.64
10125	1.53	9.18	1.53	12.24	1.53	15.30	1.53	18.36
11250	1.60	9.60	1.60	12.80	1.60	16.00	1.60	19.20

Table 12.41 continued

Excavate shafts and basements in water-bearing sand

Estimated litres of water to be pumped per minute	Pump and labour hours excavating shafts and basements in lifts of 1.5 m per cubic metre							
	First lift 0 to 1.5 m		Second lift 1.5 to 3 m		Third lift 3 to 4.5 m		Fourth lift 4.5 to 6 m	
	Pump hours	Labour hours	Pump hours	Labour hours	Pump hours	Labour hours	Pump hours	Labour hours
225	1.33	8.00	1.33	10.64	1.33	13.30	1.33	15.96
450	1.35	8.10	1.35	10.80	1.35	13.50	1.35	16.20
675	1.37	8.22	1.37	10.98	1.37	13.70	1.37	16.44
900	1.41	8.46	1.41	11.28	1.41	14.10	1.41	16.92
1350	1.45	8.70	1.45	11.60	1.45	14.50	1.45	17.40
1800	1.49	8.94	1.49	11.92	1.49	14.90	1.49	17.88
2250	1.54	9.24	1.54	12.32	1.54	15.40	1.54	18.48
2700	1.57	9.42	1.57	12.56	1.57	15.70	1.57	18.84
3150	1.60	9.60	1.60	12.80	1.60	16.00	1.60	19.20
3600	1.65	9.90	1.65	13.20	1.65	16.50	1.65	19.80
4050	1.69	10.19	1.69	13.52	1.69	16.90	1.69	20.28
4500	1.73	10.38	1.73	13.84	1.73	17.30	1.73	20.76

Excavate shafts and basements in water-bearing chalk

Estimated litres of water to be pumped per minute	Pump and labour hours excavating shafts and basements in lifts of 1.5 m per cubic metre							
	First lift 0 to 1.5 m		Second lift 1.5 to 3 m		Third lift 3 to 4.5 m		Fourth lift 4.5 to 6 m	
	Pump hours	Labour hours	Pump hours	Labour hours	Pump hours	Labour hours	Pump hours	Labour hours
225	1.25	7.50	1.25	10.00	1.25	12.50	1.25	15.00
450	1.27	7.62	1.27	10.16	1.27	12.70	1.27	15.24
675	1.29	7.74	1.29	10.32	1.29	12.90	1.29	15.48
900	1.33	7.98	1.33	10.64	1.33	13.30	1.33	15.96
1350	1.37	8.22	1.37	10.96	1.37	13.70	1.37	16.44
1800	1.41	8.46	1.41	11.28	1.41	14.10	1.41	16.92
2250	1.45	8.70	1.45	11.60	1.45	14.50	1.45	17.40
2700	1.49	8.94	1.49	11.92	1.49	14.90	1.49	17.88
3150	1.53	9.18	1.53	12.24	1.53	15.30	1.53	18.36
3600	1.57	9.42	1.57	12.56	1.57	15.70	1.57	18.84
3950	1.61	9.66	1.61	12.88	1.61	16.10	1.61	19.32
4300	1.65	9.90	1.65	13.20	1.65	16.50	1.65	19.80

Table 12.42 Earthwork support to shafts and basements in water-bearing ground

Notes: 1. The timberman and labourer hours shown are those taken in timbering per square metre of face.
2. The data shown allow for both fixing and removing the timber.
3. The timberman and labourer hours shown at the various lifts are applicable to the area of the timber in the lift to which they refer.
4. The data shown for shafts are applicable to excavations up to 16 m² in area and that for basements for areas in excess of this.
5. The total cubic metres of timber required to timber the excavations per square metre of face is as follows:

Ground water content litres of water to be pumped per minute	Shafts m³ of timber per m² of face	Basements m³ of timber per m² of face
0 to 1130	0.078	0.104
1130 to 3400	0.095	0.128
3400 to 6800	0.112	0.152
6800 to 11400	0.128	0.176

6. For purposes of estimating allow for use and waste of timber as follows:

Shafts	10 uses.
Basements	12 uses.

Earthwork support to shafts in water-bearing ground in 1.5 m lifts

Depth of excavation in metres and timberman and labourer hours timbering shafts in 1.5 m lifts. Per square metre

Estimated litres of water to be pumped per minute	First lift 0 to 1.5 m Timberman hours	Labourer hours	Second lift 1.5 to 3 m Timberman hours	Labourer hours	Third lift 3 to 4.5 m Timberman hours	Labourer hours	Fourth lift 4.5 to 6 m Timberman hours	Labourer hours
0–1325	0.43	0.43	0.65	0.65	0.96	0.96	1.08	1.08
1325–3975	0.58	0.58	0.96	0.96	1.15	1.15	1.44	1.44
3975–7950	0.72	0.72	1.08	1.08	1.44	1.44	1.80	1.80
7950–12350	0.96	0.96	1.30	1.30	1.73	1.73	2.16	2.16

Earthwork support to basements in in water-bearing ground in 1.5 m lifts

Depth of excavation in metres and timberman and labourer hours timbering basements in 1.5 m lifts. Per square metre Per cubic metre

Estimated litres of water to be pumped per minute	First lift 0 to 1.5 m Timberman hours	Labourer hours	Second lift 1.5 to 3 m Timberman hours	Labourer hours	Third lift 3 to 4.5 m Timberman hours	Labourer hours	Fourth lift 4.5 to 6 m Timberman hours	Labourer hours
0–1325	1.08	1.08	1.68	1.68	2.33	2.33	3.00	3.00
1325–3975	1.44	1.44	2.42	2.42	3.11	3.11	4.00	4.00
3975–7950	1.80	1.80	2.80	2.80	3.88	3.88	5.00	5.00
7950–12350	2.16	2.16	3.36	3.36	4.54	4.54	6.00	6.00

Table 12.43 Volume occupied by 1 metre of pipe

Notes: 1. The data shown represent the volume displaced by 1 metre of pipe and may be used for assessing the amount of excavation which has to be disposed of on account of the pipe. In assessing this quantity bulkage should be allowed for, this varying with the class of ground. See Table 12.20
2. The data shown may be used for estimating purposes in connection with uPVC, cement, cast iron, concrete and clayware pipes.

Internal diameter of pipe in millimetres	Volume occupied by 1 m of pipe in cubic metres	Internal diameter of pipe in millimetres	Volume occupied by 1 m of pipe in cubic metres
100	0.02	825	0.72
150	0.04	900	0.80
225	0.07	975	0.93
300	0.11	1050	1.07
375	0.16	1200	1.47
450	0.22	1275	1.60
525	0.32	1350	1.75
600	0.41	1500	2.18
675	0.47	1800	3.12
750	0.57		

Table 12.44 Trench excavation

Note: The table shows the cubic metres of excavation per metre of trench per 300 mm of depth for trenches of various widths. The volume of excavation in trenches per metre of trench under the various 1.5 m lifts can, therefore, readily be obtained thus:

Example. Trench width, 0.75 m; trench depth, 3.6 m.

First lift,   0 to 1.5 m. Volume of excavation per metre of trench = 5×0.222
$= 1.110 \text{ m}^3$.

Second lift, 1.5 to 3.0 m. Volume of excavation per metre of trench = 5×0.222
$= 1.110 \text{ m}^3$.

Third lift,   3.0 to 4.5 m. Volume of excavation per metre of trench = 2×0.222
$= 0.444 \text{ m}^3$.

Width of trench in metres	Cubic metres of excavation per metre of trench per 300 mm of depth	Width of trench in metres	Cubic metres of excavation per metre of trench per 300 mm of depth
0.45	0.134	1.20	0.354
0.53	0.156	1.27	0.376
0.60	0.178	1.35	0.400
0.68	0.204	1.42	0.423
0.75	0.222	1.50	0.446
0.83	0.241	1.58	0.474
0.90	0.266	1.65	0.495
0.98	0.294	1.72	0.526
1.05	0.312	1.80	0.532
1.12	0.334		

Table 12.45 Trench excavation, cubic metres of excavation per metre at various depths and widths of trench

Depth of trench in metres	Width of trench in metres, and cubic metres of excavation per metre of trench								
	0.45	0.53	0.60	0.68	0.75	0.83	0.90	0.98	1.05
0.60	0.28	0.32	0.36	0.40	0.45	0.50	0.54	0.59	0.63
0.90	0.41	0.47	0.54	0.61	0.68	0.75	0.81	0.89	0.94
1.20	0.54	0.64	0.72	0.80	0.89	1.00	1.09	1.17	1.26
1.50	0.69	0.79	0.90	1.01	1.13	1.25	1.35	1.48	1.57
1.80	–	0.95	1.08	1.22	1.34	1.50	1.62	1.78	1.88
2.10	–	–	1.26	1.40	1.57	1.75	1.90	2.06	2.20
2.40	–	–	1.44	1.60	1.78	2.00	2.18	2.34	2.52
2.70	–	–	–	1.80	2.04	2.25	2.44	2.65	2.83
3.00	–	–	–	2.00	2.25	2.50	2.70	2.94	3.15
3.30	–	–	–	2.23	2.47	2.75	2.97	3.26	3.46
3.60	–	–	–	2.40	2.68	3.00	3.24	3.56	3.76
3.90	–	–	–	–	2.91	3.25	3.52	3.89	4.08
4.20	–	–	–	–	3.10	3.50	3.80	4.12	4.40
4.50	–	–	–	–	3.35	3.75	4.08	4.40	4.72
4.80	–	–	–	–	3.56	4.00	4.36	4.68	5.04
5.10	–	–	–	–	3.82	4.25	4.62	5.00	5.35
5.40	–	–	–	–	4.08	4.50	4.88	5.30	5.66
5.70	–	–	–	–	4.29	4.75	5.14	5.59	6.00
6.00	–	–	–	–	4.50	4.98	5.40	5.88	6.30

Depth of trench in metres	Width of trench in metres, and cubic metres of excavation per metre of trench								
	1.12	1.20	1.27	1.35	1.42	1.50	1.58	1.65	1.72	1.80
0.60	0.67	0.72	0.76	0.82	0.85	0.90	0.95	0.99	1.03	1.08
0.90	1.00	1.08	1.14	1.23	1.29	1.35	1.42	1.49	1.53	1.62
1.20	1.34	1.44	1.52	1.64	1.70	1.80	1.90	1.98	2.06	2.16
1.50	1.67	1.80	1.90	2.03	2.14	2.25	2.37	2.48	2.56	2.70
1.80	2.00	2.16	2.28	2.46	2.58	2.70	2.82	2.98	3.06	3.24
2.10	2.35	2.54	2.66	2.83	2.98	3.15	3.32	3.47	3.59	3.78
2.40	2.68	2.88	3.04	3.28	3.40	3.60	3.80	3.96	4.12	4.32
2.70	3.01	3.24	3.42	3.67	3.84	4.05	4.27	4.46	4.62	4.86
3.00	3.36	3.60	3.81	4.05	4.26	4.50	4.74	4.95	5.26	5.40
3.30	3.67	3.96	4.08	4.49	4.72	4.95	5.19	5.46	5.62	5.94
3.60	4.00	4.32	4.56	4.92	5.16	5.40	5.64	5.96	6.12	6.48
3.90	4.35	4.60	4.94	5.29	5.56	5.85	6.15	6.45	6.66	7.02
4.20	4.70	5.08	5.32	5.66	5.96	6.30	6.64	6.94	7.18	7.56
4.50	5.03	5.42	5.70	6.11	6.38	6.75	7.12	7.43	7.71	8.10
4.80	5.34	5.76	6.08	6.56	6.80	7.20	7.60	7.92	8.24	8.64
5.10	5.69	6.12	6.46	6.95	7.24	7.65	8.07	8.42	8.74	9.18
5.40	6.02	6.48	6.84	7.34	7.68	8.10	8.54	8.92	9.32	9.72
5.70	6.37	6.84	7.23	7.72	8.10	8.55	9.00	9.41	9.88	10.26
6.00	6.72	7.20	7.62	8.10	8.52	9.00	9.48	9.90	10.32	10.80

13 Concrete work

In building and civil engineering work concrete is extensively used, its cost being influenced by:

1. The specification of the materials, the quality and quantity of the product and the nature of the work.
2. The method adopted of carrying out the work.

Concrete mix design

The purpose of concrete mix design is to choose and proportion the ingredients used in a concrete mix to produce economical concrete which will have the desired properties both when fresh and when hardened. The variables which can be controlled are (1) water: cement ratio; (2) maximum aggregate size; (3) aggregate grading; (4) aggregate:cement ratio; and (5) use of admixtures. Interactions between the effects of the variables complicate mix design and successive adjustments following trial mixes are usually necessary. The majority of concrete supplied to the industry is by ready-mix concrete producers who produce suitable mix designs very quickly.

In this work methods of proportioning by volume or weight are discussed, and data are shown from which to estimate the material cost of a cubic metre of concrete mixed either by volume or by weight. In estimating the cost of concreting work the estimator should carefully note the proportioning specified and whether it is by volume or by weight. He should then build up the price accordingly.

Mixing by volume

In specifying a concrete ration of mix by volume the proportioning refers to dry materials; allowance must therefore be made for any bulking in the aggregates due to their moisture content.

For estimating, bulkage may be ignored in the case of the coarse aggregates such as shingle, clean gravel, broken stone, etc., as their moisture content is extremely small. Washed sand and all-in ballast, however, bulk appreciably, due to moisture content, and where these are delivered on the site in a moist state, allowance must be made for this bulkage in computing the amount of material required per cubic metre of concrete. Experience has shown that bulkage in the case of these two materials approximates to 25 per cent in the case of washed sand and 15 per cent in that of all-in ballast, so that bulkage to this amount should be allowed for in estimating.

From the contractor's point of view it is of the utmost importance that this bulkage is allowed for, both from an estimating standpoint and in carrying out the work itself, for if such is not the case the estimated material cost of the concrete per cubic metre is too low and in carrying out the work the yield of concrete per kg of cement is less than it ought to be.

In Tables 13.2 and 13.3 for purposes of estimating the material cost of a cubic metre of concrete mixed by volume, the quantities of materials required are shown for both dry and moist materials. In those cases where washed sand or all-in ballast are used the data shown relating to moist sand and moist all-in ballast should be used in computing the material cost of the concrete per cubic metre.

Mixing by weight

As already mentioned, proportioning concrete by weight and weigh batching the materials produces concrete which is more consistent than concrete mixed by volume. By carefully grading the aggregates and rigidly controlling the water/cement ratio, taking into consideration the moisture content present in the aggregates themselves, first-quality concrete of ultimate high compressive strength is produced.

For this reason mixing by weight has become increasingly popular where large volumes of high-class concrete are required, such as in constructing airport runways, motorways, etc.

In actually carrying out the work the aggregates are carefully graded and the water/cement ratio is determined on the site by test, due allowance being made for the weight of water contained in the aggregates themselves with a view to achieving the required compressive strength and workability of the mix.

From an estimating point of view, the determination of the quantities of materials required per cubic metre of concrete when concrete is mixed by weight are best computed by the *absolute volume* method, which is illustrated in Example 2

The estimator will note that in adopting this method a knowledge of the specific gravity of the materials concerned is involved. Specific gravities of those concrete materials most commonly used are therefore given below. The specific gravities shown represent fair average values for the materials concerned, and where the actual specific gravities are unknown they may be used for estimating purposes as they stand.

Knowing the nature of the materials concerned, the proportions of mix by weight and the water/cement ratio specified, the estimator may compute the materials required per cubic metre of concrete in the manner shown.

To assist the estimator, Tables 13.4 and 13.5 set out the materials required per cubic metre of concrete mixed by weight for those proportions of mix most commonly used with appropriate water/cement ratios. The data shown relate to concrete made up of the following constituents:

- Shingle, washed sand and ordinary Portland cement.
- Broken stone, washed sand and ordinary Portland cement.
- All-in ballast and ordinary Portland cement.

The data shown in the tables are based on materials of the following specific gravities:

Shingle	2.52	All-in ballast, 2 of stone	
Broken stone	2.70	to 1 of washed sand	2.56
Washed sand	2.60	Ordinary Portland cement	3.15

The quantities of materials required per cubic metre of concrete

The manner in which to compute the quantities of materials required per cubic metre of concrete as set mixed by volume and by weight is best illustrated by example.

With this in view the work out is shown for 4:2:1 concrete mixed by volume and by weight, using shingle, washed sand and ordinary Portland cement.

Note: Where the actual weights and specific gravities of the materials concerned are unknown, the estimator is referred to the text following the Examples below, where weight/volume ratios and specific gravities of concrete materials and concrete suitable for purposes of estimating are shown.

Example 1. Mixing by Volume 4:2:1 Mix Water/Cement Ratio 0.55. For purposes of illustrating the method of computing the materials required per cubic metre of concrete as set mixed by volume, the weight of the materials used is as follows:

Shingle	= 1442 kg per m^3
Dry sand	= 1602 kg per m^3
Moist sand	= 1282 kg per m^3
Cement	= 1442 kg per m^3
Weight of 1 m^3 of shingle, sand and ordinary Portland cement concrete	= 2355 kg per m^3

To compute the materials required per cubic metre of hardened concrete, reduce to terms of kg, weight, thus:

Shingle, 1 at 1,442 kg	=	1442 kg
Dry sand, ½ at 1,602 kg	=	801 kg
Cement, ¼ at 1,442 kg	=	361 kg
Water, ⅛ at 1,442 kg	=	180 kg
		2784 kg

The materials required per cubic metre of concrete as set out are then:

$$\text{Shingle, } \frac{1442}{2784} \times 2355 \text{ kg} = 1212 \text{ kg} = \frac{1212}{1442} \text{ m}^3 = 0.841 \text{ m}^3$$

$$\text{Sand } \frac{801}{2784} \times 2355 \text{ kg} = 678 \text{ kg} = \frac{678}{1282} \text{ m}^3 = 0.529 \text{ m}^3$$

$$\text{Cement } \frac{361}{2784} \times 2355 \text{ kg} = 298 \text{ kg} = 0.298 \text{ tonnes}$$

$$\text{Water } \frac{180}{2784} \times 2355 \text{ kg} = 156 \text{ kg} = 156 \text{ litres}$$

Note: In converting the weight of sand required to cubic metres the weight of moist sand, i.e. 1282 kg per m^3 is used, since the sand is delivered in a moist condition and bulkage has therefore to be allowed for.

Example 2. Mixing Weight 4:2:1 Mix, Water/Cement Ratio 0.55. In mixing concrete by weight consideration should be given to the specific gravity of the materials concerned as they affect the bulk of the hardened concrete. Knowing these specific gravities, the materials required per cubic metre of concrete may be computed by what is known as the Absolute Volume method.

Assuming that the ratio of mix specified is 4:2:1 by weight, using shingle, sand and ordinary Portland cement, and the water/cement ratio is 0.55, the materials required per 50 kg of cement are:

Shingle	4 at 50 kg	= 200 kg
Sand	2 at 50 kg	= 100 kg
Cement	1 at 50 kg	= 50 kg
Water	0.55 × 50 kg	= 27.5 kg

The Absolute Volume of a material, i.e. the minimum volume of the material with all voids removed, is given by the equation:

$$\text{Absolute volume} = \frac{\text{Weight of the material in kg}}{\text{Specific gravity} \times \text{weight of 1 m}^3 \text{ of water}}$$

$$= \frac{\text{Weight of materials in kg}}{\text{Specific gravity} \times 1\,000 \text{ kg}}$$

For purposes of estimating, the specific gravity of the materials concerned may be taken as being:

Shingle   2.52   Cement   3.15
Sand   2.60   Water   1.00

The absolute volume of the materials is then:

$$\text{Shingle} = \frac{200}{2.52 \times 1000} = 0.079 \text{ m}^3$$

$$\text{Sand} = \frac{100}{2.60 \times 1000} = 0.039 \text{ m}^3$$

$$\text{Cement} = \frac{50}{3.15 \times 1000} = 0.016 \text{ m}^3$$

$$\text{Water} = \frac{27.5}{1000} = \frac{0.027 \text{ m}^3}{0.161 \text{ m}^3}$$

The materials required per cubic metre of concrete as set are then:

$$\text{Shingle} \quad \frac{1.0 \times 200}{0.161} = 1243 \text{ kg} = 1.243 \text{ tonnes}$$

$$\text{Sand} \quad \frac{1.0 \times 100}{0.161} = 612 \text{ kg} = 0.612 \text{ tonnes}$$

$$\text{Cement} \quad \frac{1.0 \times 50}{0.161} = 306 \text{ kg} = 0.306 \text{ tonnes}$$

$$\text{Water} \quad \frac{1.0 \times 27.5}{0.161} = 172 \text{ kg} = 172 \text{ litres}$$

Weight/volume ratios and specific gravities of concrete materials and concrete

Note: The data shown refer to average values and may be used for normal purposes of estimating where the actual weights and specific gravities of the materials concerned are unknown.

Weight/volume ratios:

Cement	$= 1.442$ tonnes per m^3
Shingle	$= 1.442$ tonnes per m^3

Broken stone:

Basalt	$= 1.506$ tonnes per m^3
Granite	$= 1.474$ tonnes per m^3
Limestone	$= 1.330$ tonnes per m^3
Whinstone	$= 1.522$ tonnes per m^3
Dry sand	$= 1.602$ tonnes per m^3
Moist washed sand	$= 1.282$ tonnes per m^3
Dry all-in ballast (2 of stone to 1 of sand)	$= 1.922$ tonnes per m^3
Moist all-in ballast (2 of stone to 1 of sand)	$= 1.682$ tonnes per m^3
Shingle and sand concrete as set	$= 2.355$ tonnes per m^3
Basalt and sand concrete as set	$= 2.499$ tonnes per m^3
Granite and sand concrete as set	$= 2.451$ tonnes per m^3
Limestone and sand concrete as set	$= 2.163$ tonnes per m^3
Whinstone and sand concrete as set	$= 2.515$ tonnes per m^3
All-in ballast concrete (2 of stone to 1 of sand)	$= 2.355$ tonnes per m^3
Water	$= 1.000$ tonnes per m^3

Specific gravity:

Basalt	$= 2.80$	Whinstone	$= 2.80$	All-in ballast	$= 2.56$
Granite	$= 2.70$	River sand	$= 2.90$	Ordinary Portland cement	$= 3.15$
Limestone	$= 2.40$	Pit sand	$= 2.50$	Ferrocrete cement	$= 3.00$
Shingle	$= 2.52$	Thames sand	$= 2.60$		

Table 13.1 The weight of broken stone knowing the weight of the stone in the solid and/or its specific gravity

Weight of stone kg per m^3	Specific gravity	Percentage voids and weight of broken stone in kg per m^3		
		40%	45%	50%
1922	1.92	1153	1.057	961
2002	2.00	1201	1105	993
2082	2.08	1249	1137	1025
2163	2.16	1298	1185	1073
2242	2.24	1345	1233	1121
2322	2.32	1394	1281	1169
2402	2.40	1442	1330	1201
2483	2.48	1490	1362	1233
2563	2.56	1538	1410	1281
2643	2.64	1586	1458	1314
2723	2.72	1633	1490	1362
2803	2.80	1682	1538	1410
2883	2.88	1730	1586	1442

Concrete tamping and vibrating plant

Modern specifications commonly call for mechanical tamping and vibrating of the concrete, this being done by using vibrating units of suitable type.

This applies in particular to concrete laid in large areas such as roads, airport runways, etc. Vibrating plant commonly used for this purpose consists of small vibrating units attached to the usual form of hand tamper, the usual practice being to attach one such unit to tampers up to 4 m in length and two on tampers whose length is in excess of this.

The units may be actuated by either small petrol engines of approximately one kilowatt, or compressed air, in which case a compressor is required, or by electricity.

The capacity of the units is such that they can comfortably tamp and vibrate up to $12\,m^3$ of concrete per hour, laid to a finished thickness of 150 mm.

The running cost of the petrol engine vibrator is extremely low as each unit consumes only about one and a half litres of petrol per hour. The electric vibrator is also inexpensive to run, provided a supply is available, which, in the case of the contractor, is not always so. The pneumatic vibrator is the most costly to use, since it requires a compressor to actuate the machine.

To consolidate concrete laid in areas, two passes of the tamper are generally required at high-speed vibration, one pass at slow speed being sufficient as a finishing pass.

Vibrating plant is also extensively used in connection with precast concrete unit work, in which case the plant is usually of the vibrating table type. The moulds containing the concrete are placed on this table, which vibrates at a controlled speed to suit the type of unit being dealt with.

The method of estimating the cost of tamping and vibrating concrete laid in areas such as roads is illustrated in Example 4 on p. 130, and in this connection it should be noted that it is calculated as an Extra Over cost on tamping manually, the labour involved in tamping being the same in each case.

In estimating the cost of vibrating precast units or other work, the amount of concrete vibrated per hour should be assessed together with the working cost of the vibrating plant, the cost of vibrating the concrete then being calculated per unit or per m^3, as the case may be.

The cost of hauling the plant and the method of allowing for its cost

The haulage cost incurred in connection with the mixing plant, attendant loading plant, transporting plant, such as locomotives, rails, wagons, etc., may be covered either:

1. By allowing for it on a lump sum basis.
2. By allowing for it in the unit cost of the concrete itself.

The cost of setting up the mixing plant, and the method of allowing for its cost

With the larger sizes of mixing plant, where mechanical transporting plant is used to transport the mixed concrete, the plant is set up on a suitable structure or platform so that the transporting plant can be loaded direct.

The type of structure erected depends on the prevailing site conditions and the type, size and weight of plant used. In the case of comparatively light plant, such as the 10/7 and 14/10 revolving drum type, a platform of built-up sleepers is usually sufficient, but for large plant from which a considerable output is required a substantial structure may be necessary. This commonly takes the form of a brick or concrete structure consisting of walls and an elevated platform.

The cost involved in setting up the plant must necessarily be allowed for, this involves both the labour cost incurred in erecting and dismantling the structure on which it is placed and the cost of any material used in its construction.

This is best done by allowing for it in the unit cost of the concrete itself, this being done in the following manner:

1. By estimating the cost of the structural work together with the labour cost involved in placing the plant on it, and on the completion of the work removing the plant and demolishing and clearing away the structure.
2. Dividing the above cost by the total cubic metres of concrete to be produced by the plant.

Data from which to estimate the cost of various forms of structures on which to set up mixing plant may be found under the relevant sections.

Water required per cubic metre of concrete and the methods of cost

In considering the amount of water required to produce 1 m³ of mixed concrete, waste and washing out the mixing plant and transporting plant must also be taken into account.

Approximately 145 litres of water are required per m³ of mixed concrete, but in actual practice, allowing for the factors mentioned, 490 litres per m³ should be allowed for in estimating the cost of the water required.

The cost of water may be allowed for in one of three ways:

1. By allowing for it on a lump sum basis in the preliminaries.
2. By allowing for it as a percentage of the estimated cost of the work.
3. By allowing for its cost in the estimated unit cost of the work itself.

The cost of mixing and placing the concrete

The cost of mixing and placing the concrete is obtained from the tables shown for mixing and placing the concrete only. The estimator should note that transporting the concrete is shown under subsequent tables.

These tables show the labour hours or, in the case of machine-mixed concrete, the plant and labour hours taken to mix and place 1 m³ of concrete in the work.

In connection with the plant hours shown it should be noted that the plant working cost or hire rate per hour must be wholly inclusive of:

1. The machine hire rate per hour, allowing for depreciation and repairs and renewals.
2. The plant driver's wages per hour.
3. The cost of the fuel and oil consumed per hour.

The estimator is referred to chapter 6, where the method of assessing the working cost of plant per hour is shown.

The cost of transporting the concrete

In estimating the cost of transporting 1 m^3 of concrete in mechanically propelled transporting plant, three factors have to be taken into consideration:

1. The time the transporting plant stands while it is loaded.
2. The time the transporting plant takes to haul the concrete from the mixing plant to where it is off-loaded, off-load and return to the mixing plant per load.
3. The cubic metres of concrete hauled per load.

The estimator should note that the time the plant stands while 1 m^3 of concrete is loaded into it is governed by the output of the mixing plant, i.e. the time taken to load 1 m^3 of concrete into the transporting plant is the mixing plant hours per cubic metre of concrete as shown in the mixing and placing only tables.

For purposes of estimating the cost of transporting concrete by mechanically propelled vehicles Table 13.17 shows the time taken to haul 1 m^3 of concrete to various lengths of haul, using lorries and dumpers, the time taken to haul a set of wagons, using locomotives on 0.6 m and 1.44 m track, and the time taken to haul a wagon or wagons using tractors. Table 13.16 sets out the volume of concrete hauled per load in various types and sizes of transporting plant. By the correct use of the data shown in these tables in conjunction with the data shown in the tables for mixing and placing the concrete only, the estimator can calculate the cost of mixing, transporting and placing the concrete per cubic metre for all combinations of mixing plant, class of work concreted, type of transporting plant used and length of haul.

In considering the haulage cost of materials in vehicles such as lorries, dumpers, etc., it should be noted that this cost must be based on the actual number of vehicles used, as distinct from the theoretical number, because the vehicles are only kept constantly employed when the time of loading is equal to the time of haul.

The computation of the actual number of vehicles required is given by the following equation:

$$\text{No. of vehicles required} = \frac{\text{(Time to load vehicle)} + \text{(Time of haul)}}{\text{Time to load vehicle}}$$

the actual number used being to the nearest round number of vehicles in excess of this, e.g.:

Time to load vehicle = 0.20 h
Time of haul   = 0.14 h

$$\text{No. of vehicles required} = \frac{0.20 + 0.14}{0.20} = 1.70$$

Actual number of vehicles used = 2

This example shows that consideration should be given to the capacity of the vehicles used in conjunction with the length of haul, otherwise a little over one vehicle may be required and in actual fact two would have to be used. This applies in general to those cases where the time of loading is greatly in excess of the time of haul, and the estimator should note that in certain cases it might be more economical to use only one vehicle instead of two.

The method of estimating the cost of various classes of concreting is best illustrated by example, and to do so typical examples are shown. The method of estimating the Extra Over cost of tamping concrete with vibrating hand screeds is also shown.

In the case of all the examples, the cost of hauling the plant and the cost of water are not included, it being assumed that these are allowed for elsewhere in the estimate. The alternative ways in which those elements of the cost may be allowed for are dealt with in chapter 3.

For purposes of the examples shown it is assumed that the concrete in each case is concrete of a 4:2:1 by volume mix of 38 to 12 mm shingle, washed sand and cement, the following being the rates quoted for the respective materials delivered on the site:

38 to 12 mm shingle	£13.00 per m^3
Washed sand	£12.00
Cement	£75.00 per tonne

The material cost of concrete per cubic metre

The estimated material cost of 4:2:1 concrete per cubic metre is obtained from Table 13.2, thus:

38 to 12 mm shingle	0.84 m^3 at £13.00 per m^3	= £10.92 per m^3
Washed sand allowing for bulkage	0.53 m^3 at £12.00 per m^3	= £ 6.36 per m^3
Cement	0.304 tonnes at £75.00 per tonne	= £22.80 per m^3
The estimated material cost of the concrete		= £40.08 per m^3

Example 1.   Estimate the cost per cubic metre of mixing, wheeling and placing 4:2:1 concrete in a building foundation, using a 7/5 concrete mixer of the revolving drum type, the concrete being wheeled to position in single-wheel handbarrows.

For purposes of this example the following are assumed:

1. The working cost of hire rate of the mixing plant, inclusive of fuel and oil = £3.00 per h.
2. Labour rate = £6.00 per h.
3. The material cost of the concrete = £40.08 per m^3.

The estimated cost of the concrete is built up as follows:

The cost of mixing and placing the concrete per cubic metre

Referring to Table 13.6 cost of mixing and placing concrete is given by:

Concrete mixer, including fuel and oil =	0.65 h at £3.00 per h	= £ 1.95 per m^3
Labour	= 3.90 h at £6.00 per h	= £23.40 per m^3
Estimated cost of mixing and placing the concrete		= £25.35 per m^3

The cost of wheeling the concrete per cubic metre

Referring to Table 13.15 the labour hours taken to wheel concrete in single-wheeled barrows is given as 1.07 hours per cubic metre.

The estimated cost of wheeling the concrete = 1.07 h at £6.00 per h
= £6.42 per m^3

The total cost of supplying, mixing, wheeling and placing the concrete then is:

Mixing and placing the concrete	= £25.35 per m^3
Wheeling the concrete	= £6.42 per m^3
Material cost of the concrete	= £40.08 per m^3
Estimated cost of the concrete	= £71.85 per m^3

Example 2.   Estimate the cost per cubic metre of mixing, transporting, placing and tamping 4:2:1 concrete in a new concrete road, the concrete being laid to a finished thickness of 150 mm. The mixing plant used is a 14/10 concrete mixer of the drum and hopper type, the plant being loaded by a 0.2 m³ dragline. The concrete is transported from the mixing plant to where it is to be deposited in the work in a 4 tonne dumper fitted with mechanical tipping gear.

For purposes of this example the following are assumed:

1. The average length of haul from the mixing plant is 1.5 kilometre.
2. The estimated cost of setting up the mixing plant on a sleeper platform is £80, and the total volume of concrete to be placed in the work is 1000 m³.
3. The working cost or hire rate of the mixing plant, including driver, fuel and oil = £9.00 per h.
4. The working cost or hire rate of the 0.2 m³ dragline, including driver, fuel and oil = £8.50 per h.
5. The working rate cost or hire rate of a 4 tonne dumper, including driver, fuel and oil = £10.00 per h.
6. The labour rate = £6.00 per h.
7. The material cost of the concrete = £40.08.

The estimated cost of the concrete is built up in the following manner:

The cost of setting up the mixing plant per cubic metre of concrete
The estimator, having assessed the cost of setting up the mixing plant at £80 and knowing that the total volume of concrete to be produced by the plant is 1000 m³, calculates the cost of setting up the plant per cubic metre of concrete thus:

$$\text{Plant setting up cost per m}^3 \text{ of concrete } = \frac{£80}{1000} \text{ m}^3$$

$$= £0.08 \text{ per m}^3$$

The cost of mixing, and placing the concrete per cubic metre
Referring to Table 13.6 the cost of mixing and placing the concrete per cubic metre is given by:

14/10 concrete mixer, including driver,
    fuel and oil   = 0.36 h at £8.00 per h = £ 2.88 per m³
Labour   = 2.88 h at £6.00 per h = £17.28 per m³

    Estimated cost of mixing and placing the concrete   = £20.16 per m³

The cost of transporting the concrete per cubic metre
From Table 13.6.

Hours taken by 14/10 concrete mixer to produce 1 m³ of concrete = 0.36 h.

From Table 13.16.

The cubic metres of concrete hauled per load in a 4 tonne dumper = 1.80 m³.

From Table 13.17.

The time taken by the lorry to haul one load of concrete to the point of off-loading 1.5 kilometres away, off-load and return for the next load = 0.27 h.

Time taken to load lorry　　　 = 0.23 × 1.80 = 0.41 h.
Time of haul　　　　　　　　　　　　　= 0.27 h.

$$\text{Number of lorries required} = \frac{(\text{Time to load}) + (\text{Time to haul})}{\text{Time to load}}$$

$$= (0.41 + 0.27)/0.41$$

$$= 1.7 \text{ lorries}$$

Actual number of lorries used　= 2

The cost of transporting the concrete per cubic metre is therefore built up as follows:

Transport cost per h = 2 dumpers at £10.00 per h = £20.00
Output of 14/10 concrete mixing plant = 0.36 h per m^3

Estimated cost of hauling the concrete per m^3 = 0.23 × £20.00　= £4.60 per m^3

The total cost of supplying, mixing, transporting and placing the concrete per cubic metre then is:

Setting up the mixing plant　　　　= £　0.08 per m^3
Mixing and placing the concrete　= £20.16 per m^3
Transporting the concrete　　　　　= £　4.60 per m^3
Material cost of the concrete　　　 = £40.08 per m^3

　Estimated cost of the concrete　= £64.92 per m^3

Concrete laid 150 mm thick = Cost of one cubic metre × 0.150
　　　　　　　　　　　　　　 = £64.92 × 0.150 = £9.74

Note: The estimated cost of the concrete per square metre shown does not include incidental items such as fixing side forms and expansion joints, curing, bullnosing edges, etc., for which see Table 13.23.

Example 3.　Estimate the cost per cubic metre of mixing, transporting, placing and tamping 4:2:1 concrete in a new concrete road, the concrete being laid to a finished thickness of 150 mm. The mixing plant used is a 14/10 mixer of the drum and hopper type, the plant being loaded by a 0.2 m^3 dragline. The concrete is transported from the mixing plant to where it is placed in the work in Decauville wagons, hauled by locomotive on 0.6 m gauge track.

For purposes of this example the following are assumed:

1. The average length of haul from the mixing plant is 0.75 kilometres.
2. The total volume of concrete to be placed in the work is 1000 m^3.
3. The estimated cost of setting up the mixing plant is £80.
4. The working cost or hire rate of the 14/10 mixer, including driver, fuel and oil = £8.00 per h.
5. The working cost or hire rate of the 0.2 m^3 dragline, including driver, fuel and oil = £8.50 per h.

6. The working cost or hire rate of locomotive, including the hire rate of wagons and track = £8.00 per h.
7. The labour rate = £6.00 per h.
8. The material cost of the concrete = £40.08 per m³.

The estimated cost of the concrete is built up as follows:

The cost of setting up the mixing plant per cubic metre of concrete
The cost of setting up the plant on a suitable structure so that the skips or wagons may be loaded direct having been assessed at £80 and the total cubic metres of the concrete to be produced by the mixing plant as being 1000 m³, the cost of 'setting up' the plant is:

$$\text{Plant setting up cost per m}^3 \text{ of concrete} = \frac{£80}{1000} = £0.08 \text{ per m}^3$$

The number of skips or wagons required and the m³ of concrete hauled per set
By correctly assessing the number of skips or wagons per set in conjunction with the output of the mixing plant and the length of haul, the cubic metres of concrete hauled per set is obtained thus:

Referring to Table 23.29, 1 14/10 concrete mixer loaded by dragline produces 1 m³ of concrete in 0.23 h.

Referring to Table 13.17, the time taken by the locomotive in hauling an average length of 0.75 km is 0.40 h.

$$\text{In 0.40 h the concrete mixer produces } \frac{0.40}{0.23} = 1.74 \text{ m}^3 \text{ of concrete}$$

Sufficient skips or wagons per set must be employed to transport 1.74 m³ of concrete.

The cost of laying and removing the rail track per cubic metre of concrete
The average length of haul being 0.75 km, the track to be laid is 0.75 km, plus, say, 30 m for turnouts, a total of 0.78 km.

Referring to chapter 21, the labour hours taken to lay and remove the track are:

Laying the track	= 0.36 h per m
Removing the track (multiplier 0.80) = 0.80 × 0.36	= 0.29 h per m
Total	= 0.65 h per m

The cost of laying and removing the track = 780m at 0.65 h per m = 507.0 labour hours at £6.00 per hour = £3042.00

The total m³ of concrete to be placed = 1000 m³

$$\text{Cost of track per m}^3 \text{ of concrete} = \frac{£3042.00}{1000}$$
$$= £3.042 \text{ per m}^3$$

Note: The rail track may be used on other operations in connection with the work, such as the preparatory excavation. The cost of laying and renewing it may be apportioned to the various operations if desired, but from an estimating point of view is best allowed for under a single operation, as is done in the case of this example. This makes for both simplicity and accuracy, which is to be desired. If the rail track is used in connection with excavation, it is usual to allow for the cost of laying and removing it wholly under this item.

The cost of mixing, placing and manually tamping the concrete per cubic metre

Referring to Table 21.29, the cost of mixing, placing and tamping the concrete is given by:

14/10 concrete mixer,
including driver, fuel and oil   = 0.23 h at £8.00 per h = £1.840 per m³
0.2 m³ dragline, including
driver, fuel and oil   = 0.23 h at £8.50 per h = £1.955 per m³
Labour   = 2.04 h at £6.00 per h = £12.240 per m³

Estimated cost of mixing, placing and tamping the concrete = £14.195 per m³

The cost of transporting the concrete per cubic metre

In transporting concrete by locomotive and skips or wagons on 0.6 m rail track, allowance must be made for one man attending on the locomotive and also for the labour employed in maintaining the track. For purposes of this example one man may be allowed for track maintenance.

The transporting cost per cubic metre of concrete is estimated as follows:

Locomotive time taken to haul one set of skips or wagons containing 1.80 m³ of concrete = 0.40 h

Labour in attendance on locomotive and track = 2 men at 0.40 h = 0.80 h

The estimated cost of transporting 1.80 m³ of concrete is:

Hire of locomotive, wagons and track, 0.40 h at £8.00 per h = £3.20
Labour   0.80 h at £6.00 per h = £4.80

Total   = £8.20

$$\text{Cost of transporting the concrete per cubic metre} = \frac{£8.00}{1.80 \text{ m}^3}$$

$$= £4.444 \text{ per m}^3$$

The total cost of supplying, mixing, transporting, placing and manually tamping the concrete per cubic metre then is:

Setting up the mixing plant   = £0.080 per m³
Laying and removing the rail track   = £3.042 per m³
Mixing, placing and tamping the concrete   = £14.195 per m³
Transporting the concrete   = £4.444 per m³
Material cost of the concrete (£40.08 per m³) = £40.080 per m³

Estimated cost of the concrete   = £61.841 per m³

$$150 \text{ mm concrete} = \frac{£61.841}{1000} \times 150 = £9.276 \text{ per m}^2$$

Note: The estimated cost of the concrete per square metre shown does not include for incidental items such as fixing side forms and expansion joints, curing bullnosing edges, etc. for which see Table 21.33.

Example 4.   Calculate the Extra Over cost per square metre, for tamping concrete laid in a new road, using petrol-driven vibrating units and also pneumatically actuated vibrating units attached to normal timber hand tampers.

For purposes of this example the following is assumed:

1. The price of the petrol-driven vibrating units is £1600 each.
2. The price of pneumatic-driven vibrating units is £1500 each.
3. The life of the units is five years, working 1,200 h per annum.
4. The petrol consumption of the petrol-driven unit is 1.5 litre per h.
5. The oil consumption of the petrol-driven unit is 0.005 litres per h.
6. The cost of the petrol is £0.600 per litre and oil £0.500 per litre.
7. 2 petrol type or pneumatic units are used on a single tamper for carrying out the work.
8. The hire or working cost per hour of a compressor to work the pneumatic units is £6.00 per hour inclusive.
9. The mixer hours per cubic metre of concrete are as shown in Example No. 2, i.e. 0.36 h per cubic metre. This means that the vibrating plant and the compressor hours per cubic metre of concrete are also 0.36 h, since the concrete is mixed and placed at this rate of output.

Tamping with petrol-driven vibrating units

Cost of one petrol-driven vibrating unit = £1600
Add 5 per cent per annum for five years = £ 400
                                           _____
                                           £2000

Allowing a 5 year life, cost per annun = $\dfrac{£2000}{5}$ = £400

Allowing 1200 working hours per annum, plant cost per h = $\dfrac{£400}{1200}$ = £0.333 per h

Allow for repairs and renewals, say, 20 per cent = £0.066 per h
Allow for petrol, 1.5 litres at £0.600 per litre = £0.900 per h
Allow for oil, 0.005 litres at £0.500 per litre = £0.003 per h
                                                   _____
  Working cost of 1 unit per h   = £0.969 per h
  Working cost of 2 units on
  the one tamper per h = 2 × £0.969   = £1.938 per h

Since the concrete mixer hours per cubic metre of concrete is 0.36 h, the working cost of vibrating plant per cubic metre of concrete = 0.36 h at £1.938 per h = £0.698 per m³
The Extra Over cost of tamping concrete 150 mm thick with petrol-driven vibrating units, is

$\dfrac{£0.698}{1000}$ × 150 = £0.105 per m².

Tamping with pneumatic type vibrating units

Cost of one pneumatic unit   = £1500
Add 5 per cent per annum for 5 years   = £375
                                                   = £1875

Allowing a 5 year life, cost per annum   = $\dfrac{£1875}{5}$ = £375

Table 13.2 The quantities by volume of materials per cubic metre of concrete, using shingle and broken stone as coarse aggregates, sand and ordinary Portland cement

Notes: 1. The table shows the materials required per cubic metre of hardened concrete as set, using shingle and broken stone as coarse aggregates with dry sand, and moist sand which bulks 25 per cent, due to its moisture content, with appropriate water/cement ratios.
2. The data shown relating to cement refer to ordinary Portland cement weighing 1.44 tonnes per cubic metre.

Nominal mix by volume, Dry materials			Appropriate water/cement ratio by weight	Shingle or clean gravel 40% voids						Broken stone 45% voids					
				Dry sand			Moist sand 25% bulkage			Dry sand			Moist sand 25% bulkage		
Coarse aggregate	Sand	Cement		Coarse agg. *m³*	Sand *m³*	Cement *tonnes*	Coarse agg. *m³*	Sand *m³*	Cement *tonnes*	Coarse agg. *m³*	Sand *m³*	Cement *tonnes*	Coarse agg. *m³*	Sand *m³*	Cement *tonnes*
2	1	1	0.35	0.72	0.36	0.521	0.72	0.45	0.521	0.76	0.38	0.552	0.76	0.48	0.552
2	1½	1	0.40	0.64	0.48	0.469	0.64	0.60	0.469	0.66	0.49	0.485	0.66	0.61	0.485
3	1½	1	0.45	0.80	0.40	0.389	0.80	0.50	0.389	0.82	0.41	0.405	0.82	0.51	0.405
3	2	1	0.50	0.72	0.48	0.355	0.72	0.60	0.355	0.75	0.50	0.369	0.75	0.63	0.369
4	2	1	0.55	0.84	0.42	0.304	0.84	0.53	0.304	0.88	0.44	0.320	0.88	0.55	0.320
4	2½	1	0.60	0.78	0.48	0.287	0.78	0.60	0.287	0.81	0.50	0.299	0.81	0.63	0.299
5	2½	1	0.65	0.88	0.44	0.253	0.88	0.55	0.253	0.90	0.45	0.264	0.90	0.56	0.264
5	3	1	0.70	0.82	0.48	0.237	0.82	0.60	0.237	0.85	0.51	0.249	0.85	0.64	0.249
6	3	1	0.75	0.88	0.44	0.215	0.88	0.55	0.215	0.92	0.46	0.227	0.92	0.58	0.227
8	4	1	0.80	0.90	0.45	0.167	0.90	0.56	0.167	0.96	0.48	0.173	0.96	0.60	0.173
10	5	1	0.85	0.84	0.42	0.125	0.84	0.53	0.125	0.90	0.45	0.131	0.90	0.56	0.131

Table 13.3 The quantities by volume of materials per cubic metre of concrete, using all-in ballast and ordinary Portland cement

Notes: 1. The table shows the materials required per cubic metre of hardened concrete as set, using dry all-in ballast and moist all-in ballast which bulks 15 per cent, due to its moisture content, with appropriate water/cement ratios.
2. The data shown relating to all-in ballast refer to material containing 2 parts of stone to 1 part of sand.
3. The data shown relating to cement refer to ordinary Portland cement, weighing 1,440 kg per m^3.

Nominal mix by volume. Dry materials		Appropriate water/cement ratio by weight	Dry all-in ballast		Moist all-in bulkage 15% bulkage	
All-in ballast	Cement		All-in ballast cubic metres	Cement Tonnes	All-in ballast cubic metres	Cement Tonnes
6	1	0.60	1.26	0.304	1.45	0.304
7	1	0.65	1.28	0.275	1.47	0.275
8	1	0.70	1.30	0.244	1.49	0.244
9	1	0.75	1.32	0.214	1.52	0.214
10	1	0.80	1.33	0.199	1.53	0.199
12	1	0.85	1.35	0.167	1.55	0.167

Table 13.4 The quantities by weight of material per cubic metre of concrete, using shingle and broken stone as coarse aggregates with washed sand and ordinary Portland cement

Notes: 1. The table shows the materials required per cubic metre of hardened concrete as set with appropriate water/cement ratios.
2. The specific gravity of the materials on which the data are based is as follows:

Shingle or clean gravel = 2.52   Sand = 2.60
Broken stone = 2.70   Ordinary Portland cement = 3.15

Nominal mix, by weight. Dry materials			Appropriate water/cement ratio by weight	Shingle or clean gravel 40% voids			Broken stone 45% voids		
Coarse agg.	Sand	Cement		Coarse aggregate tonnes	Sand tonnes	Cement tonnes	Coarse aggregate tonnes	Sand tonnes	Cement tonnes
2	1	1	0.35	1.09	0.55	0.55	1.12	0.56	0.56
2	1½	1	0.40	0.96	0.72	0.48	0.99	0.75	0.493
3	1½	1	0.45	1.17	0.59	0.391	1.21	0.60	0.404
3	2	1	0.50	1.08	0.72	0.36	1.12	0.75	0.373
4	2	1	0.55	1.24	0.65	0.312	1.28	0.64	0.320
4	2½	1	0.60	1.16	0.83	0.291	1.20	0.76	0.300
5	2½	1	0.65	1.28	0.64	0.256	1.33	0.67	0.267
5	3	1	0.70	1.20	0.72	0.240	1.25	0.76	0.251
6	3	1	0.75	1.31	0.65	0.216	1.36	0.68	0.227
8	4	1	0.80	1.36	0.69	0.173	1.43	0.72	0.179

Table 13.5 The quantity by weight of materials per cubic metre of concrete, using all-in ballast and ordinary Portland cement

Notes: 1. The table shows the materials required per cubic metre of hardened concrete as set with appropriate water/cement ratios.
2. The specific gravity of the materials on which the data are based is as follows:

      All-in ballast   = 2.56
      Ordinary Portland cement = 3.15

Nominal mix by weight Dry materials		Appropriate water/cement ratio by weight	All-in ballast tonnes	Cement tonnes
All-in ballast	Cement			
6	1	0.60	1.87	0.312
7	1	0.65	1.89	0.280
8	1	0.70	1.92	0.248
9	1	0.75	1.96	0.217
10	1	0.80	2.0	0.203
12	1	0.85	2.07	0.172

Allowing 1,200 working hours per annum, unit cost per hour $= \dfrac{£375}{1200}$

                                                        $= £0.313$ per h

Allow for repairs and renewals, 20 per cent $= £0.063$

  Cost of 1 unit per hour $= £0.376$ per h

Cost of 2 units on tampers per hour $= 2 \times £0.376 = £0.752$ per h
Add working cost of compressor £6.00 per h $= £6.000$ per h
Working cost of vibrating plant per hour $= £6.752$ per h

Since the concrete mixer hours per cubic metre of concrete is 0.36 h, the working cost of the vibrating plant per cubic metre of concrete = 0.36 h at £6.752 per h = £2.431 per m^3.

The Extra Over cost of tamping concrete 150 mm thick with pneumatic vibrating units per square metre is

$\dfrac{£2.431}{1000} \times 150 = £0.365$ per m^2.

The treatment of concrete surfaces with silicate of soda

Concrete surfaces are treated with silicate of soda in order to add to the life of the concrete. Its application tends to lessen the amount of water which the concrete absorbs in passing from the wet to the dry state and also increases the resistance of the surface to abrasion.

The method of applying the solution is as follows:

1. The silicate of soda is mixed with water in the proportion of four parts of water to one part of silicate of soda, this being suitable for normal dense concrete, such as concrete roads. For concrete of a porous nature a greater quantity of silicate would be necessary, the amount depending on the porosity of the concrete.

2. The solution is applied to the surface by means of a water-can, sprayer or water-cart, when it is spread evenly with a soft brush or mop.
3. Three applications are generally necessary to complete the treatment, each application having thoroughly dried out before the next is applied. It is usual to carry out the applications at 24-hourly intervals.
4. Before applying the first coat, care should be taken to see that the surface of the concrete is clean, free from grease, dust, etc., and as dry as possible.

This treatment is particularly suitable for concrete roads.

One litre of silicate of soda mixed in the proportion of 4 parts of water to 1 part of silicate of soda will cover 6 m^2 one coat, the labour hours involved in mixing and brushing or mopping it being 0.04 hour per m^2. A water-cart will comfortably cover 280 m^2 per hour, i.e. 0.0038 h per m^2.

Assuming the labour rate is £6.00 per hour, the hire rate of the water-cart £1.50 per hour, and the cost of the silicate of soda £0.50 per litre, the cost of a single application per square metre then is:

Material cost of silicate of soda $\dfrac{£0.50}{6}$ = £0.083 per m^2

Labour cost, mixing and applying 0.04 h at £6.00 = £0.240 per m^2
Hire of water-cart, 0.0038 h at £1.50 = £0.006 per m^2

Total cost of application = £0.329 per m^2

Table 13.6 Mixing and placing concrete in small areas, foundations and narrow trenches

Notes: 1. The data shown are for mixing and placing the concrete only per cubic metre, and do not allow for wheeling or transporting, for which see Tables 13.15, 13.16 and 13.17
  2. For fixing bar or sheet reinforcement, see Table 13.18, 13.19 and 13.20

Description	*Mixing and placing concrete, using concrete mixing plant of the sizes shown. Per cubic metre*					
	7/5		10/7		14/10	
Concrete						
Mixing and placing only. (The hours shown do not allow for wheeling or transporting. See notes above.)	*Mixing plant hours*	*Labour hours*	*Mixing plant hours*	*Labour hours*	*Mixing plant hours*	*Labour hours*
Concrete in small areas such as paths, garage washdowns, yards, etc.	0.60	3.60	0.37	2.59	0.33	2.64
Concrete in foundations to chambers, manholes, etc.	0.70	4.20	0.48	3.36	0.39	3.12
Concrete in mass foundations.	0.56	3.36	0.44	3.08	0.36	2.88
Concrete in narrow trenches such as foundations to buildings, beds and backing to kerb and channels, etc.	0.65	3.90	0.49	3.43	0.40	3.20

Table 13.7 Mixing, hoisting and placing concrete in floors, landings and roofs up to a height of 3 m

Notes: 1. The data shown in the tables below are applicable to concrete joisted up to a height of 3 m, a table of multipliers being shown for heights in excess of this.
2. The data shown do not allow for wheeling or transporting, for which see. Tables 13.15, 13.16 and 13.17
3. For erecting and striking formwork, see text below
4. For fixing bar and sheet reinforcement, see Tables 13.18, 13.19 and 13.20
5. For finishing concrete surfaces, see Table 13.21
6. For sundry works to concrete, see Table 13.23
7. If the concrete is reinforced use a multiplier of 1.10 with the plant and labour hours shown.

Description *Mixing, hoisting and placing concrete up to a height of 3 m. Use table of multipliers shown for heights in excess of this. The hours shown do not allow for wheeling or transporting. See notes above*	Mixing, hoisting and placing concrete, using concrete mixing plant of the sizes shown and hoisting plant. Per cubic metre								
	7/5 mixer			10/7 mixer			14/10 mixer		
	Mixing plant hours	*Hoisting plant hours*	*Labour hours*	*Mixing plant hours*	*Hoisting plant hours*	*Labour hours*	*Mixing plant hours*	*Hoisting plant hours*	*Labour hours*
Concrete in floors, landings and roofs up to 100 mm thick.	0.96	0.96	5.76	0.80	0.80	5.60	0.67	0.67	5.36
Exceeding 100 mm and not exceeding 150 mm thick.	0.88	0.88	5.28	0.72	0.72	5.04	0.60	0.60	4.80
Exceeding 150 mm thick and not exceeding 300 mm thick.	0.81	0.81	4.86	0.67	0.67	4.69	0.56	0.56	4.48

Table 13.8 Multipliers for mixing, hoisting and placing concrete in floors, landings and roofs, to heights in excess of 3 m

Height to which concrete is hoisted in metres	*Plant and labour hour multipliers*
3	1.00
6	1.05
9	1.10
12	1.15
15	1.20

Table 13.9 Mixing, hoisting and placing concrete in walls up to a height of 3 m

Notes: 1. The data shown are applicable to concrete placed at a height of 3 m, a table of multipliers being shown for heights in excess of this.
2. The data shown do not allow for wheeling or transporting, for which see Tables 13.15, 13.16 and 13.17
3. For erecting and striking formwork, see text below
4. For fixing bar and sheet reinforcement, see. Tables 13.18, 13.19 and 13.20
5. For finishing concrete surfaces, see Table 13.21
6. For concreting walls to slope, use a multiplier of 1.05 with the plant and labour hours shown.
7. For concreting walls to curve, use a multiplier of 1.10 with the plant and labour hours shown.
8. If the concrete is reinforced, use a multiplier of 1.10.

Description	*Mixing, hoisting and placing concrete, using concrete mixing plant of the sizes shown and hoisting plant. Per cubic metre*								
Concreting, mixing, hoisting and placing up to a height of 3 m. For heights in excess of this use the table of multipliers shown. The hours shown do not allow for wheeling or transporting. See notes above	*7/5 mixer*			*10/7 mixer*			*14/10 mixer*		
	Concrete mixing plant hours	*Hoisting plant hours*	*Labour hours*	*Concrete mixing plant hours*	*Hoisting plant hours*	*Labour hours*	*Concrete mixing plant hours*	*Hoisting plant hours*	*Labour hours*
Concreting walls:									
Not exceeding 150 mm thick.	0.80	0.80	5.60	0.67	0.67	5.36	0.57	0.57	5.13
Exceeding 150 mm and not exceeding 225 mm thick.	0.69	0.69	4.83	0.59	0.59	4.72	0.49	0.49	4.41
Exceeding 225 mm and not exceeding 300 mm thick.	0.61	0.61	4.27	0.51	0.51	4.08	0.44	0.44	3.96
Exceeding 300 mm thick.	0.53	0.53	3.71	0.44	0.44	3.52	0.37	0.37	3.33

Table 13.10 Multipliers for mixing, hoisting and placing concrete in walls to heights in excess of 3 m

Height to which concrete is hoisted in metres	*Plant and labour hour multipliers*
3	1.00
6	1.05
9	1.10
12	1.15
15	1.20

Table 13.11 Mixing, hoisting and placing concrete in beams, braces, columns, etc. Up to a height of 3 m

Notes: 1. The data shown in the tables below are applicable to concrete hoisted up to a height of 3m, a table of multipliers being shown for heights in excess of this.
2. The data shown do not allow for wheeling or transporting, for which see. Tables 13.15, 13.16 and 13.17
3. For erecting and striking formwork, text below.
4. For fixing bar and sheet reinforcement, see Tables 13.18, 13.19 and 13.20
5. For finishing concrete surfaces, see Table 13.21
6. For sundry works to concrete see Table 13.23
7. If the concrete is reinforced use a multiplier of 1.10 with the plant and labour hours shown.

| Description | Mixing, hoisting and placing concrete, using concrete-mixing plant of the sizes shown and mechanical hoisting plant. Per cubic metre | | | | | |
| | 7/5 | | | 10/7 | | |
Concreting, mixing, hoisting and placing up to a height of 3 m. For heights in excess of this use the table of multipliers shown. The hours shown do not allow for wheeling or transporting	*Concrete mixing plant hours*	*Hoisting plant hours*	*Labour hours*	*Concrete mixing plant hours*	*Hoisting plant hours*	*Labour hours*
Concrete in beams, girder casings, plinths, projections, strings and similar. Cross-sectional area:						
0 to 0.025 m²	1.25	1.25	8.75	1.07	1.07	8.56
Exceeding 0.025 and up to 0.05 m²	1.07	1.07	7.49	0.91	0.91	7.28
Exceeding 0.05 and up to 0.075 m²	0.93	0.93	6.51	0.80	0.80	6.40
Exceeding 0.075 and up to 0.1 m²	0.88	0.88	6.16	0.75	0.75	6.00
Exceeding 0.1 m²	0.84	0.84	5.88	0.70	0.70	5.60
Concrete in braces. Cross-sectional area:						
0 to 0.025 m²	1.47	1.47	10.29	1.27	1.27	10.16
Exceeding 0.025 and up to 0.05 m²	1.33	1.33	9.31	1.13	1.13	9.04
Exceeding 0.05 and up to 0.075 m²	1.13	1.13	8.16	0.99	0.99	7.92
Exceeding 0.075 and up to 0.1 m²	1.08	1.08	7.56	0.93	0.93	7.44
Exceeding 0.1 m²	0.94	0.94	6.58	0.88	0.88	7.04
Concrete in columns, piers and pilasters. Cross-sectional area:						
0 to 0.025 m²	1.33	1.33	9.31	1.15	1.15	9.20
Exceeding 0.025 and up to 0.05 m²	1.24	1.24	8.68	1.03	1.03	8.24
Exceeding 0.05 and up to 0.075 m²	1.01	1.01	7.07	0.87	0.87	6.96
Exceeding 0.075 and up to 0.1 m²	0.95	0.95	6.65	0.81	0.81	6.48
Exceeding 0.1 m²	0.91	0.91	6.37	0.79	0.79	6.32

Table 13.12 Multipliers for concreting beams, columns, braces, etc. to heights in excess of 3m

Height to which the concrete is hoisted in metres	*Plant and labour hour multipliers*
3	1.00
6	1.05
9	1.10
12	1.15
15	1.20

Ready-mixed concrete

The overall cost of labour and material of concrete shows very little difference whether mixed on site by contractor's labour and machines or by using a ready-mixed concrete delivery service. The additional cost of ready-mixed delivery service being offset by the saving of labour time on site. Much will depend on site conditions and site management. If the new building is on a confined site facing on to a busy commercial thoroughfare with little or no storage space at the back, and several reinforced concrete floors and stair flights, then ready-mixed concrete service is the best solution. On the other hand on a large open site there is plenty of storage space and room for lorry turn round, the agent or foreman can arrange the concrete mixing programme from hour to hour independent of ready-mix delivery times.

Labourer times for wheeling and depositing ready-mixed concrete are shown in Table 13.13.

Table 13.13 Labourer hours for wheeling and depositing ready-mixed concrete

	Labourer hours
Unreinforced concrete	
Foundations exceeding 300 mm thick	3.00
Floors and oversite concrete 100 mm thick	6.00
Floors and oversite concrete 150 mm thick	5.50
Floors and oversite concrete 300 mm thick	4.80
Reinforced concrete including extra time tamping and vibrating concrete around reinforcement	
Foundations exceeding 300 mm thick	4.00
Suspended floors or roofs not exceeding 3 m above ground floor or pavement level 100 mm thick	7.50
Suspended floors or roofs not exceeding 3 m above ground floor or pavement level 150 mm thick	7.00
Suspended floors or roofs not exceeding 3 m above ground floor or pavement level 300 mm thick	6.50
Columns not exceeding 3 m above ground floor or pavement sectional area not exceeding $0.05\,m.^2$	10.00
Columns not exceeding 3 m above ground floor or pavement sectional area not exceeding $0.05–0.10\,m.^2$	9.00
Columns not exceeding 3 m above ground floor or pavement sectional area not exceeding $0.10\,m.^2$	8.00
Beams not exceeding 3m above ground floor or pavement	
Sectional area not exceeding $0.05.\,m.^2$	9.00
Sectional area not exceeding $0.05–0.10.\,m.^2$	8.00
Sectional area not exceeding $0.10\,m.^2$	7.00
Work at heights greater than 3m above ground or pavement level	
Add the extra labourer hours to the time shown in the last twelve items	2.00
Hoisting 3–6 m high	2.60
Hoisting 6–9 m high	3.20
Hoisting 12–15 m high	3.80
Hoisting 15–18 m high	4.40
Hoisting exceeding 18 m high	5.00

Table 13.14 Precast concrete unit work mixing and placing concrete in steel or timber moulds

Notes: 1. The data shown are for mixing and placing the concrete only, and do not allow for wheeling or
transporting, for which see Tables 13.15, 13.16 and 13.17
2. The hours shown allow for:
 • Mixing and placing in the mould.
 • Finishing off the unit after its removal from the mould.
 • Erecting, striking, cleaning and oiling the moulds.
3. For timber moulds for precast concrete unit work, see Table 13.33
4. For fixing reinforcement, see Tables 13.18, 13.19 and 13.20
5. If the precast units are reinforced, use a multiplier of 1.10 with the plant and labour hours shown.

| | Mixing and placing concrete, using concrete-mixing plant of the sizes shown. Per cubic metre | | | |
| | 7/5 | | 10/7 | |
Description *Concreting, mixing and placing. (The hours shown do not allow for wheeling or transporting. See notes above.)*	*Mixing plant hours*	*Labour hours*	*Mixing plant hours*	*Labour hours*
Concrete in moulds such as fence posts etc:				
Cross-sectional area up to 0.015 m^2	1.20	24.50	1.05	23.75
Cross-sectional area exceeding 0.015 and up 0.027 m^2	1.15	22.25	0.95	20.75
Concrete in moulds such as kerb, channel, precast edging, plain lintels, sills, etc.:				
Cross-sectional area up to 0.007 m^2	1.45	24.50	1.25	22.75
Cross-sectional area exceeding 0.007 and up to 0.015 m^2	1.15	18.75	1.00	18.00
exceeding 0.015 and up to 0.022 m^2	0.85	14.25	0.75	13.50
exceeding 0.022 and up to 0.030 m^2	0.68	10.65	0.57	9.12
exceeding 0.030 and up to 0.038 m^2	0.53	8.40	0.45	7.20
exceeding 0.038 and up to 0.045 m^2	0.45	7.35	0.40	6.40
Concrete in moulds such as paving slabs etc.:				
Slabs, 50 mm thick	1.60	34.65	1.35	32.00
Slabs, 75 mm thick	1.30	27.60	1.10	26.15
Concrete in moulds such as wall blocks or similar:				
Blocks, 75 mm thick	1.00	9.60	0.88	8.25
Blocks, 100 mm thick	0.79	7.35	0.67	6.40
Blocks, 150 mm thick	0.64	5.96	0.55	5.35

Table 13.15 Wheeling concrete in barrows

Types of barrows used	*Labour hours wheeling concrete up to a 22 m run. Per cubic metre*
Steel barrows, single-wheel type 0.08 m^3	1.07
Two-wheel steel hand-tipping concrete barrows 0.16 m^3	0.80

Table 13.16 Cubic metres of concrete hauled per load by vehicles

Note: In hauling concrete in dumpers the weight of load hauled should not exceed that which the vehicle is designed to carry. The data shown take this into account.

Transporting plant used	Volume of concrete hauled per load. Cubic metres
1 tonne dumper	0.45
2 tonne dumper	0.90
3 tonne dumper	1.35
4 tonne dumper	1.80
5 tonne dumper	2.25
6 tonne dumper	2.70

Dumpers, Decauville wagons, etc.	Such vehicles are rated on a volume basis. For purposes of estimating, the concrete they contain per load may be taken as being 85 per cent of their rated capacity. For example, a 1.5 m^3 dumper hauls 85 per cent of 1.5 m^3 = 1.2 m^3 of concrete per load.

Table 13.17 Travelling time taken per load of concrete from the mixing plant to the point of off-loading, off-load and return

Notes: 1. The hours shown are those taken per round trip. They do not allow for the time the vehicle stands while it is being loaded.
2. The hours shown for haulage by lorries are based on lorries fitted with mechanical gear and assume that the vehicles run on roads or reasonably firm surfaces.
3. The hours shown for haulage by dumper or by locomotive and skips or wagons are the hours taken per set of wagons, allowing for off-loading time and, in the case of locomotive and skips or wagons, for the normal shunting time involved.

Length of haul (one way) from mixing plant to point of off-loading in km	Lorries fitted with mechanical tipping gear	Dumpers	Caterpillar tractor-drawn wagons	Locomotive and wagons on 0.6 m gauge Decauville track	Locomotive and wagons on 1.4 m gauge track
0.2	0.11	0.13	0.22	0.22	0.25
0.6	0.14	0.19	0.28	0.28	0.32
0.75	0.18	0.26	0.38	0.40	0.46
1.5	0.27	0.40	0.56	0.62	0.68
5	0.54	–	–	–	–
8	0.79	–	–	–	–

Type of haulage plant used and plant hours per load

Table 13.18 Fix sheet reinforcement in various classes of work

Notes: 1. The hours shown are for fixing only and do not include for hoisting for which see Table 13.20
2. In computing the amount of reinforcement required allowance should be made for laps.

Description — Fixing in the work. (The hours do not allow for hoisting. See notes above.)	Weight of reinforcement in kg and labour hours fixing. Per square metre							
	0.45	0.90	1.35	1.80	2.25	2.70	3.15	3.60
Reinforcement to beams	0.13	0.17	0.20	0.24	0.28	0.31	0.35	0.38
Reinforcement to columns	0.18	0.22	0.25	0.29	0.33	0.36	0.40	0.43
Reinforcement to floors and slabs	0.02	0.03	0.05	0.06	0.07	0.08	0.10	0.11
Reinforcement to girder casings	0.22	0.26	0.31	0.35	0.41	0.46	0.50	0.55
Reinforcement to roads	0.01	0.02	0.03	0.05	0.06	0.07	0.08	0.10
Reinforcement to roofs	0.05	0.07	0.10	0.12	0.14	0.17	0.19	0.22
Reinforcement to walls	0.12	0.16	0.19	0.23	0.26	0.30	0.34	0.37
Circular cutting. Labour hours per metre	0.10	0.18	0.21	0.28	0.31	0.38	0.42	0.49
Straight cutting	0.07	0.11	0.14	0.18	0.21	0.24	0.28	0.31

Table 13.19 Fixing bar reinforcement

Notes: The hours shown are for cutting, bending and fixing in the work only and do not allow for hoisting, for which see table below.

Description — Cut, bend and fix in the work. (The hours shown do not allow for hoisting. See notes above.)	Size of bars and steel fixing and labourer hours cutting, bending and fixing, per tonne							
	up to 6 mm		7 to 12 mm		13 to 19 mm		over 19 mm	
	Steel fixer hours	Labourer hours	Steel fixer hours	Labourer hours	Steel fixer hours	Labourer hours	Steel fixer hours	Labourer hours
Straight round bars								
To beams, floors, roofs and walls	32.00	32.00	22.00	22.00	16.00	16.00	12.00	12.00
To braces, columns, sloping roofs and battered walls	64.00	64.00	46.00	46.00	32.00	32.00	20.00	20.00
Bent round bars								
To beams, floors, roofs and walls	42.00	42.00	32.00	32.00	26.00	26.00	22.00	22.00
To braces, columns, sloping roofs and battered walls	74.00	74.00	56.00	56.00	42.00	42.00	30.00	30.00
Straight indented or square bars								
To beams, floors, roofs and walls	36.00	36.00	24.00	24.00	18.00	18.00	14.00	14.00
To braces, columns, sloping roofs and battered walls	68.00	68.00	48.00	48.00	34.00	34.00	22.00	22.00
Bent indented or square bars								
To beams, floors, roofs and walls	48.00	48.00	36.00	36.00	30.00	30.00	26.00	26.00
To braces, columns, sloping roofs and battered walls	80.00	80.00	60.00	60.00	46.00	46.00	34.00	34.00

Table 13.20 Hoisting bar or sheet reinforcement

Notes: 1. The hours shown are the additional labour hours taken per tonne to hoist the reinforcement to the heights shown.
2. For sheet reinforcement use the data shown for the nearest bar diameters, generally, columns Nos. 1 and 2 being applicable.

Heights to which the reinforcement is hoisted in metres	Diameter of bars in millimetres and labourer hours hoisting the reinforcement, per tonne			
	up to 6 mm	7 to 12 mm	13 to 19 mm	over 19 mm
3	3.60	3.20	2.80	2.40
6	4.80	4.20	3.60	3.00
9	6.00	5.20	4.40	3.60
12	7.20	6.20	5.20	4.20
15	8.40	7.20	6.00	4.80

Table 13.21 Surface treatment of concrete

Note: The data shown are for work carried out to walls. For surface treatment to floors and ceilings use the table of multipliers shown.

Description	Material	Unit	Labour hours
Bush hammer surfaces	–	m²	2.40
Cement wash surfaces	0.29 kg cement	m²	0.12
Hack face for key	–	m²	0.60
Paint surfaces	0.23 kg paint	m²	0.48
Polish surfaces after rubbing down with stone and sand	–	m²	1.93
Remove excrescences from face	–	m²	0.30
Remove skin and expose aggregate while green	–	m²	0.60
Rub down surfaces with stone and sand	–	m²	0.96
Stop holes	–	m²	0.60
Studded roller finish	–	m²	0.07
Tool surface	–	m²	5.76
Trowel surface	–	m²	0.48

Table 13.22 Multipliers for Table 13.21

Description	Labour hour multipliers
Work to walls	1.00
Work to floors	0.75
Work to ceilings	2.00

Table 13.23 Sundry work to concrete

Description	Unit	Labour hours
Fix expansion jointing	m²	1.30
Fix steel side forms – straight and to slow sweep	m²	1.50
Fix steel side forms – to quick sweep	m²	2.00
Grout in rag bolts up to 25 mm diameter in brickwork, including punching hole	Each	0.70
Grout in rag bolts up to 25 mm diameter in concrete, including punching hole	Each	1.20
Grout in rag bolts only up to 25 mm diameter	Each	0.30
Mix and place concrete in small quantities by hand, such as to bases of fence-posts, etc.	m³	8.75
Round edges with bullnose trowel	m	0.14
Trowel surfaces	m²	0.48

Table 13.24 Fixing precast hollow floor and roof slabs

Note: The data shown are for fixing in the work at a height not in excess of 3 m. For fixing at heights in excess of this use the table of multipliers shown.

	Fix slabs per square metre	
Description	Slab fixer hours	Labourer hours
Hollow slabs, 100 mm in depth	0.25	0.25
Hollow slabs, 125 mm in depth	0.31	0.31
Hollow slabs, 150 mm in depth	0.36	0.36
Hollow slabs, 175 mm in depth	0.42	0.42

Table 13.25 Multipliers for where the height is in excess of 3 m

Height of the work in metres	Multipliers
3.0	1.0
6.0	1.1
9.0	1.2
12.0	1.3
15.0	1.4

Formwork

Formwork to floors, roofs, walls, etc. In estimating the cost of formwork the following factors have to be taken into consideration:

1. The nature of the work on which the formwork has to be used.
2. Whether the formwork can be used once only or whether it may be used several times in repetition work.
3. That if repetition formwork is used the subsequent erection costs are less than the first.
4. Materials cost – timber, steel and special plastics.

Formwork used once only. In estimating the cost of the material in formwork which can be used once only, the usual practice is to charge the whole of the material, making a reasonable allowance for the value of any timber recovered on the completion of the work.

In order to illustrate the method of computing the material cost of formwork an example is shown, and the estimator must himself assess what he thinks is a fair recovery value of the timber. In the case of intricate shuttering with much fitting to angles, etc. this recovery value may be very small; on the other hand, if the work is simple flat work it may be as large as 75 per cent.

Example.   Calculate the cost of the material required for formwork per square metre using plywood boarding, the work being such that it can only be used once, and an assessment having been made that the recovery value of the timber in the formwork on the completion of the work is 75 per cent of its original value.

It will be assumed that the original cost of the materials and the quantities required per square metre of formwork are:

Plywood, at £9.00 per m^2	= £ 9.00 per m^2
Adjustable props and braces, 0.015 m^3 at £200.00 per m^3	= £ 3.00 per m^2
	= £12.00 per m^2
Allow for wastage 6 per cent	= £ 0.72 per m^2
	= £12.72 per m^2
Deduct recovery value assessed at 75 per cent	= £ 9.54 per m^2
	= £ 3.18 per m^2
Allow for nails: first and only use, 0.5 kg at £4.00	= £ 2.00
Total cost of material in formwork on which to base the estimate	= £ 5.18 per m^2

Repetition formwork. In simple straightforward work such as in walls, roofs, etc. formwork can generally be so constructed that it may be used over and over again. There is a limit to the number of times it can be used, however, as after continued use due to nailing, knocking it about, winding, etc. it would be to the detriment of the work to continue using it. Mould oil should be applied for each use.

For purposes of estimating, where repetition formwork is used, the maximum number of uses should be taken as being six. If used less than this a recovery value may be assessed, this value depending on the nature of the work. The estimator, knowing the type of work which has to be constructed, must assess what timber will be required and what recovery value, if any, it has as a percentage of the original cost, bearing in mind the number of times it will be used.

In order to illustrate the method of computing the cost of the material used in formwork on repetition work the following example is shown.

Example.   Calculate the cost of the material required for formwork per m², the work being such that it is used six times.

Note: The recovery value of timber for estimating purposes in this case will be nil, since six uses will be made of the formwork. Were the number of uses less than six, an assessment would be made of the recovery value of the timber and allowance made for this as shown in the example above.

It will be assumed that the original cost of the materials and the quantities required per square metre of formwork are:

Plywood at £9.00 per m²	= £ 9.00 per m²
Adjustable props and braces, 0.015 m³ at £200.00 per m³	= £ 3.00 per m²
	= £12.00 per m²
Allow for wastage 6 per cent	= £ 0.72 per m²
	= £12.72 per m²
Allow for nails: first use, 0.5 kg at £4.00	= £ 2.00 per m²
Allow for nails for 5 subsequent uses,	
5 at 0.125 kg = 0.625 kg at £4.00	= £ 2.50 per m²
Allow for mould oil: 6 uses @ £1.00	= £ 6.00
Total cost of material in formwork	= £23.22 per m²
Allowing 6 uses, cost per use = $\dfrac{£23.22}{6}$	= £ 3.87 per m²

The erection cost of repetition formwork. In erecting formwork in repetition work, provided there is no extensive alteration required after the first use, the subsequent erection costs are less. In order to tabulate this for estimating purposes the tables show multipliers for use with the main tables to allow for all numbers of uses up to and including six.

The Tables for Use in Estimating Formwork. In the main tables shown for estimating the cost of formwork the timber is assumed to be plywood. The carpenter and labourer hours shown allow for all off-loading, handling and erecting. The number of props per m² is stated for each table, these being a fair average for normal work, and the props are taken as being 100 × 75 mm up to 1.8 m, 100 × 100 mm from 1.8 to 3.6 m, and 150 × 150 mm from 3.6 m to 6.0 m long.

Table 13.26 Formwork to beams, lintels and walls

Notes: 1. The timber in formwork is taken as being plywood, with 100 × 75 mm props up to 1.8 m, 100 ×
100 mm 1.8 to 3.6 m and 150 × 150 mm 3.6 to 6.0 m long, each prop supporting 1m².
2. In calculating the timber required for formwork allow 6 per cent for waste.
3. Nails: allow 0.4 kg per m² for first use and 0.1 kg for each subsequent use.
4. If walls require formwork on both faces, both must be measured.

Erect and strike formwork	Formwork per m². First use	
	Carpenter hours	Labourer hours
Beams up to 0.6 m deep at heights up to 3.0 m	1.20	1.20
Beams up to 0.6 m deep at heights 3.0 to 3.6 m	1.25	1.25
Beams up to 0.6 m deep at heights 3.6 to 4.2 m	1.30	1.30
Beams up to 0.6 m deep at heights 4.2 to 4.8 m	1.35	1.35
Beams up to 0.6 m deep at heights 4.8 to 5.4 m	1.40	1.40
Beams up to 0.6 m deep at heights 5.4 to 6.0 m	1.50	1.50
Walls – vertical face up to 1.5 m	0.60	1.20
Walls – vertical face 1.5 to 3.0 m	0.72	1.44
Walls – vertical face 3.0 to 4.5 m	0.84	1.68
Walls – vertical face 4.5 to 6.0 m	0.96	1.92

Table 13.27 Multipliers for formwork to beams and walls

	Multipliers
Formwork used once	1.00
Formwork used twice, per use	0.85
Formwork used three times, per use	0.80
Formwork used four times, per use	0.78
Formwork used five times, per use	0.76
Formwork used six times, per use	0.75
Formwork used seven times or more, per use	0.70
Walls built to batter	1.30
Walls built circular to large sweep	1.50
Walls built circular to quick sweep	2.00
Beams over 0.6 m and up to 0.9 m deep	1.25
Beams over 0.9 m and up to 1.2 m deep	1.50

Table 13.28 Formwork to braces, columns, piers, pilasters and stanchions

Notes: 1. The timber in formwork is taken as plywood.
2. In calculating the timber required for formwork allow 6 per cent for waste.
3. Nails: allow 0.2 kg per m² for first use and 0.025 kg for each subsequent use.

	Formwork per square metre. First use	
Erect and strike formwork	*Carpenter hours*	*Labourer hours*
Square or rectangular braces, columns, piers, pilasters and stanchions at heights up to 3.0 m	1.08	1.08
Square or rectangular braces, columns, piers, pilasters and stanchions at heights 3.0 to 3.6 m	1.14	1.14
Square or rectangular braces, columns, piers, pilasters and stanchions at heights 3.6 to 4.2 m	1.26	1.26
Square or rectangular braces, columns, piers, pilasters and stanchions at heights 4.2 to 4.8 m	1.44	1.44
Square or rectangular braces, columns, piers, pilasters and stanchions at heights 4.8 to 5.4 m	1.68	1.68
Square or rectangular braces, columns, piers, pilasters and stanchions at heights 5.4 to 6.0 m	2.00	2.00
Column bases	2.40	2.40
Column heads	3.60	3.60

Table 13.29 Multipliers for formwork to braces, columns, piers, pilasters and stanchions

Description	*Multipliers*
Formwork used once	1.00
Formwork used twice, per use	0.94
Formwork used three times, per use	0.87
Formwork used four times, per use	0.85
Formwork used five times, per use	0.84
Formwork used six times, per use	0.82
Formwork used seven times or more	0.78
Formwork to circular shape	3.00
Formwork to shaped sides, e.g. six-sided	1.60
Formwork to slopes	1.40

Table 13.30 Formwork to floors and roofs

Notes: 1. The timber in formwork is taken as plywood.
2. Props are taken as being 100 × 75 mm up to 1.8 mm, 120 × 120mm 1.8 to 3.6 m and 150 × 150mm from 3.6 to 6.0 m long, each prop propping 2 m².
3. In calculating the timber required for formwork allow 6 per cent for waste.
4. Nails: allow 0.5 kg per m² for first use and 0.12 kg for each subsequent use.

	Formwork per square metre. First use	
Erect and strike formwork	*Carpenter hours*	*Labourer hours*
Horizontal flat formwork at heights up to 3.0m	0.90	0.90
Horizontal flat formwork at heights 3.0 to 3.6m	0.95	0.95
Horizontal flat formwork at heights 3.6 to 4.2m	1.00	1.00
Horizontal flat formwork at heights 4.2 to 4.8m	1.05	1.05
Horizontal flat formwork at heights 4.8 to 5.4m	1.10	1.10
Horizontal flat formwork at heights 5.4 to 6.0m	1.20	1.20

Table 13.31 Multipliers for formwork to floors and roofs

Description	*Multipliers*
Formwork used once	1.00
Formwork used twice, per use	0.80
Formwork used three times, per use	0.73
Formwork used four times, per use	0.70
Formwork used five times, per use	0.68
Formwork used six times, per use	0.66
Formwork used seven times or more	0.60
Formwork to slope not over 45° from horizontal	1.25

Table 13.32 Steel panel and prefabricated formwork

Note: Labours shown allow for all strutting and propping and the formwork is taken as being prepared ready for use. The labour hours shown are for erecting and striking.

	Formwork per square metre	
Erect and strike formwork	*Carpenter or erector hours*	*Labourer hours*
Beams and lintels at heights up to 3.0 m	0.66	0.66
Beams and lintels at heights 3.0 to 3.6 m	0.70	0.70
Beams and lintels at heights 3.6 to 4.2 m	0.74	0.74
Beams and lintels at heights 4.2 to 4.8 m	0.82	0.82
Beams and lintels at heights 4.8 to 5.4 m	0.88	0.88
Beams and lintels at heights 5.4 to 6.0 m	0.96	0.96
Columns and stanchions at heights up to 3.0 m	0.86	0.86
Columns and stanchions at heights 3.0 to 3.6 m	0.91	0.91
Columns and stanchions at heights 3.6 to 4.2 m	1.01	1.01
Columns and stanchions at heights 4.2 to 4.8 m	1.15	1.15
Columns and stanchions at heights 4.8 to 5.4 m	1.35	1.35
Columns and stanchions at heights 5.4 to 6.0 m	1.58	1.58
Floors and roofs at heights up to 3.0 m	0.58	0.58
Floors and roofs at heights 3.0 to 3.6 m	0.60	0.60
Floors and roofs at heights 3.6 to 4.2 m	0.64	0.64
Floors and roofs at heights 4.2 to 4.8 m	0.68	0.68
Floors and roofs at heights 4.8 to 5.4 m	0.74	0.74
Floors and roofs at heights 5.4 to 6.0 m	0.82	0.82
Walls – vertical face up to 1.5 m	0.48	0.96
Walls – vertical face 1.5 to 3.0 m	0.58	1.16
Walls – vertical face 3.0 to 4.5 m	0.67	1.34
Walls – vertical face 4.5 to 6.0 m	0.77	1.54

The mould cost per cubic metre of precast concrete units

In calculating the mould cost per m^3 of precast unit, the procedure is as follows:

1. Calculate the cost of the mould.
2. Assess the number of precast units which are to be produced from the mould and, knowing the volume of each unit, the m^3 of concrete produced.
3. Divide the cost of the mould by the m^3 of precast concrete units produced from it.

Estimating data are shown, giving the joinery hours taken to make timber moulds for some of the most commonly used precast concrete units such as fence posts, paving slabs, etc. This may be used as a guide in computing the joiner hours taken in constructing moulds for work of a similar type.

In the case of plastics or all-metal moulds such as steel or aluminium, the estimator may obtain prices for these from firms who specialise in producing such a commodity.

Example.   For purposes of this example the volume of the concrete unit concerned is taken as being 0.03 m³ and 50 precast units are cast from it.

Cost of mould.

125mm × 50mm timber, 0.016 m³ at £200.00 per m³	= £ 3.200
Plywood at £9.00	= £ 9.000
50mm × 25mm timber, 0.005 m³ at £200.00 per m³	= £ 1.000
	£13.200
Add 10 per cent for waste	= £ 1.320
	£14.520
Screws	= £ 2.000
Bolts	= £ 5.000
	£21.520
Joiner, say 6.00 h at £7.00 per h	= £42.000
Total cost of one mould	= £63.520

Total volume of concrete cast per mould = 50×0.03 m³ = 1.50 m³

Cost of mould of precast unit $\dfrac{£63.520}{1.5}$ = £42.347 per m³

Table 13.33 Timber moulds for precast concrete units

Note: For moulds lined with sheet aluminium or zinc use a labour multiplier of 1.25 with the joiner hours shown.

Type of precast concrete unit	Description of mould	Joiner hours per mould
Kerb		
250 mm × 100 mm × 1 m long	Single timber moulds with wooden bottom and sides	8.80
250 mm × 150 mm × 1 m long	Single timber moulds with wooden bottom and sides	9.30
300 mm × 150 mm × 1 m long	Single timber moulds with wooden bottom and sides	9.90
Lintels		
112 mm × 75 mm × 1 m long	Timber mould in 'nest' of 4 lintels	13.20
112 mm × 225 mm × 1.60 m long	Timber mould in 'nest' of 4 lintels	15.40
150 mm × 225 mm × 2 m long	Timber mould in 'nest' of 4 lintels	17.60
Paving slabs		
0.60 m × 0.60 m × 50 mm	Single timber moulds with wooden bottom and sides	7.15
0.75 m × 0.75 m × 50 mm	Single timber moulds with wooden bottom and sides	7.80
1 m × 0.75 m × 50 mm	Single timber moulds with wooden bottom and sides	8.40
Posts		
100 mm × 100 mm × 1.30 m long	Single timber moulds with wooden bottom and sides	6.60
100 mm × 125 mm × 1.60 m	Single timber moulds with wooden bottom and sides	7.10
100 mm × 125 mm × 2.00 m	Single timber moulds with wooden bottom and sides	8.20
Slabs		
1.30 m × 1 m × 50 mm	Single timber moulds with no bottoms. Unit cast on smooth concrete surface	2.40
2 m × 1 m × 50 mm	Single timber moulds with no bottoms. Unit cast on smooth concrete surface	3.60

14 Masonry

Brick and block walling

The data shown for estimating the cost of brickwork deals with the standard size of bricks, viz.

$$215 \times 102.5 \times 65 \, mm$$

The data is based on the standard thickness of joint i.e. 10 mm this being the difference between two widths of 'headers' and the length of a 'stretcher'.

The number of bricks laid per bricklayer per hour

Considerable controversy exists as to how many bricks a bricklayer can lay per hour, and so far as estimating is concerned it is the 'fair average' number laid per hour which must be considered. In this connection it should be noted that the bricklayer is not constantly laying bricks.

A considerable portion of his time is taken up in carrying out other work as the bricklaying proceeds such as bedding door frames, sills, windows, etc. Time is also spent in plumbing to quoins and angles. Thus, while actually laying bricks the bricklayer may do so at a rate in the region of 100 per hour, the average rate of laying the actual bricks might only be in the region of 60 when the time spent on the other labours is taken into account.

The class of work built also has a distinct bearing on the number of bricks laid per hour, for considerably more bricks can be laid per hour in a low wall where no scaffolding is required than in a high one built from scaffolding where extensive plumbing of angles is entailed. The human element also enters into it, for some bricklayers lay more bricks in similar work in the same period of time than others.

From the foregoing it will be seen that for purposes of estimating, the estimator must assess the output per bricklayer per hour on the class of work being tendered for, and in doing so must take into account the extent of scaffolding required and the amount of plumbing to angles, bedding of door frames, windows, etc. involved. Such outputs are obtained from actual experience of similar work carried out and for estimating purposes the data in Table 14.1 may be used:

Table 14.1 Output per bricklayer hour

Type of brickwork	Output per bricklayer per hour/bricks
Ordinary brickwork. Half brick wall.	50
Ordinary brickwork. One brick wall.	60
Ordinary brickwork. One and a half brick wall.	70
Brick facings.	40

In the tables shown for estimating purposes the data is based on the outputs given in Table 14.1. If the estimator considers that more or less bricks than shown may be laid per bricklayer per hour, having carefully taken into consideration the nature of the work which has to be carried out and possibly having knowledge of the capabilities of the bricklayers who will carry it out, the bricklayer and labourer hours shown in the table may be adjusted as follows.

Assuming the table for one brick thick brickwork shows an output of 60 bricks per bricklayer per hour, and the estimator considers that 80 bricks per hour would be laid, the bricklayer and labourer hours shown in the table may be multiplied by 60/80 i.e. 0.75 this 0.75 being the labour multipliers for use with the hours shown in the table for an assessed output of 80 bricks per hour in lieu of 60.

Allowance for waste in bricks

In laying bricks of a reasonably hard and strong nature there is little waste, however, in the case of softer bricks the waste is greater, therefore an allowance of $2\frac{1}{2}$ to 10 per cent should be made to allow for loss through breakage.

Mortars for brickwork

Brickwork is generally carried out with cement sand mortar; cement lime sand mortar and cement sand with plasticiser.

Lime is hydrated lime powder to BS 890.

Portland cement mortar consists of cement to BS 12, sand to BS 1200 Table 1, water, and for brickwork is usually used in the proportion of 3 of sand to 1 of cement.

The labour hours shown in the brickwork tables do not include for 'knocking up' or mixing the mortar. In computing the cost of a cubic metre of mortar, therefore, its cost must be wholly inclusive of the material involved and also the labour or plant and labour cost of 'knocking it up' or mixing.

Scaffolding

In constructing brickwork to height, two factors have to be taken into consideration:

1. The cost of bricklaying increases with the height as more cost is involved in getting the bricks and mortar to the bricklayer.
2. Scaffolding has to be erected in order that the bricklayer can reach the work.

Regarding (1), the data shown in the main estimating tables for brickwork may be used as they stand for work carried out to a height of 12 m above ground level. In carrying out brickwork to heights in excess of this, multipliers are shown for use with the data shown in the main estimating tables. These multipliers allow for additional cost of carrying out the brickwork itself at height, but do not allow for the cost of the erection and dismantling of scaffolding.

Regarding (2), scaffolding is often allowed for by making a percentage addition to the labour and material cost of the brickwork itself, this generally being in the neighbourhood of $1\frac{1}{2}\%$. This rule of thumb method of allowing for the cost of scaffolding is to be deprecated as the ratio of the labour cost to the material cost is variable as is the thickness of the work. It is therefore best allowed for by assessing the cost of erecting and dismantling it per square metre when it may then be allowed for in one or other of the following ways:

1. Its cost being allowed for per square metre of brickwork in the cost of brickwork itself.
2. Knowing the cost of the scaffolding per square yard and the total square metres of brickwork to be built by allowing for it on a 'lump sum' basis in the preliminaries.

Data from which to estimate the cost of erecting and dismantling scaffolding are shown (see Table 6.3).

In connection with the cost of the scaffolding it should be noted that this is used for trades other than bricklayer, such as slater, tiler, roof carpenter, painter, external renderer, etc. Allowing for it under Preliminaries is the most simple way of dealing with it as apportioning its cost to the various trades is almost impossible and would, in any case, only tend to make the estimation of the cost of the work as a whole more complicated.

Note: Any loading and haulage cost involved in connection with the scaffolding should not be overlooked and should also be allowed for.

Fletton or common brickwork

Fletton or common brickwork is carried out in half brick walls, skins of cavity walls, one brick walls, one and half brick walls and thickness in excess of this.

All brickwork is quoted in square metres, the thickness of the work being stated.

Note: The hours shown in the tables from which to estimate the cost of stock brickwork do not include the following:

- The cost of mixing the mortar.
- The cost of erecting and dismantling scaffolding.
- The cost of pointing the brickwork.

The cost of these must be allowed for by referring to the data for mortars Tables 12.2 to 14.6, pointing Table 14.14, and scaffolding Table 6.3.

The following examples show the method of estimating the cost of Fletton or common brickwork. For purposes of illustration, examples are shown for brickwork carried out with single frogged Fletton bricks, $215 \times 102.5 \times 65$ mm, built in three-to-one cement mortar.

No sum is allowed for the cost of water used in connection with the mortar, it being assumed that all the water required for the contract as a whole has been computed and allowed for on a lump sum or other basis.

The cost of scaffolding is allowed for in Preliminaries.

The cost of carrying out the pointing to the brickwork is not included. Examples showing the method of estimating the cost of this are on pp. 15.8

The examples shown for Fletton or common brickwork are based on the following assumed rates:

Bricklayer's rate	= £ 7.00 per h
Labourer's rate	= £ 6.00 per h
Fletton bricks delivered on the site	= £150.00 per 1000
Cement delivered on the site	= £100.00 per tonne
Sand delivered on the site	= £ 16.00 per m^3
Hollow wall ties delivered on the site	= £ 0.10 each

The data used in the various examples are found by referring to Table 14.7 and tables shown for mortar, Fletton or common brickwork carried out with 3-to-1 cement mortar, and $215 \times 102.5 \times 65$mm bricks.

The Cost of 3-to-1 Cement Mortar per cubic metre.

Cement	0.52 tonnes at £100.00 per tonne	= £ 52.00
Sand	1.10m^3 at £16.00 per m^3	= £ 17.60
Labour mixing	9.20 h at £6.00 per h	= £ 55.20
Total cost per m^3		= £124.80

Note: The cost of the water required is taken as being allowed for on the contract as a whole, on a lump sum or other basis.

Example 1.   Calculate the cost per m^2 of Fletton brickwork in half-brick walls in 3-to-1 cement mortar where no scaffolding is required.

65 bricks at £150.00 per 1000		= £ 9.75
Mortar 0.022 m^3 at £115 60 per m^3		= £ 2.54
Bricklayer 1.15 h at £7.00 per h		= £ 8.05
Labourer 1.15 h at £6.00 per h		= £ 6.90
Total labour and material cost per m^2 exclusive of pointing		= £27.24

Example 2.   Calculate the cost per square metre of one brick wall in Fletton brickwork in 3-to-1 cement mortar.

128 bricks at £150.00 per 1000		= £19.20
Mortar 0.055 m^3 at £115.60 per m^3		= £ 6.36
Bricklayer 1.90 h at £7.00 per h		= £13.30
Labourer 1.90 h at £6.00 per h		= £11.40
Total labour and material cost per m^2 exclusive of pointing		= £50.26

Brickwork facings

In laying facing bricks the bricklayer lays less facings per hour than ordinary bricks, and, in the case of good-class facing work, additional labour cost is involved as the facing bricks are selected or hand picked. The facing work may be carried out in various bonds, each of which involves a different number of facings per square metre.

The pointing of brick facings is usually included in the item billed for the facings, but for purposes of estimating, pointing is shown separately; pointing in various forms can then be dealt with in building up the cost of facing work.

The estimator should note that brickwork facings are billed as Extra Over ordinary brickwork generally or entirely in facings in some such form as the following:

1. Extra Over ordinary brickwork in gauged mortar for facing bricks as specified in English bond.
2. Extra Over ordinary brickwork in cement mortar in stretching bond for facing with facing bricks as specified.
3. Facing bricks as specified to outer walls of hollow brick walls.

This Extra Over or additional cost is made up of:

1. The additional cost of the facing bricks over and above the cost of ordinary bricks
2. Any additional labour cost involved in selecting or hand picking the facing bricks.

3. The additional bricklayer and labourer hours taken in laying the facing bricks in the work as compared with ordinary brickwork.

The method of calculating this Extra Over cost may be worked out on a basis of one or other of the following:

- Per m^2 of facing work.
- Per 1000 facing bricks.

In the table shown for brick facings, the bricklayer and labourer hours refer to the Extra Over or additional hours taken over ordinary brickwork to lay 1 m^2 of facing bricks in the work, the number of facing bricks required per square metre for the various bonds being shown.

The hours shown are exclusive of pointing.

To illustrate the method of estimating the Extra Over cost of facing work, examples are shown, and in the case of each it is assumed that the work is carried out in English bond. The cost of pointing is not included, examples of how to estimate the cost of this being shown below.

The examples shown are based on the following rates:

Bricklayer's rate	= £ 7.00 per h
Labourer's rate	= £ 6.00 per h
Facing bricks delivered on the site	= £200.00 per 1000
Ordinary bricks delivered on the site	= £150.00 per 1000

Example 1.   Calculate the Extra Over cost per m^2 for hand-picked facings over ordinary brickwork, exclusive of pointing, the bricks used being 215 × 102.5 × 65mm and the work carried out in English bond.

Facing bricks delivered on the site	= £200.00 per 1000
Ordinary bricks delivered on the site	= £150.00 per 1000
Extra Over cost of facing bricks	= £ 50.00 per 1000

Referring to Table 14.10 for brick facings the number of facing bricks required per m^2 in English bond is 92, and the Extra Over bricklayer and labourer hours per m^2 for facing work in English bond are 0.72 and 1.15 hours respectively.

The total Extra Over cost of the facing work over ordinary brickwork per square metre is:

$$\text{Extra Over cost of facing bricks } \frac{92}{1000} \times £50.00 = £4.60 \text{ per m}^2$$

Bricklayer 0.72 h at £7.00 per h	= £ 5.04 per m^2
Labourer 1.15 h at £6.00 per h	= £ 6.90 per m^2
Extra Over cost of facing work	= £16.54 per m^2

Note: The above is exclusive of pointing.

Example 2.   Calculate the Extra Over cost per 1000 facing bricks over ordinary bricks for hand picking and laying the facings in the work, exclusive of pointing, the facing bricks used being 215 × 102.5 × 65mm, laid in English bond.

For the purpose of this example the same basic rates as used in Example 1 are taken.

This method of dealing with the Extra Over cost of facings over ordinary brickwork is similar to the method illustrated in Example 1 only it is based on 1000 facing bricks and not 1 m^2 or

92 bricks. Referring to Example 1, it will be seen that the additional hours taken by the bricklayer and labourer in laying 92 facings is:

Bricklayer   0.72 h per m^2 or per 92 facing bricks
Labourer   1.15 h per m^2 or per 92 facing bricks

The additional bricklayer and labourer hours per 1000 facing bricks are:

Bricklayer hours $\dfrac{1000}{92} \times 0.72 = 7.83$ h per 1000 facing bricks

Labourer hours $\dfrac{1000}{92} \times 1.15 = 12.50$ h per 1000 facing bricks

The Extra Over cost of facings over ordinary bricks per 1000 facing bricks, therefore, is:

Cost of 1000 facing bricks at £200.00 per 1000	= £200.00
Cost of 1000 ordinary bricks at £150.00 per 1000	= £150.00
Additional cost of 1000 facing bricks	= £ 50.00
Additional bricklayer hours 7.83 at £7.00 per h	= £ 54.81
Additional labourer hours 12.50 at £6.00 per h	= £ 75.00
Extra Over cost per 1000 facing bricks	= £179.81

Single face work in English bond requires 92 facing bricks per m^2.

The Extra Over cost of facing per m^2 is: $\dfrac{92}{1000} \times £179.81$

$$= 0.092 \times £179.81$$

$$= £16.54 \text{ per m}^2$$

Note: The above is exclusive of pointing and the result is the same as in Example 1.

Example 3.   Calculate the cost of facing with best picked Fletton or common bricks per m^2 in English bond, exclusive of pointing, using 215 × 102.5 × 65mm Fletton or common bricks.

Note: In this case the whole of the work is carried out with Fletton or common bricks, and the Extra Over is only the additional bricklayer and labourer hours involved in selecting and laying the bricks. Referring to Example 1, it will be seen that the extra bricklayer and labourer hours taken to hand pick and lay facings in English Bond single-faced work, exclusive of pointing, per m^2 are:

Bricklayer	= 0.72 per m^2
Labourer	= 1.15 per m^2
The Extra Over cost per m^2 is:	
Bricklayer 0.72 h at £7.00 per h	= £5.04 per m^2
Labourer 1.15 h at £6.00 per h	= £6.90 per m^2
	£11.94 per m^2

Note: The above is exclusive of pointing.

Pointing brickwork

Pointing brickwork may be carried out as the work proceeds or as a separate operation. In the former case the joints are 'flushed' or 'struck' by the bricklayer as the work of bricklaying goes on, while in the latter case the joints are raked out before the mortar has set and the actual pointing carried out as a separate operation after the mortar has set. This latter method is more costly to execute. The actual pointing may be carried out with lime or coloured cement mortars being selected to give different ranges of colour.

A form of pointing which is now commonly used in pointing brick facings is the 'ironed' or 'rubbed' joint. This method of pointing is carried out as follows:

In laying the bricks the joints are well mortared and struck or flushed with the trowel as the work proceeds. Prior to the mortar in the joint setting it is rubbed or ironed in with a piece of round bar or tubing 9mm in diameter, slightly bent at the end. By continuous rubbing a very pleasing joint is obtained and it is possible to polish the joints by this means.

In the case of new brickwork carried out at height, the pointing is carried out from scaffolding erected for the purpose of building the brickwork itself. Since the cost of this scaffolding is allowed for in the cost of the brickwork or on a lump sum basis, no allowance need be made for it in the cost of the pointing. If, however, pointing has to be carried out to old brickwork, and scaffolding has to be erected and dismantled for the specific purpose of carrying this out, its cost must be allowed for and included in the estimate for the pointing. The estimator is referred to where the data are shown from which the cost of erecting and dismantling scaffolding can be found. Any hire charge and haulage cost in connection with the scaffolding should also be taken into consideration.

To illustrate the method of estimating the cost of pointing brickwork, examples are shown for pointing carried out in English bond with $215 \times 102.5 \times 65$ mm bricks.

Examples showing the method of estimating the cost of pointing brickwork. For purposes of the examples shown it is assumed that the pointing is carried out to brickwork in English bond in 3-to-1 cement mortar, the following being the assumed labour and material rates:

Bricklayer's rate   £   7.00 per h
Labourer's rate   £   6.00 per h
Cement   £100.00 per tonne, delivered on the site
Sand   £ 16.00 per m^3, delivered on the site

The cost of 3-to-1 lime mortar per cubic metre.

Cement	0.52 tonnes at £100.00 per tonne	= £ 52.00
Sand	1.10 m^3 at £16.00 per m^3	= £ 17.60
Labour mixing	6.75 h at £6.00 per h	= £ 40.50
Total cost per m^3		= £110.10

Note: The cost of the water required is taken as being allowed for on the contract as a whole, on a lump sum or other basis.

Example 1.   Calculate the cost per square metre of raking out and pointing existing brickwork in English bond with 3-to-1 cement mortar with a neat weathered joint, the bricks used being $215 \times 102.5 \times 65$ mm.

From Table 14.14 the cost of pointing is given by:

Mortar	0.0018 m^3 at £110.10 per m^3	= £ 0.20
Bricklayer	1.25×0.96 h at £7.00 per h	= £ 8.40
Labourer	1.25×0.96 h at £6.00 per h	= £ 7.20
Labour and material cost per m^2		= £15.80

Example No. 2. Calculate the cost per square metre of raking out and pointing facings in English bond as the work proceeds, the bricks used being 215 × 102.5 × 65 mm, built in 3-to-1 cement mortar.

Note: Referring to Table 14.15 it will be seen that for this type of pointing a multiplier of 0.60 is used with the bricklayer and labourer hours shown in the pointing table.

From Table 14.14 the cost of pointing is given by:

Mortar	0.0018 m^3 at £110.10 per m^3		= £0.20
Bricklayer	1.25 × 0.60 × 0.96 = 0.72 h at £7.00 per h		= £5.04
Labourer	1.25 × 0.60 × 0.96 = 0.72 h at £6.00 per h		= £4.32
Labour and material cost per m^2			= £9.56

Table 14.2 Mortar for brickwork walling mortars

Notes: 1. The data shown are based on the following:

Cement weighing   1442 kg per m^3
Lime weighing   720 kg per m^3

2. The data shown relating to sand refer to *dry sand*. If washed sand containing moisture is used a multiplier of 1.25 should be used in conjunction with the quantities of sand shown in order to allow for bulkage due to the moisture content.
3. Allow the following plant and labour hours for mixing per cubic metre of mortar:

Nature of mortar	Plant hours per m^3	Labour hours per m^3
Sand, lime	1.30	2.6
Sand, cement, lime	1.40	2.8
Sand, cement	1.50	3.0
Ash, lime	1.75	3.5

Table 14.3 Lime mortar

Note: Allow for mixing by hand 6.75 labour hours per cubic metre.

Mix		Material required per cubic metre of mortar	
Sand	Lime	Sand m^3	Lime tonnes
1	1	0.73	0.53
1½	1	0.93	0.44
2	1	1.00	0.36
2½	1	1.07	0.31
3	1	1.10	0.27
3½	1	1.15	0.25
4	1	1.20	0.20

Table 14.4 Ash mortar

Note: Allow for mixing by hand 7.50 h per metre.

Mix		Material required per cubic metre of mortar	
Ashes	Lime	Ashes m^3	Lime tonnes
1	1	1.16	0.83
1½	1	1.38	0.57
2	1	1.50	0.45
2½	1	1.58	0.39
3	1	1.66	0.35
3½	1	1.91	0.33
4	1	2.08	0.32

Table 14.5 Cement lime mortar

Note: Allow for mixing by hand 8.0 labour hours per cubic metre.

Mix			Material required per cubic metre of mortar		
Sand	Cement	Lime	Sand m^3	Cement tonnes	Lime tonnes
2	1	1	0.73	0.53	0.27
3	1	1	0.93	0.45	0.23
4	1	1	1.00	0.37	0.19
5	1	1	1.07	0.32	0.16
6	1	1	1.10	0.27	0.13
7	1	1	1.15	0.24	0.12
8	1	1	1.20	0.21	0.11

Table 14.6 Cement mortar

Note: Allow for mixing by hand 9.2 labour hours per cubic metre.

Mix		Material required per cubic metre of mortar	
Sand	Cement	Sand m^3	Cement tonnes
1	1	0.73	1.05
1½	1	0.93	0.89
2	1	1.00	0.72
2½	1	1.07	0.61
3	1	1.10	0.52
3½	1	1.15	0.44
4	1	1.20	0.38

Table 14.7 Fletton or common brickwork

Notes: 1. The data shown are for brickwork carried out with 10 mm joints in cement mortar to heights not in excess of 12 m. For work to heights in excess of this see table of multipliers, Table 14.8.
2. The number of bricks laid per bricklayer per hour on which the data are based is shown in each case. If this number be assessed at more or less than that shown, the bricklayer and labourer hours may be adjusted thus: If the output tabulated is 60 bricks per hour and the assessed output is 80 bricks per hour, use a labour multiplier with the bricklayer and labour hours shown of 60/80 = 0.75.
3. The data shown do not allow for knocking up the mortar. See mortar tables, Tables 14.2 to 14.6.
4. The data shown do not allow for pointing for which see Table 14.14.
5. The data shown are for brickwork carried out in cement mortar. If the brickwork is carried out in gauged mortar use a multiplier of 0.80 with the brickwork and labourer hours shown.
6. The labour hours shown do not allow for scaffolding, for which see Table 6.3.
7. For hollow or cavity walls allow for fixing wall ties: bricklayer, 0.02 h per tie, labourer, 0.02 h per tie.
8. The data shown includes 10 per cent waste for the number of bricks required in a square metre.

Bricks, 215 × 102.5 × 65 mm

| Description | | Number of bricks laid per | Number | Mortar required cubic metres | | | Brick- | |
Rough brickwork not pointed	Unit	bricklayer per hour	of bricks required	No frogs	Single frogs	Double frogs	layer hours	Labourer hours
Half-brick walls, 103 mm work	m²	50	65	0.022	0.027	0.032	1.44	1.44
Two half-brick hollow walls, 270 mm work	m²	50	128	0.044	0.054	0.064	2.88	2.88
Solid brick walls, 215 mm work	m²	60	128	0.052	0.064	0.076	2.40	2.40
Brick and a half walls, 328 mm work	m²	70	191	0.073	0.091	0.108	3.10	3.10

Table 14.8 Fletton or common brickwork

Table of multipliers for use with the bricklayer and labourer hours shown in the Fletton or common brickwork tables for carrying out work at height and various types of work.

	Bricklayer and labourer hour multipliers
Brickwork built at height ranging from ground level to 12 m	1.00
Brickwork exceeding 12 m but not exceeding 18 m	1.20
Brickwork exceeding 18 m but not exceeding 24 m	1.30
Brickwork exceeding 24 m but not exceeding 30 m	1.40
Brickwork built in arches	2.50
Brickwork built to batter	1.50
Brickwork built in cement mortar	1.20
Brickwork built in chimneys	2.00
Brickwork built to flat sweeps	1.50
Brickwork built in heavy engineering bricks	1.25
Brickwork built in flues	1.80
Brickwork built to quick sweeps	2.00
Brickwork built overhand	1.10
Brickwork built in large chambers below ground level	1.10
Brickwork built in small chambers below ground level, manholes, etc.	1.20
Brickwork built in raising old walls	1.25
Brickwork built in sewers	3.20
Brickwork built in tapered walls	2.00
Brickwork built in underpinning	2.25

Table 14.9 Fletton or common brickwork

	Unit	Material	Bricklayer hours	Labourer hours
Beam filling, per 112 mm width	m	–	0.40	0.40
Bedding wall plates, 122 mm wide	m	–	0.34	0.34
Brick noggin on edge	m²	36 bricks 0.02 m³ mortar	0.96	0.96
Brick noggin on flat	m²	50 bricks 0.03 m³ mortar	1.20	1.20
Cutting and pinning landing edges, chase girth 150 mm	m	–	1.33	1.33
Cutting and pinning landing edges, chase girth 225 mm	m	–	2.00	2.00
Cutting and pinning landing edges, chase girth 300 mm	m	–	2.65	2.65
Cutting and pinning landing edges, chase girth 450 mm	m	–	3.30	3.30
Cutting toothing and bonding to old walls	m²	–	6.50	6.50
Cutting and fitting brickwork to joists, etc. per 112 mm width of cutting	m	–	0.83	0.83
Chases – horizontal, 75 mm girth	m	–	0.66	0.66
Chases – horizontal, 150 mm girth	m	–	1.00	1.00
Chases – horizontal, 225 mm girth	m	–	1.32	1.32
Chases – horizontal, 300 mm girth	m	–	1.65	1.65
Chases – horizontal, 450 mm girth	m	–	1.90	1.90
Chases – raking, 75 mm girth	m	–	1.32	1.32
Chases – raking, 150 mm girth	m	–	2.00	2.00
Chases – raking, 225 mm girth	m	–	2.64	2.64
Chases – raking, 300 mm girth	m	–	3.30	3.30
Chases – raking, 450 mm girth	m	–	4.00	4.00
Chases – vertical, 75 mm girth	m	–	1.00	1.00
Chases – vertical, 150 mm girth	m	–	1.50	1.50
Chases – vertical, 225 mm girth	m	–	2.00	2.00
Chases – vertical, 300 mm girth	m	–	2.50	2.50
Chases – vertical, 450 mm girth	m	–	3.00	3.00
Flues – form, parge and core:				
cross-sectional area 155 cm²	m	0.013 m³ mortar	1.00	1.00
cross-sectional area 155 cm²	m	0.021 m³ mortar	1.16	1.16
cross-sectional area 155 cm²	m	0.029 m³ mortar	1.32	1.32
cross-sectional area 155 cm²	m	0.046 m³ mortar	1.68	1.68
cross-sectional area 155 cm²	m	0.052 m³ mortar	1.80	1.80
Hoop iron bond, single	m	1 m	0.11	0.11
Oversailing course, per course	m	–	0.17	0.17
Projections to aprons	m²	1 m² of 112 mm	1.70	1.70
Projections to friezes	m²	brickwork and mortar		
		as required	1.20	1.20
Projections to pilasters	m²	–	1.80	1.80
Projections to plinths	m²	–	0.95	0.95
Reveals – cutting and waste	m	3.5 bricks for waste	0.40	0.40
Rounded angles, 150 mm girth	m	–	1.32	1.32
Rounded angles, 225 mm girth	m	–	1.65	1.65
Rounded angles, 300 mm girth	m	–	2.00	2.00
Rounded angles, 375 mm girth	m	–	2.30	2.30
Rounded angles, 450 mm girth	m	–	2.65	2.65
Splays, 225 mm wide	m	3.5 bricks for waste	0.66	0.66
Splays, 350 mm wide	m	5 bricks for waste	1.00	1.00
Splays, 450 mm wide	m	7 bricks for waste	1.32	1.32
Squints and birdsmouths with purpose made bricks	m	13 bricks for squints	0.40	0.40
		7 bricks for birdsmouths		
Squints and birdsmouths with ordinary bricks	m	3.5 bricks for waste	1.10	1.10
Wedging to underpinning	m²	11 bricks for waste	2.75	2.75

Table 14.10 Brick facings

Notes: 1. The table shows the number of facing bricks required for facing work carried out in various bonds and also the additional or Extra Over bricklayer and labour hours taken to lay the facing bricks over ordinary brickwork per square metre.
2. The data refer to good-class facing work with selected hand-picked facing bricks. For seconds facings with no selection, the labourer hours may be taken as being the same as those shown in the table for bricklayer.
3. The data shown refer to facing work carried out in sandfaced or multicoloured facings. For facings other than this, use the following multipliers with the bricklayer and labourer hours shown:

   Facing work in Staffordshire and similar bricks, 1.50
   Facing work in glazed bricks, 2 00.

4. The data shown are exclusive of pointing, for which see Table 14.14

	Size of bricks, number of bricks required and the Extra Over bricklayer and labourer hours over ordinary brickwork in layer facing bricks per square metre		
	215 × 102.5 × 65 mm		
Type of Work	*No. facing bricks reqd.*	*Bricklayer hours*	*Labourer hours*
English bond, single-faced work	92	0.90	1.44
Flemish bond, single-faced work	81	0.79	1.27
103 mm or single-faced cavity work	65	0.60	0.96
215 mm solid and 270 mm cavity work	128	1.20	1.93
English garden wall bond, single-faced work	78	0.74	1.20
Flemish garden wall bond, single-faced work	68	0.70	1.12

Table 14.11 Multipliers for brickwork facings

	Bricklayer and labourer hour multipliers
Brickwork built in arches	5.50
Brickwork built in bands	1.50
Brickwork built to batter	1.25
Brickwork built to flat sweeps	1.50
Brickwork built to quick sweeps	2.00
Brickwork built to flush quoins	1.20
Brickwork built to projecting quoins	1.50
Brickwork built to small panels	2.60
Brickwork built to large panels	1.80
Brickwork built to overhand	1.25

Table 14.12 Brickwork facings

	Unit	Material	Bricklayer hours	Labourer hours
Angles, chamfered – purpose made	m	13½ bricks	0.82	0.82
Angles, chamfered – cut and rubbed	m	3½ bricks for waste	1.65	1.65
Angles, moulded – purpose made	m	13½ bricks	0.82	0.82
Aprons	m²	–	2.70	2.70
Aprons, returns to	m	–	0.66	0.66
Aprons, shape edges of	m	–	0.82	0.82
Birdsmouths – purpose made	m	7 bricks	0.82	0.82
Birdsmouths – cut and rubbed	m	5 bricks	1.65	1.65
Copings – bricks on edge per 225 mm width	m	13½ bricks 0.002 m³ mortar	0.66	0.66
Copings – brick on flat, per 225 mm width	m	10 bricks 0.002 m³ mortar	0.50	0.50
Cutting, circular	m	3½ bricks for waste	1.09	1.09
Cutting, horizontal	m	3½ bricks for waste	0.66	0.66
Cutting, raking	m	3½ bricks for waste	0.82	0.82
Cutting, vertical	m	3½ bricks for waste	0.39	0.39
Cutting against mouldings 150 mm girth	Each	1 brick for waste	0.36	0.36
Cutting against mouldings 300 mm girth	Each	2 bricks for waste	0.75	0.75
Cutting holes for pipes up to 75 mm diameter per 112 mm thickness of wall	Each	–	1.00	1.00
Cutting holes for pipes over 75 mm but not exceeding 150 mm diameter per 112 mm thickness of wall	Each	–	1.40	1.40
Cutting holes for pipes over 150 mm but not exceeding 225 mm diameter per 112 mm thickness of wall	Each	–	2.00	2.00
Cuttings and rubbings to groin points	m	10 bricks for waste	3.00	3.00
Cuttings and rubbings to ribs per 112 mm thickness	m	3½ bricks for waste	1.50	1.50
Dentil course up to 50 mm projection	m	–	0.33	0.33
Fair cutting and fitting facings to joists, etc. per 112 mm width of cutting	m	7 bricks for waste	1.64	1.64
Moulded course per course to 56 mm projection	m	–	0.20	0.20
Moulded angles – purpose made	m	14 bricks for waste	0.82	0.82
Notches in strings per 56 mm notch in one course	Each	m	0.60	0.60
Perforations in strings per 56 mm notch in one course	Each	m	1.50	1.50
Oversailing per course	m	m	0.20	0.20
Pilasters up to 350 mm width	m	m	1.64	1.64
Plinth course, per course to 50 mm projection	m	m	0.10	0.10
Plumbing to external angles	m	m	0.33	0.33
Reveals – cutting and waste	m	3½ bricks for waste	0.82	0.82
Squint quoins – purpose made	m	14 bricks for waste	0.82	0.82
Squint quoins – cut and rubbed	m	7 bricks for waste	1.64	1.64
Tile creasings to 225 mm walls	m	17 tiles 0.002 m³ mortar	0.82	0.82
Tile creasings to 350 mm walls	m	23 tiles 0.003 m³ mortar	1.00	1.00
Toothing and bonding to old walls	m	3½ bricks for waste	1.09	1.09

Table 14.13 Rubbed and gauged facings

	Unit	Material	Bricklayer hours	Labourer hours
Mouldings, cut and fix single course	m	7 bricks	5.00	5.00
External angle to single course	m	–	1.64	1.64
Internal angle to single course	m	–	2.50	2.50
Stops to single course	Each	–	0.60	0.60
Niche heads	m²	130 bricks	75.00	75.00
External angle of arch to niche head	m	–	4.00	4.00
Facing to niches	m²	110 bricks	27.00	27.00
External angle of niche facing	m	–	2.00	2.00
Sills to niches	m²	–	6.50	6.50
Pilasters	m²	–	5.40	5.40
Cutting and bonding pilasters to facing	m	–	2.00	2.00
Throatings	m	–	0.40	0.40
Weatherings	m²	–	1.10	1.10
Cleaning down	m²	–	0.30	0.30

Table 14.14 Pointing brickwork

Notes: 1. The bricklayer and labourer hours shown are for raking out the joints and pointing as a separate operation with a neat weathered joint.
2. The hours shown are for pointing with lime mortar. If cement mortar is used a labour multiplier of 1.25 should be used with the hours shown.
3. The hours shown refer to 9 mm joints.

Raking out the joints as the work proceeds and pointing as a separate operation with a weathered joint. Size of bricks, mortar required, bricklayer and labourer hours per m²

	215 × 102.5 × 65 mm		
Type of bond	Mortar required m³	Bricklayer hours	Labourer hours
English bond	0.0021	1.20	1.20
Flemish bond	0.0020	1.14	1.14
Stretcher bond	0.0019	1.02	1.02
Header bond	0.0026	1.20	1.20
English garden bond	0.0020	1.12	1.12
Flemish garden bond	0.0019	1.08	1.08

Table 14.15 Multipliers for pointing brickwork

Description	Bricklayer and labourer hour multipliers
Raking out the joints as the work proceeds and pointing with a neat weathered joint	0.60
Raking out joints only before the mortar is set	0.20
Raking out joints only in perished lime mortar	0.70
Raking out joints only in perished cement mortar	0.90
Pointing only in a coloured mortar with a neat weathered joint	0.40
Ironing joints as the work proceeds before mortar has set*	0.60
Tuck pointing	2.30

* This form of pointing consists of rubbing or ironing the joints with a 6 mm diameter bent steel bar

Table 14.16 Sundry works

	Unit	Material	Bricklayer hours	Labourer hours
Break down only brick walls built in lime mortar	m³	–	–	2.11
Break down only brick walls built in cement mortar	m³	–	–	3.16
Break down only stone walls built in lime mortar	m³	–	–	2.63
Break down only stone walls built in cement mortar	m³	–	–	3.94
Note. Add loading and transporting if required				
Brick on edge steps or sills	m	13½ bricks 0.002 m³ mortar	0.66	0.66
Brick on edge steps or sills radiating	m	13½ bricks 0.002 m³ mortar	1.32	1.32
Fair ends to the above	Each	–	0.25	0.25
Cement fillets to roof joints	m	0.001 m³ mortar	0.20	0.20
Cement wash	m²	0.03 kg cement	0.12	0.12
Clean and stack old bricks lime mortar	per 1000	–	–	18.00
Damp proof course in bitumen horizontal	m²	Allow 5% for waste	0.22	0.22
Damp proof course in bitumen vertical	m²	Allow 5% for waste	0.54	0.54
Damp proof course in sheet lead horizontal	m²	Allow 5% for waste	0.32	0.32
Damp proof course in sheet lead vertical	m²	Allow 5% for waste	0.65	0.65
Damp proof course in two courses of slates horizontal	m²	2.4 m² of slates	2.70	2.70
Damp proof course in two courses of slates vertical	m²	0.015 m³ mortar	5.40	5.40
Damp proof course in two courses of bricks horizontal	m²	72 bricks, 0.030 m³ mortar	3.25	3.25
Fix air bricks, 225 × 150 mm	Each	0.003 m³ mortar	0.20	0.20
Fix caps to pillars	m²	0.03 m³ mortar	2.16	2.16
Fix chimney pots, 300 mm diameter	Each	0.003 m³ mortar	0.60	0.60
Fix chimney bars	Each	–	0.10	0.10
Fix copings, 230 cm² in cross section	m	0.005 m³ mortar	0.82	0.82
Fix copings, 308 cm² in cross section	m	0.006 m³ mortar	0.86	0.86
Fix copings, 388 cm² in cross section	m	0.007 m³ mortar	0.89	0.89
Fix copings, 465 cm² in cross section	m	0.009 m³ mortar	0.96	0.96
Fix door or window frames	m²	–	0.88	0.88
Fix sinks on metal supports	Each	–	2.25	2.25
Hack brickwork for key	m²	–	–	0.96
Hack concrete for key	m²	–	–	1.20
Lime whiting	m²	0.25 kg lime	0.12	0.12
Pointing to door or window-frames	m	0.002 m³ mortar	0.36	0.36
Rake, wedge and point lead flashings	m	0.003 m³ mortar	0.90	0.90
Rake, wedge and point step flashings	m	0.003 m³ mortar	1.00	1.00
Reinforcement				
Render chimney backs	m²	0.015 m³ mortar	0.36	0.36
Set and fix mantel register grates, 600 mm	Each	24 bricks 0.015 m³ mortar	2.20	2.20
Set and fix mantel register grates, 900 mm	Each	32 bricks 0.020 m³ mortar	2.60	2.60
Set and fix register grate, 900 mm	Each	32 bricks 0.020 m³ mortar	2.00	2.00
Tiles 2 courses in cement for sills	m	13½ tiles 0.01 m³ mortar	1.70	1.70
Unload and stack bricks	per 1000	–	–	3.00
Wallplates bedding 112 mm wide	m	–	0.33	0.33
Form cavity	m	–	0.10	0.10

Table 14.17 Hollow wall ties

Notes: 1. Ties are usually fixed at 5 or 7 per m² and are sold by the kg.
2. The data shown are for wrought galvanised ties with twisted centre and with section forked or swallow-tailed ends.

Dimensions of tie in mm	Number of ties per 50 kg
200 × 18 × 3	510
200 × 18 × 5	340
200 × 25 × 3	390
200 × 25 × 5	260
212 × 18 × 3	480
212 × 18 × 5	320
212 × 25 × 3	360
212 × 25 × 5	240
225 × 18 × 3	460
225 × 18 × 5	310
225 × 25 × 3	330
225 × 25 × 5	220

Table 14.18 Hoop iron bond

Gauge S.W.G.	Width in mm	kg per 100 m	metres per kg
13	50	95	1.3
14	43	70	1.6
15	37	54	2.2
16	31	39	3.0
17	28	32	4.0
18	25	24	4.8

Table 14.19 Partition block work in walls

	Thickness of slab in mm	Unit	Mortar required m³	Bricklayer hours	Labourer hours
Concrete solid blocks	75	m²	0.008	0.72	0.72
	100	m²	0.010	0.84	0.84
Hollow blocks	75	m²	0.009	0.84	0.84
	100	m²	0.010	0.96	0.96
	125	m²	0.012	1.08	1.08
	150	m²	0.014	1.20	1.20

Table 14.20 Fixing copings, string courses and similar work

Description	Unit	Mortar required m^3	Bricklayer hours	Labourer hours
Fix coping 225 mm wide	m	0.008	0.83	0.83
Fix coping 350 mm wide	m	0.012	1.10	1.10
Fix coping 450 mm wide	m	0.016	1.15	1.15
Fix coping 600 mm wide	m	0.019	1.32	1.32
Fix enriched work	m^3	0.090	25.20	25.20
Fix moulded work	m^3	0.090	21.80	21.80
Fix plain work	m^3	0.090	14.40	14.40
Fix splayed work	m^3	0.090	24.90	24.90
Fix 75 mm string courses	m	0.003	0.40	0.40
Fix 150 mm string courses	m	0.006	0.53	0.53
Fix 225 mm string courses	m	0.009	0.66	0.66
Fix caps to 350 mm pillars	Each	0.100	0.40	0.40
Fix caps to 450 mm pillars	Each	0.150	0.60	0.60
Fix caps to 600 mm pillars	Each	0.200	0.80	0.80

Centering

Centering is used in the construction of work such as brick arches, vaulting, etc., and in estimating its cost several factors have to be taken into consideration:

1. The thickness of the ribs required depends on the width of the span and the weight of the work which has to be carried by the centering.
2. The thickness of the lagging depends on the distance apart at which the ribs are placed.

The data shown for estimating purposes are based on the rib and lagging thicknesses shown, the centres at which the ribs are placed being stated. This allows for substantial heavy work, but should the work be of a comparatively light nature the distance between the ribs may be increased. The estimator should bear in mind, however, that with centering work it is wise to err on the side of safety, so far as strength is concerned, as any movement or collapse of the centering results in destroyed work, which is a costly matter.

Table 14.21 Span and the rib and lagging thicknesses, together with the distance apart at which the ribs are set

Span in metres	Rib thickness in mm	Distance between ribs in mm	Lagging thickness in mm
0 to 1.5	25	300	25
1.5 to 3.0	38	375	38
3.0 to 4.5	50	450	42
4.5 to 6.0	62	525	50

In order that the cost of centering work may be estimated for variable widths of span, heights, and rib and lagging thicknesses, the following tables are given:

Table 14.22 Data from which to estimate the cost of erecting and striking one rib or one unit of built up ribs for various widths of span data, for flat, segmental and semicircular soffits.

Table 14.23 Data for propping or supporting the ribs in position to various heights, per rib or per unit of built up ribs.

Table 14.24 Data for fixing the lagging to the ribs for laggings of various thicknesses and for cutting, notching, etc.

Table 14.25 This table consists of multipliers for use with the data shown in the previous three Tables for dealing with centering fixed to slope ranging from 0 to 30°, 30 to 45° and 45 to 60°.

In estimating the cost of centering the most simple way of doing so is as follows:

1. Assess the total number of ribs required, bearing in mind the centres at which they are placed.
2. Assess the total square metres of lagging required.

Having done this, it is a simple matter to compute the amount of timber required and, by using the data shown in the tables, estimate the cost of the centering as a whole.

In this connection the estimator should note the following:

- The data shown allow for erecting and striking the centering.
- The data shown in Tables 14.22 and 14.23 are per rib.
- The cost of the timber, nails, dogs, etc., must be taken into account. A value may be allowed for any timber recovered if the nature of the work is such that this is merited.

Table 14.22 Fixing ribs

Notes: 1. The data shown are for erecting and striking single ribs or a unit of built-up ribs.
    2. For propping or strutting and supporting the ribs, see Table 14.23
    3. For dogs, nails, etc., allow 15 kg per cubic metre of timber.

| | Span in m, thickness of ribs in mm, and carpenter and labourer hours erecting and striking per rib or per unit of built-up ribs | | | | | | | |
| | Span 0 to 1.5 m 25 mm ribs 300 m centres | | Span 1.5 to 3 m 37 mm ribs 375 mm centres | | Span 3 to 4.5 m 50 mm ribs 450 mm centres | | Span 4.5 to 6 m 62 mm ribs 525 mm centres | |
Erect and strike ribs	*Carp. hours*	*Lab. hours*	*Carp. hours*	*Lab. hours*	*Carp. hours*	*Lab. hours*	*Carp. hours*	*Lab. hours*
Flat soffits: single rib	0.10	0.10	0.15	0.15	0.20	0.20	0.25	0.25
Two ribs: built up	0.20	0.20	0.30	0.30	0.40	0.40	0.50	0.50
Three ribs: built up	0.30	0.30	0.45	0.45	0.60	0.60	0.75	0.75
Segmental soffits: single rib	0.20	0.20	0.30	0.30	0.40	0.40	0.50	0.50
Two ribs: built up	0.40	0.40	0.60	0.60	0.80	0.80	1.00	1.00
Three ribs: built up	0.60	0.60	0.90	0.90	1.20	1.20	1.50	1.50
Semicircular soffits: single rib	0.25	0.25	0.40	0.40	0.50	0.50	0.60	0.60
Two ribs: built up	0.50	0.50	0.80	0.80	1.00	1.00	1.20	1.20
Three ribs: built up	0.75	0.75	1.20	1.20	1.50	1.50	1.80	1.80

Table 14.23 Propping and supporting ribs

Notes: 1. The data shown are for propping and supporting the ribs to the heights shown and include for plates, etc.
2. The data shown are per rib for the spans and heights shown.

Propping and supporting ribs per rib	Span in metres and carpenter and labourer hours propping and supporting ribs per rib							
	0 to 1.5 m		*1.5 to 3 m*		*3 to 4.5 m*		*4.5 to 6 m*	
	Carp. hours	*Lab. hours*	*Carp. hours*	*Lab. hours*	*Carp. hours*	*Lab. hours*	*Carp. hours*	*Lab. hours*
Height 0 to 1.5 m	0.20	0.20	0.30	0.30	0.45	0.45	0.60	0.60
Height 1.5 to 3 m	0.30	0.30	0.60	0.60	0.85	0.85	1.10	1.10
Height 3 to 4.5 m	0.40	0.40	0.80	0.80	1.25	1.25	1.60	1.60
Height 4.5 to 6 m	0.50	0.50	1.00	1.00	1.65	1.65	2.10	2.10

Table 14.24 Fix lagging

Notes: 1. For waste in timber allow 5 per cent.
2. For nails allow 0.5 kg per m^2.

Fix and remove lagging	Unit	Span in metres, size of lagging and carpenter and labourer hours nailing, lagging, etc.							
		Span 0 to 2 m 50 × 25 mm Lagging		*Span 2 to 4 m 50 m × 37 mm Lagging*		*Span 4 to 6 m 50 × 442 mm Laggging*		*Span 6 to 8 m 50 × 50 mm Lagging*	
		Carp. hours	*Lab. hours*	*Carp. hours*	*Lab. hours*	*Carp. hours*	*Lab. hours*	*Carp. hours*	*Lab. hours*
Fix lagging, close nailed	m^2	0.77	0.77	1.00	1.00	1.21	1.21	1.43	1.43
Fix lagging, nailed with 12 mm gaps	m^2	0.66	0.66	0.88	0.88	1.10	1.10	1.32	1.32
Fix lagging, nailed with 25 mm gaps	m^2	0.55	0.55	0.77	0.77	1.00	1.00	1.21	1.21
Cutting – circular	m	0.66	–	0.82	–	1.00	–	1.10	–
Cutting – raking	m	0.40	–	0.50	–	0.60	–	0.70	–
Cutting – to groins	m	1.00	–	1.32	–	1.56	–	1.98	–
Cutting – intersections	m	0.82	–	1.00	–	1.15	–	1.32	–
Notching for keys	Each	0.25	–	0.30	–	0.35	–	0.40	–

Table 14.25 Multipliers for use with tables 14.22, 14.23 and 14.24

Description	Multipliers
Centering fixed on the level	1.00
Centering fixed on slope 0° to 30° from the horizontal	1.50
Centering fixed on slope 30° to 45° from the horizontal	2.00
Centering fixed on slope 45° to 60° from the horizontal	2.50

15 Woodwork

This chapter deals with the labours of a carpenter and a joiner in fixing only in the work.

The manufacture or making of joinery is not dealt with. The data in the tables assume that this has been manufactured and delivered on the site ready for building in the work, and allows for off-loading and fixing.

The estimating data shown refer to softwoods, such as deal, yellow pine, spruce, etc. and a table of multipliers is included for use with the data shown for fixing hardwoods.

The fixing of ironmongery is also included in this section.

Allowance for waste. This is done by adding a percentage to the cost of the timber per cubic metre, per metre or per square metre, as the case may be. These percentages vary, those shown in the table being a fair average and suitable for estimating purposes. It will be seen that they are shown under these headings:

1. Cubed unwrought timber in carcase work.
2. Flooring, boarding, etc.
3. Mouldings, skirtings, grounds, etc.

Table 15.1 Percentage additions to the cost of timber to allow for waste

Description	Percentage addition to add to the cost of timber to allow for waste
Unwrought cubed timber in carcase work	7
Flooring, boarding, etc.	5
Mouldings etc.	5

Table 15.2 Decking

Note: Allow 5 per cent for waste in timber.

Description	Unit	Labour hours
Decking, close nailed, 100 × 38 mm	m^2	1.45
Decking, close nailed, 100 × 50 mm	m^2	1.75
Decking, close nailed, 100 × 75 mm	m^2	2.50
Decking with 25 mm gaps, 100 × 38 mm	m^2	1.30
Decking with 25 mm gaps, 100 × 50 mm	m^2	1.50
Decking with 25 mm gaps, 100 × 75 mm	m^2	2.30

Table 15.3 Shores

Note: Allow 5 per cent for waste.

	Labour hours fixing shores per m³			
Description	50 × 225 to 100 × 225	100 × 225 to 150 × 200	150 × 200 to 150 × 300	over 150 × 300
Fix only:				
Dead shores	18.20	12.60	11.50	10.50
Flying shores	39.20	34.60	30.80	28.00
Raking shores	24.50	21.00	19.25	17.50
Remove shores	12.25	10.50	9.62	8.75

Table 15.4 Timber work in bridges, centering to large spans, gantries and stands

Notes: 1. Add 5 per cent for waste in timber.
2. For nails, dogs, etc., for cube work, allow 16 kg per m³.
3. For nails for decking allow:

For decking 37 mm thick 0.54 kg per m².
For decking 50 mm thick 0.8 kg per m².
For decking 75 mm thick 1.6 kg per m².

	Cross section of timber and labour hours fixing per m³			
Description	50 × 225 to 100 × 225	100 × 225 to 150 × 300	150 × 300 to 250 × 250	250 × 250 to 300 × 300
Timber in bridges, heavy centering, gantries and stands:				
up to 3 m high	17.5	14.0	12.6	10.5
3 to 4.5 m high	18.2	14.7	13.3	11.2
4.5 to 6.0 m high	19.65	15.75	14.35	11.95
6.0 to 9.0 m high	21.7	15.7	15.75	13.3

Table 15.5 Work to floors and partitions

Notes: 1. Allow for waste in timber 7 per cent.
2. Nails: for joists allow 0.15 kg per m; for herringbone strutting allow 0.08 kg per m; for solid strutting allow 0.07 kg per m.

Description	Unit	Carpenter hours	Labourer hours
50 mm joists on plates, 100 and 125 mm	m	0.05	0.05
50 mm joists on plates, 150 and 175 mm	m	0.07	0.07
50 mm joists on plates, 200 and 225 mm	m	0.10	0.10
75 mm joists on plates, 100 and 125 mm	m	0.12	0.12
75 mm joists on plates, 150 and 175 mm	m	0.14	0.14
75 mm joists on plates, 200 and 225 mm	m	0.16	0.16
50 mm joists with no plates, 100 mm and 125 mm	m	0.06	0.06
50 mm joists with no plates, 150 mm and 175 mm	m	0.09	0.09
50 mm joists with no plates, 200 mm and 225 mm	m	0.12	0.12
75 mm joists with no plates, 100 mm and 125 mm	m	0.15	0.15
75 mm joists with no plates, 150 mm and 175 mm	m	0.17	0.17
75 mm joists with no plates, 200 mm and 225 mm	m	0.20	0.20
Framed floors or partitions up to 6000 mm² in cross-sectional area	m	0.06	0.06
Framed floors or partitions over 6000 mm² in cross-sectional area	m	0.05	0.05
Trim ends of joists up to 150 mm deep	Each	0.20	0.20
Trim ends of joists over 150 mm and up to 225 mm deep	Each	0.30	0.30
Herringbone strutting	m	0.50	0.50
Solid strutting	m	0.33	0.33
Wall plates	m	0.05	0.05

Table 15.6 Floorboarding

Notes: 1. Allow for waste in timber 5 per cent.
2. When using tongued and grooved boards, allow 10 per cent for waste and tongue width.
3. Brads: allow for 75 mm boards 0.6 kg per m².
   allow for 100 mm boards 0.5 kg per m².
   allow for 125 mm boards 0.4 kg per m².
   allow for 150 mm boards 0.35 kg per m².
4. The hours shown allow for cutting, cramping, nailing and puttying nail-holes.

	Lay and nail boards, carpenter and labourer hours per square metre			
	Square edge or rough boarding		Tongued and grooved boarding	
Description	Carpenter hours	Labourer hours	Carpenter hours	Labourer hours
Floorboarding 75 mm wide up to 19 mm thick	0.50	0.25	0.56	0.28
Floorboarding 100 mm wide up to 19 mm thick	0.48	0.24	0.52	0.26
Floorboarding 125 mm wide up to 19 mm thick	0.44	0.22	0.48	0.24
Floorboarding 150 mm wide up to 19 mm thick	0.42	0.21	0.46	0.23

Table 15.7 Multipliers for floorboarding of different thicknesses

Description	Multipliers
Floorboards up to 19 mm thick	1.00
Floorboards up to 25 mm thick	1.10
Floorboards up to 32 mm thick	1.30
Floorboards up to 38 mm thick	1.60
Floorboards up to 44 mm thick	1.80
Floorboards up to 50 mm thick	2.00

Table 15.8 Sundry works to flooring

Description	Unit	Carpenter hours	Labourer hours
Circular cutting, per 25 mm thickness	m	0.66	–
Raking cutting, per 25 mm thickness	m	0.50	–
Border to hearth, 75 × 25 mm mitred and glued	m	0.40	0.40
Nosings to floors	m	0.66	–
Mitres	Each	0.15	–
Returned ends	Each	0.15	–

Table 15.9 Work to roofs

Notes: 1. Allow for waste in timber 7 per cent.
2. Nails: allow for nails to rafters, plates, purlins, joists, ridge boards, 3.9 kg per m³.
Allow for nails to tilting fillets, rolls and drips, 0.15 kg per m.

Description	Unit	Carpenter hours	Labourer hours
Ceiling beams, 150 × 75 mm	m	0.20	0.10
Ceiling joists, 150 × 50 mm	m	0.10	0.05
Ceiling struts, 150 × 50 mm	m	0.12	0.06
Collars, 125 × 50 mm	m	0.12	0.06
Purlins, 150 × 100 mm	m	0.24	0.12
Rafters – not framed, 125 × 50 mm	m	0.10	0.05
Rafters – hip and valley, 125 × 50 mm	m	0.16	0.08
Ridge boards, 175 × 25 mm	m	0.12	0.06
Trusses – framed, 150 × 100 mm	m	0.50	0.25
Wall plates, 100 × 75 mm	m	0.10	0.10
Rolls, 50 mm to flat roofs	m	0.33	0.33
Drips, 50 mm to flat roofs	m	0.80	0.80
Cleats for purlins	Each	0.80	0.40
Tilting fillets, 50 × 37 mm	m	0.20	0.20
Cutting rafters	Each	0.04	0.02
Shaped and wrot. ends to 100 × 50 mm rafters	Each	0.24	0.12
Shaped and wrot. ends to 150 × 50 mm rafters	Each	0.30	0.15

Table 15.10 Work to trusses

Description	Unit	Carpenter hours	Labourer hours
Truss, Hydro-Nail, 45 degree pitch			
Span 5 m	Each	0.90	0.30
Span 7 m	Each	1.20	0.40
Span 9 m	Each	1.50	0.50

Table 15.11 Hoisting roof trusses

Height in metres	Labourer hours per cubic metre
9	1.75
18	2.10
27	2.45
36	2.80
45	3.15

Table 15.12 Roof boarding

Notes: 1. Allow for waste in timber 5 per cent.
2. Brads: allow for 100 mm boards 0.5 kg per m^2.
   allow for 125 mm boards 0.4 kg per m^2.
   allow for 150 mm boards 0.3 kg per m^2.
3. For boarding laid diagonally use a labour multiplier of 1.25.
4. For boarding laid to ceilings use a labour multiplier of 1.80

Description	Unit	Square edge or rough boarding		Tongued and grooved or weathered boarding	
		Carpenter hours	Labourer hours	Carpenter hours	Labourer hours
Boards 100 mm wide × 19 mm	m^2	0.46	0.23	0.48	0.24
Boards 100 mm wide × 25 mm	m^2	0.50	0.25	0.52	0.26
Boards 125 mm wide × 19 mm	m^2	0.42	0.21	0.44	0.22
Boards 125 mm wide × 25 mm	m^2	0.46	0.23	0.48	0.24
Boards 150 mm wide × 19 mm	m^2	0.40	0.20	0.42	0.21
Boards 150 mm wide × 25 mm	m^2	0.44	0.22	0.46	0.23

Table 15.13 Sundry works to roofs

Description	Unit	Carpenter hours	Labourer hours
Barge boards 31 mm thick	m²	1.65	1.65
Barge boards 31 mm thick to dormers	m²	2.75	2.75
Boarding to soffits	m²	0.88	0.88
Fascias	m	0.13	0.13
Fascias to dormers	m	0.27	0.27
Gutter boards	m²	2.20	2.20
Gutter sides	m²	1.32	1.32
Gutter, mitres to	Each	0.06	0.06
Gutter, intersections to	Each	0.20	0.20
Gutter, gussets	Each	0.10	0.10
Hip and valley boards laid flat on rafters	m²	1.10	1.10
Mouldings, applied, up to 300 mm² in cross-sectional area	m	0.13	0.13
Mouldings, applied, up to 300 mm² to dormers	m	0.26	0.26

Architraves, cornices, mouldings, skirtings and grounds

Note: In fixing architraves and mouldings, two methods may be adopted:

- They may be nailed to grounds;
- They may be plugged direct to walls.

The grounds themselves are fixed to walls. Plugging direct is generally carried out by means of plugs of Rawlplug or other proprietary type.

The cost of fixing a ground per metre plugged to walls depends on the number of grounds fixed per metre. For practical purposes this is generally carried out at about 0.5 m centres.

To assist the estimator a table is shown giving the carpenter hours plugging for various gauges of screws to different depths. It should be noted that, depending on the material which has to be plugged, the hours shown may vary somewhat, but the table shown may be used as it stands for normal brickwork or concrete. In plugging through tiles or similar work special care must be taken, and for such plugging a multiplier of 1.50 should be used with the carpenter hours shown in the table.

Table 15.14 Plugging to walls

Gauge number of screw	Depth of hole, and carpenter hours per plug				
	12 mm	25 mm	37 mm	50 mm	62 mm
6	0.08	0.12	–	–	–
8	0.10	0.14	–	–	–
10	0.12	0.16	0.20	0.24	–
12	0.14	0.18	0.22	0.26	–
14	–	0.20	0.24	0.28	–
16	–	0.22	0.26	0.30	0.34
18	–	0.24	0.28	0.32	0.36

Table 15.15 Grounds

Notes: 1. Allow for waste in timber 5 per cent.
     2. Nails: allow for single grounds 0.04 kg per m.
              allow for double grounds 0.08 kg per m.
              allow for framed grounds 0.12 kg per m.
     3. The data shown for grounds plugged to walls are based on plugging carried out at 0.5 m centres.

Description	Unit	Carpenter hours	Labourer hours
Grounds – single: nailed to partitions	m	0.07	0.07
Grounds – single: nailed to studs	m	0.05	0.05
Grounds – single: plugged to walls	m	0.23	0.23
Grounds – double: nailed to partitions	m	0.10	0.10
Grounds – double: nailed to studs	m	0.07	0.07
Grounds – double: plugged to walls	m	0.40	0.40
Grounds – double, framed: nailed to partitions	m	0.17	0.17
Grounds – double, framed: nailed to studs	m	0.13	0.13
Grounds – double, framed: plugged to walls	m	0.53	0.53

Table 15.16 Architraves, cornices, mouldings, skirtings, etc.

Notes: 1. Allow for waste in timber 5 per cent.
     2. Nails: allow for moulds 125 mm wide 0.06 kg per m.
              allow for moulds 150 mm wide 0.08 kg per m.
              allow for moulds 225 mm wide 0.09 kg per m.
              allow for moulds 300 mm wide 0.11 kg per m.
              allow for narrow moulds up to 75 mm wide 0.03 kg per m.
              allow for cornices 0.15 kg per m^2.
     3. The data shown for moulds etc., plugged to walls are based on plugging carried out at 450 mm centres.

Description	Unit	Nailed to grounds		Plugged to walls	
		Carpenter hours	Labourer hours	Carpenter hours	Labourer hours
Mouldings – single widths					
up to 75 mm wide picture rails etc.	m	0.05	0.05	0.20	0.20
up to 100 mm wide dados etc.	m	0.07	0.07	0.27	0.27
up to 150 mm wide skirtings etc.	m	0.08	0.08	0.33	0.33
up to 225 mm wide skirtings etc.	m	0.10	0.10	0.40	0.40
Mouldings – two widths, built up					
125mm architraves, skirtings, etc.	m	0.08	0.08	0.40	0.40
150mm architraves, skirtings, etc.	m	0.10	0.10	0.45	0.45
225mm architraves, skirtings, etc.	m	0.12	0.12	0.55	0.55
Cornices					
Built up in two widths	m^2	1.30	1.30	–	–
Built up in three widths	m^2	2.00	2.00	–	–
Angle brackets	Each	0.30	0.30	–	–
Cornice bracketing, including grounds and plugging	m^2	4.40	4.40	–	–

Table 15.17 Battening

Notes: 1. Allow for waste in timber 3 per cent.
   2. Allow for nails 0.09 kg per m.
   3. The data shown for battening plugged to walls are based on plugging carried out at 450 mm centres.

Description	Unit	Carpenter hours	Labourer hours
Battens, nailed to partitions	m	0.04	0.04
Battens, nailed to rafters	m	0.02	0.02
Battens, nailed to felted roofs	m	0.02	0.02
Battens, nailed to studs	m	0.02	0.02
Battens, plugged to walls	m	0.23	0.23

Table 15.18 Wall and ceiling linings

Notes: 1. Allow 5 per cent for waste to wall and ceiling linings.
   2. The labour hours shown for wall and ceiling linings are for nailing to timber or grounds fixed ready to receive them.
   3. Nails: allow for nails to non-asbestos sheets, plaster, fibre and wall boards, 0.09 kg per m^2.

Description	Unit	Carpenter hours	Labourer hours
Non-asbestos flat sheets fixed to ceilings	m^2	0.43	0.43
Non-asbestos flat sheets fixed to walls	m^2	0.33	0.33
Plaster board fixed to ceilings	m^2	0.40	0.40
Plaster board fixed to walls	m^2	0.30	0.30
Wall board or fibre board fixed to ceilings	m^2	0.36	0.36
Wall board or fibre board fixed to walls	m^2	0.29	0.29
Circular cutting to asbestos sheets	m	0.40	0.40
Circular cutting to plaster board	m	0.23	0.23
Circular cutting to fibre board	m	0.20	0.20
Raking cutting to non-asbestos sheets	m	0.20	0.20
Raking cutting to plaster board	m	0.14	0.14
Raking cutting to fibre board	m	0.10	0.10

Table 15.19 Door frames and window linings

Note: The labours are for fixing only.

Description	Unit	Nailed to grounds		Plugged to walls	
		Carpenter hours	Labourer hours	Carpenter hours	Labourer hours
Door and window linings and frames, 75 × 50 mm	m	0.05	0.05	0.20	0.20
Door and window linings and frames, 75 × 75 mm	m	0.07	0.07	0.20	0.20
Door and window linings and frames, 100 × 50 mm	m	0.08	0.08	0.23	0.23
Door and window linings and frames, 100 × 75 mm	m	0.10	0.10	0.25	0.25
Door and window linings and frames, 100 × 100 mm	m	0.12	0.12	0.27	0.27
Door and window linings and frames, 125 × 75 mm	m	0.13	0.13	0.27	0.27
Door and window linings and frames, 125 × 100 mm	m	0.15	0.15	0.33	0.33
Door beading	m	0.10	0.10	–	–
Iron dowels to frames	Each	0.06	0.06	–	–

Table 15.20 Windows and skylights

Notes: 1. Labours are for fixing only.
2. For fixing window linings, see Table 15.19
3. The fixing of metal or wood window frames is generally done by a bricklayer, but they are also shown under this section.

Description	Unit	Carpenter hours	Labourer hours
Window frames, metal or wood, fixed only (usually fixed by a bricklayer)	m²	1.10	1.10
Window frames, build in lugs or holdfast	Each	0.04	0.04
Window frames, bed and point one side	m	0.27	0.27
Window frames, bed and point two sides	m	0.40	0.40
Window			
Boards and bearers	m²	2.75	2.75
Nosings to window boards	m	0.20	0.20
Housing ends of window linings to boards	Each	0.20	0.20
Notched and returned ends to window boards	Each	0.12	0.12
Mitres to window boards	Each	0.10	0.10
Returned ends and mitres to window boards	Each	0.12	0.12
Skylights, fix only	m²	1.65	1.65
Frames to skylights	m²	1.65	1.65
Intersections to ridges and hips of skylight	Each	0.18	0.18
Linings and fascias to skylights	m²	3.30	3.30
Mitres to ridges and hips of skylight	Each	0.06	0.06
Rolls to ridges and hips of skylight	m	0.20	0.20

Table 15.21 Doors and gates

Note: The labours are for fixing only, exclusive of ironmongery.

Description	Unit	Thickness in mm and carpenter and labourer hours fixing only, per m²							
		35 mm		*44 mm*		*54 mm*		*63 mm*	
		Carp. hours	Labourer hours	Carp. hours	Labourer hours	Carp. hours	Labourer hours	Carp. hours	Labourer hours
Doors – house type	m²	0.33	0.33	0.44	0.44	0.55	0.55	0.66	0.66
Gates	m²	0.55	0.55	0.66	0.66	0.77	0.77	0.88	0.88

Table 15.22 Stairs, complete with handrails, balusters or panels

Notes: 1. The labours shown are for fixing only. The stairs are the usual housing type, 1 m wide, purchased ready-made for fitting, complete with outer strings, panels or balusters, handrails and newel posts.
   2. The labour hours shown are for fixing complete with all strings, newel posts, balusters, panels and handrails.

Description	Erect stairs per 300 mm rise	
	Carpenter hours	Labourer hours
Erect straight stairs, 1 m wide	0.60	1.20
Add for half landing	1.80	3.60
Add for one set of winders	2.20	4.40
Add for two sets of winders	2.80	5.60

Table 15.23 Sundry carpentry works

Note: The labours shown are for fixing only.

Description	Unit	Carpenter hours	Labourer hours
Bench or table legs, 75 × 75 mm	m	0.40	0.40
Bench or table legs, 100 × 100 mm	m	0.60	0.60
Bearers to tables	m	0.50	0.50
Bench or table-tops, 19 mm boards	m²	1.10	1.10
Bench or table-tops, 25 mm boards	m²	1.32	1.32
Draining boards	m²	2.75	2.75
Cupboard fronts to recesses, exclusive of fitting doors	m²	1.10	1.10
Dressers, 1.8 m long	Each	1.50	1.50
Dressers, 2.1 m long	Each	1.70	1.70
Dressers, 2.4 m long	Each	2.00	2.00
Kitchen cabinets, 1.2 m long	Each	1.00	1.00
Kitchen cabinets, 1.5 m long	Each	1.30	1.30
Kitchen cabinets, 1.8 m long	Each	1.50	1.50
Shelves in cupboard recesses	m²	1.54	1.54
Shelving – lattice type	m²	1.32	1.32
Shelving – plain, 150 mm wide, fixed	m	0.14	0.14
Shelving – plain, 225 mm wide, fixed	m	0.17	0.17
Shelving – plain, 300 mm wide, fixed	m²	0.66	0.66
Shelf bearers fixed to previously fixed grounds or plugs	m	0.33	0.33
Shelf bearers fixed to previously fixed grounds or plugs, including plugging	m	0.60	0.60
W.C. backboards	Each	0.70	0.70

Table 15.24 Multipliers for work other than softwood

Nature of wood	Multipliers
Ash	1.50
Beech	1.60
Birch	1.40
Chestnut	1.30
Elm	1.60
Larch	1.20
Oak	1.70
Pitch pine	1 20

Table 15.25 Ironmongery

Note: As for carpentry, the labours shown are for fixing to softwoods. For fixing to hardwoods use table of multipliers shown in Table 15.24

Description	Unit	Carpenter hours
Bolts		
Barrel bolts, 100 mm	Each	0.40
Barrel bolts, 150 mm	Each	0.50
Barrel bolts, 200 mm	Each	0.60
Brass flush bolts, 100 mm	Each	1.30
Brass flush bolts, 150 mm	Each	1.50
Panic bolts, arm pattern	Each	1.40
Panic bolts, top and bottom action	Each	2.80
Door checks and springs		
Check or spring hinge – single	Each	1.80
Check or spring hinge – double	Each	2.50
Plain hinge	Each	1.00
Spring attached to lock rail	Each	1.20
Springs attached to top rail	Each	2.00
Fasteners		
Buttons	Each	0.12
Casement	Each	0.60
Casement stays	Each	0.40
Cupboard	Each	0.33
Hasp and staples – light type	Each	0.25
Hooks and hinges for gates	Per set	1.80
Sash	Each	0.40
Strap hasps and staples	Each	1.00
Hinges		
Butt hinges, steel or brass, 50 mm	Per pair	0.60
Butt hinges, steel or brass, 75 mm	Per pair	0.80
Butt hinges, steel or brass, 100 mm	Per pair	1.10
Cast rising butt hinges, 100 mm	Per pair	1.25
Cast rising butt hinges, 125 mm	Per pair	1.50
Double-jointed butt hinges, 50 mm	Per pair	0.90
Double-jointed butt hinges, 75 mm	Per pair	1.20
Double-jointed butt hinges, 100 mm	Per pair	1.60
Gate hinges, heavy	Per pair	2.50
Tee or cross garnet hinges, 150 mm	Per pair	0.40
Tee or cross garnet hinges, 200 mm	Per pair	0.50
Tee or cross garnet hinges, 250 mm	Per pair	0.60
Tee or cross garnet hinges, 300 mm	Per pair	0.70
Tee or cross garnet hinges, 350 mm	Per pair	0.80
Tee or cross garnet hinges, 400 mm	Per pair	0.90
Tee or cross garnet hinges, 450 mm	Per pair	1.00
Latches		
Gate latch	Each	0.90
Indicator W.C. latch	Each	1.40
Night latch – ordinary type	Each	1.00
Night latch – cylinder type	Each	1.60
Norfolk	Each	0.90
Rim	Each	0.80
Suffolk	Each	0.90

Table 15.25 continued

Description	Unit	Carpenter hours
Locks		
Drawback, 150 mm	Each	1.00
Drawback, 200 mm	Each	1.20
Mortice lock	Each	2.00
Norfolk lock	Each	2.20
Padlocks	Each	0.30
Rim lock	Each	1.00
Suffolk lock	Each	2.30
Sundries		
Finger plates	Each	0.25
Handles – door	Each	0.50
Handles – drawer	Each	0.30
Hooks – coat	Each	0.12
Hooks – hat and coat	Each	0.15
Letter plates, including perforation	Each	1.75
Number plates	Each	0.10

16 *Structural steelwork*

Steelwork is purchased by the tonne. To the purchase price of the steelwork are added those charges necessary for fabrication. Fabrication includes all operations up to and including delivery to site.

Fabricated steelwork is then ready for erection. Erection includes all operations subsequent to fabrication.

The labour constants are given for use with a crane.

Although there may be only one item for erection of structural steel framing measured in the bills of quantities, the estimator may well wish to do this synthesis in some detail and the labour constants can be used for this purpose.

Table 16.1 New work

| | Steelwork fixed per tonne (Weight up to 40 kg/m) | | | | | | | |
| | Plain steelwork Unjointed | | Plain steelwork Jointed | | Compound beams and stanchions | | Roof trusses | |
Description	Fitter hours	Labour hours	Fitter hours	Labour hours	Fitter hours	Labour hours	Fitter hours	Labour hours
Fit and erect at ground floor level.	4.80	4.80	14.60	14.60	19.60	19.60	–	–
Fit and erect 3 m above ground level.	10.60	10.60	21.20	21.20	28.20	28.20	38.80	38.80
Fit and erect 6 m above ground level.	11.20	11.20	–	–	–	–	–	–
Fix and erect 15 m above ground level.	12.80	12.80	25.40	25.40	33.80	33.80	46.60	46.60

Table 16.2 Alteration work

| | Steelwork fixed per tonne (Weight up to 40 kg/m) | | | | | | | |
| | Plain steelwork Unjointed | | Plain steelwork Jointed | | Compound beams and stanchions | | Roof trusses | |
Description	Fitter hours	Labour hours	Fitter hours	Labour hours	Fitter hours	Labour hours	Fitter hours	Labour hours
Fit and erect at ground floor level.	11.60	11.60	36.60	36.60	48.00	48.00	–	–
Fit and erect 3 m above ground level.	23.00	23.00	46.00	46.60	61.40	61.40	84.20	84.20
Fit and erect 6 m above ground level.	24.20	24.20	48.40	48.40	64.20	64.20	88.60	88.60
Fix and erect 15 m above ground level.	27.60	27.60	55.20	55.20	73.60	73.60	100.00	100.00

Table 16.3 Sundry work

Description	Unit	Smith hours	Labourer hours
Cutting girders etc.	Per 25 mm cross-sectioned area	0.25	0.25
Drilling holes in the shop for M12 bolt.	Each	0.08	0.08
Drilling holes in the shop for M20 bolt.	Each	0.10	0.10
Drilling holes in the shop for M24 bolt.	Each	0.12	0.12
Drilling holes in position for M12 bolt.	Each	0.20	0.20
Drilling holes in position for M20 bolt.	Each	0.25	0.25
Drilling holes in position for M24 bolt.	Each	0.30	0.30
Fix M12 bolts.	Each	0.08	0.08
Fix M20 bolts.	Each	0.10	0.10
Fix M24 bolts.	Each	0.12	0.12
Fix rivets in position 12mm diameter.	Each	0.20	0.20
Fix distance pieces.	Per kg	0.10	0.10

Table 16.4 Fixing universal beams, channel iron, etc.

Description	Height above ground in metres and labour hours fixing per tonne					
Fix UB: channel iron and similar in the work	Ground Level	3	6	9	12	15
Fix plain joists etc.	20.00	22.00	24.00	26.00	28.00	30.00
Fix cross beams etc. with angle cleats. Jointed work.	40.00	44.00	48.00	52.00	56.00	60.00

Table 16.5 Sharpen tools etc.

Description	Unit	Smith hours	Labourer hours
Cut and bend chimney bars, plain.	Each	0.30	0.30
Cut and bend chimney bars, with ends caulked.	Each	0.60	0.60
Cutting iron dowels for door frames.	Each	0.02	0.02
Forge gutter brackets.	Each	0.12	0.12
Forge hip strap.	Each	0.50	0.50
Forge pipe hooks for 50 mm pipe.	Each	0.12	0.12
Sharpen chisels – light up to 12 mm diameter.	Each	0.15	0.15
Sharpen chisels – heavy over 19 mm and up to 25 mm diameter.	Each	0.20	0.20
Sharpen drills for pneumatic hammers.	Each	0.75	0.75
Sharpen points, medium over 12 mm and up to 19 mm such as navvy picks.	Each	0.12	0.12
Sharpen tines of scarifier.	Each	0.20	0.20

Table 16.6 Sitework and metalwork – sundry works

Notes: 1. The data shown does not include for cutting holes, fixing ragbolts or building in.
  2. Allow for sundry materials necessary in the way of bolts, washers, ragbolts, screws, etc.

Description	Unit	Fixer hours	Labourer hours
Fix brackets to walls, up to 9 kg weight.	Each	0.30	0.30
Fix handrails 50mm galvanised tube.	m	0.82	0.82
Fix hydrant boxes.	Each	0.30	0.30
Fix ladders to walls, fire escape type.	m	0.82	0.82
Fix roof ventilators 150 mm diameter.	Each	1.00	1.00

Fire protection – structural steelwork

Several products are available for the fire protection of buildings. Quelfire Ltd. produce fire stop seals, fire stop pillows, fire protection compounds, ceramic fibre fire stops, flexible fire resistant seals and intumescent coatings. For intumescent coatings the weight of the steel not the girth is the critical factor.

Table 16.7 Fire protection of structural steelwork – example

152 × 152 mm Universal Column, 4 sides exposed, 1 hour fire protection		
Weight of column kgs mass/m	Intumescent coating required litres per square metre	Microns (D.F.T.)
37	2.46	1475
30	4.05	2425
23	5.64	3375

17 Roofing

Scaffolding

Scaffolding is required in connection with roofing work, but it should be noted that the same scaffolding is used by other trades such as bricklayer, carpenter, external renderer, etc.

It is usual to allow for the scaffolding either as a lump sum in the preliminaries or under brickwork and blockwork, in which case there is no need to allow for it under Roofing.

If, however, scaffolding has to be specially erected and dismantled in order to carry out specific works in connection with roofing, its cost must be included.

Profiled fibre cement sheet roofing

Table 17.1 Profiled fibre cement sheet roofing

Notes: 1. The data shown are for fixing of 10 m².
2. The data shown allow for 53 galvanised roofing nails and washers per 10 m² of roof using:

  63 mm roofing nails weighing 0.018 kg each.
  Washers weighing 0.006 kg.
  Weight of nails per 10 m² as fixed = 0.78 kg.
  Weight of washers per 10 m² as fixed = 0.27 kg.

3. For labour hours for nailing more or less than 53 nails per 10 m², see Table 17.3
4. Allow 5 per cent for waste on sheets and 5 per cent on nails and washers.
5. The fixer hours are generally those of a carpenter.

	Fixer and labourer hours fixing sheets per 10 m² of roof					
	Sheets 5.5 mm thick 72 mm pitch width = 0.782 m Width as fixed, allowing for lap = 0.648 m			Sheets 6 mm thick 72 mm pitch width = 1.086 m Width as fixed, allowing for lap = 1.016 m		
Length of sheets in metres	*No. of sheets required per 10 m² of roof*	*Fixer hours*	*Labourer hours*	*No. of sheets required per 10 m² of roof*	*Fixer hours*	*Labourer hours*
1.53	11.29	2.59	2.59	7.16	2.33	2.33
1.83	9.24	2.51	2.51	5.88	2.26	2.26
2.13	7.82	2.42	2.42	4.98	2.16	2.16
2.45	6.77	2.33	2.33	4.28	2.10	2.10
2.75	6.00	2.24	2.24	3.79	2.02	2.02
3.05	5.36	2.16	2.16	3.39	1.94	1.94

Table 17.2 Multipliers for profiled fibre cement sheet roofing

Description	Fixer and labourer hour multipliers
Fixing to pitch not in excess of 45°	1.00
Fixing to pitch exceeding 45° and not exceeding 75°	1.25
Fixing to pitch exceeding 75° to vertical	1.50
Fixing in small areas	2.25
Fixing to steel angle irons with galvanised hook bolts	1.50
Fixing curved sheets	1.25
For jointing with mastic prior to nailing	2.00

Table 17.3 Sundry work to profiled fibre cement sheet roofing

Description	Unit	Fixer hours	Labourer hours
Cutting – straight or raking	m	0.17	0.17
Cutting – circular	m	0.33	0.33
Fix apron pieces of flushing pieces	m	0.20	0.20
Fix barge pieces	m	0.23	0.23
Fix corner pieces to vertical work	m	0.17	0.17
Fix hips	m	0.33	0.33
Fix ridges – two-piece	m	0.26	0.26
Allow labour for nailing in excess of or less than 53 nails per 10 m^2	Per nail	0.01	0.01

Table 17.4 Corrugated galvanised sheet roofing with 996 mm sheets with 50 mm side lap and 150 mm vertical lap

Notes: 1. The data shown are for fixing 10 m^2.
2. The data shown allow for 53 galvanised roofing nails and washers per 10 m^2 of roof, using:

   112 mm roofing nails weighing 0.018 kg each.
   Washers weighing 0.06 kg each.
   Weight of nails per 10 m^2 as fixed = 0.78 kg.
   Weight of washers per 10 m^2 as fixed = 0.27 kg.

3. Allow 5 per cent for waste on nails and washers and 2½ per cent on sheets.
4. The fixer hours are generally those of a carpenter.

Length of sheets in metres	No. of sheets required per 10 m^2 of roof	Thickness of sheets and fixer and labourer hours fixing sheets per 10 m^2 of roof					
		0.70 mm		0.90 mm		1.20 mm	
		Fixer hours	Labourer hours	Fixer hours	Labourer hours	Fixer hours	Labourer hours
1.50	7.29	2.48	2.48	2.23	2.23	2.00	2.00
1.80	6.08	2.40	2.40	2.16	2.16	1.92	1.92
2.40	4.56	2.22	2.22	2.00	2.00	1.79	1.79
3.00	3.65	2.05	2.05	1.85	1.85	1.64	1.64

Table 17.5 Multipliers for fixing corrugated sheet roofing to ordinary 50 mm lap

Description	Fixer and labourer hour multipliers
Fixing to pitch not in excess of 45°	1.00
Fixing to pitch exceeding 45° and not exceeding 75°	1.25
Fixing to pitch exceeding 75° to vertical	1.50
Fixing in small areas	2.25
Fixing to angle irons with galvanised hook bolts	1.50
Fixing curved sheets	1.25
Jointing with mastic prior to nailing	2.00

Table 17.6 Sundry work to corrugated sheet roofing

Note: Allow for tar for tarring roofs 4.5 m² per litre. Allow for preservative compositions for treating roofs 18 m² per litre. Allow for sand for sanding roofs 290 m² per m².

Description	Unit	Fixer hours	Labourer hours
Circular cutting	m	0.66	0.66
Raking or straight cutting	m	0.33	0.33
Fix galvanised ridge with 100 mm lap	m	0.17	0.17
Allow for nails and washers in excess of or less than 53 per 10 m²	Per nail	0.01	0.01
Brooming down	m²	–	0.04
Wire brushing	m²	–	0.12
Sanding roofs	m²	–	0.06
Tarring roofs one coat	m²	–	0.18
Treating roofs with preservative compositions	m²	–	0.14

Table 17.7 Felt roofing

Notes: 1. The data shown are per 10 m².
2. The data shown are based on using felt 1 m × 10 m long with 50 mm laps, nailed with 25 mm clout nails.
3. Allow 2½ per cent waste on felt and 5 per cent on nails.
4. The fixer hours are generally those of a carpenter.

Description	Material required per 10 m² of roof				Labour fixing	
	Felt square metres	Strips metres	25 mm clout nails kg	50 mm clout nails kg	Fixer hours	Labourer
Felt nailed and fixed without strips	10	–	0.30	–	0.30	0.10
Felt nailed and fixed with strips	10	27	0.30	0.25	0.40	0.12
Felt nailed and with laps jointed in mastic – without strips	11	–	0.30	–	0.50	0.15
Felt nailed and with laps jointed in mastic – with strips	11	27	0.30	0.25	0.60	0.20

Table 17.8 Multipliers for felt roofing

Description	Fixer and labourer hour multipliers
Fixing felt to pitch not in excess of 45°	1.00
Fixing felt to pitch exceeding 45° and not exceeding 75°	1.25
Fixing felt to pitch exceeding 75° to vertical	1.50
Fixing felt in small areas	2.25

Table 17.9 Sundry work to felt roofing

Note: Allow for tar for tarring roofs 2 m² per litre. Allow for preservative compositions for treating roofs 4 m² per litre. Allow for sand 200 m² per m³.

Description	Unit	Fixer hours	Labourer hours
Circular cutting	m	0.11	0.33
Raking or straight cutting	m	0.05	0.17
Brooming down roofs prior to painting or tarring	m²	–	0.32
Tarring roofs one coat	m²	–	0.24
Treating roofs with preservative compositions	m²	–	0.16
Sanding to roofs after tarring	m²	–	0.06

Tiled roofing

Plain Tiling with 270 × 165 mm Tiles

Plain tiles for roofing can be obtained in a variety of shades and colours.

The battens used are generally 2.5 m long, these being nailed to the rafters; 50 mm cut nails are commonly used for this purpose.

The weight of these nails may be taken as being 0.003 kg each. The battens are usually sawn 30 × 18 mm, 50 × 25 mm or 75 × 25 mm.

It is usual to nail the tiles to the battens about every fifth course, using 38 mm copper or zinc nails, weighing 0.002 kg each. The specification may call for more extensive nailing, such as every fourth course, in which case the extra labour in nailing and the material cost of the nails must be included. A 100 mm gauge is almost universally adopted for tiling.

Note. 1. The lap is obtained by deducting twice the gauge from the length of the tile, thus the lap with 100 mm gauge is: 270 mm − (2 × 100 mm) = 70 mm
2. The gauge is obtained by deducting the lap from the lengths of the tile and dividing by two.

The data shown for estimating the cost of fixing battens and tiles are based on the rafters being fixed at 400 mm centres and the tiles nailed every fifth course, but a table of multipliers is given for use with the main estimating table to allow for the additional labour involved in nailing, should closer nailing be specified. The additional cost of the extra nails should also be included.

Table 17.10 Plain tiling to roofs with 270 × 165 mm tiles

Notes: 1. The tiles are taken as being nailed every fifth course with two 31 mm nails, the battens being at 400 mm
centres.
2. The hours shown are per 10 m^2.
3. For sundry work to tiling, see Table 17.13
4. For battening for tiling, see Table 17.11
5. Allow 5 per cent for waste on tiles and nails.

Gauge to which tiles are laid in millimetres	Number of tiles required per 10 m^2 of roof	Tile nails required per 10 m^2 of roof kg	Tiler and labourer hours fixing tiles per 10 m^2 of roof	
			Tiler hours	Labourer hours
86	684	0.63	5.94	5.94
100	599	0.54	5.40	5.40

Table 17.11 Battening per 10 square metres of plain tiling

Notes: 1. Battens are taken as being nailed with 50 mm cuts, the rafters being at 400 mm centres.
2. The data shown are for fixing battens per 10 m^2 of roof.
3. Allow 5 per cent for waste on battens and nails.
4. The battens may be fixed by a carpenter or tiler, and are usually 30 × 18 mm.

Gauge of tiles in millimetres	Battens required per 10 m^2 of roof m	Batten nails required per 10 m^2 of roof 50 mm nails kg	Carpenter or tiler and labourer hours fixing battens per 10 m^2 of roof	
			Carpenter h	Labourer h
86	113	1.00	3.00	1.50
100	100	0.81	2.60	1.30

Table 17.12 Multipliers for plain tiling and battening

Description	Tiler, carpenter and labourer hour multipliers
Fixing battens and tiles to pitch not in excess of 45°	1.00
Fixing battens and tiles to pitch exceeding 45° and not exceeding 75°	1.25
Fixing battens and tiles to pitch exceeding 75° to vertical	1.50
Tiling in small areas	2.25
Tiles nailed every fifth course	1.00
Tiles nailed every fourth course	1.20
Tiles nailed every third course	1.40
Tiles nailed every second course	1.60
Tiles nailed every course	1.80

Table 17.13 Sundry work to plain tiling

Description	Unit	Tiler hours	Labourer hours
Cement fillets	m	0.20	0.20
Double eaves course bedded in mortar	m	0.10	0.10
Double eaves course not bedded	m	0.07	0.07
Fix felt underlays	m^2	0.11	0.11
Hips and valleys cut from plain tiles (both sides measured)	m	0.33	0.33
Pointing verges (0.002 m^3 mortar)	m	0.20	0.20
Ridging	m	0.40	0.40
Raking, cutting to gables	m	0.50	0.50
Torching to underside of tiling	m^2	0.32	0.32

Interlocking tiles

Interlocking tiles can be obtained in a variety of colours and shades and in various sizes, the most commonly used being:

380 × 205 mm, requiring 170 tiles per 10 m^2 of roof
375 × 260 mm, requiring 130 tiles per 10 m^2 of roof
335 × 260 mm, requiring 160 tiles per 10 m^2 of roof

The tiles generally have projecting lugs by which they are suspended from the roof battens, and are laid with the same vertical lap throughout. The horizontal lap may be varied, thus obviating the necessity of cutting to the ridge. The horizontal lap is usually 75 mm, while the vertical lap is in the region of 19 mm.

The data shown for interlocking tiles are based on the 380 × 205 mm tiles laid to a 19 mm vertical and a 75 mm horizontal lap. The net covering area is, therefore, 312 × 185 mm, and the number of tiles required per 10 m^2 is 170.

To find the cost of tiling with tiles other than the 380 × 205 mm, knowing the number of tiles required per 10 m^2 of roof, a multiplier may be evolved for use with the data shown for the appropriate type of tile. Thus, if tiles are being used which require 130 per 10 m^2, such as in the case of 375 × 260 mm tiles, the multiplier to use with the tiling and battening hours shown in the table is 130/170 = 0.77.

Italian tiles and Spanish tiles

Italian and Spanish type tiles have a decided charm and provide an effective covering. They consist of two tiles, one forming a roll and the other a channel. The tiles are tapered and in fixing them battens are required fixed on the flat to support the channel and on edge to support the rolls.

The tiles are available in various sizes, the most commonly used being:

Italian Tiles:
335 mm long, fixed at 300 mm centres, requiring 120 pairs of tiles per 10 m^2 of roof.
405 mm long, fixed at 275 mm centres, requiring 108 pairs of tiles per 10 m^2 of roof.

Spanish Tiles:
350 mm long, fixed at 262 mm centres, requiring 135 pairs of tiles per 10 m^2 of roof.

The data shown for estimating the cost per 10 m^2 of this type of tiling are based on Italian tiles 335 mm long, laid to an 85 mm lap, requiring 120 pairs per 10 m^2 of roof. If the estimator has to estimate the cost of fixing tiles other than this, he may evolve a multiplier for use with the data shown, knowing the number of tiles required per square of roof for the particular case. Thus if the tiles to be used are Spanish tiles requiring 135 pairs of tiles per 10 m^2 of roof the labour multiplier is

$$\frac{135}{120} = 1.12.$$

Pantiles

Pantiles may be obtained in various sizes. Those most commonly used are 338 × 245 mm, laid to a 75 mm lap, the number of tiles required per 10 m^2 of roof being 180. The battens used are generally 38 × 25 mm.

The data shown for fixing pantiles are based on this size, but if the estimator has to estimate the cost of fixing tiles of a size other than this, he can, knowing the number of tiles required per 10 m^2 of roof, evolve a multiplier for use with the tiling and battening hours shown in the main estimating table. Thus, if the number of tiles per 10 m^2 of roof is 158, the labour multiplier for use with the data shown in the table is 158/180 = 0.88.

Allowance for waste. Five per cent should be allowed for waste in battening, pantiles and batten nails.

Roman tiles

Roman tiles may be obtained in single or double tile type to a variety of colours and tints. Various sizes of tiles are available, such as:

420 × 330 mm, the tiles required per 10 m^2 of roof being 95.
380 × 275 mm, the tiles required per 10 m^2 of roof being 134.
360 × 260 mm, the tiles required per 10 m^2 of roof being 147.

The data shown for estimating purposes are based on the 420 × 330 mm tile laid to a 75 mm horizontal lap and with a 38 mm vertical lap, the tiling being nailed every third course, the number of tiles required per 10 m^2 of roof being 95. If tiles other than this are used the estimator may evolve a labour multiplier for use with the tiling and battening hours shown in the main estimating table. Thus, if the tiles used are such that 134 per 10 m^2 of roof are required the multiplier is

$$\frac{134}{95} = 1.41.$$

Table 17.14 Tiling to roofs – various

Notes: 1. The hours shown are per 10 m².
2. For tile nails for Italian tiling allow 1.8 kg of nails per 10 m².
3. For tile nails for Roman tiling allow 0.11 kg of nails per 10 m².
4. For sundry work to tiling, see Table 17.16
5. Allow 5 per cent for waste on battening, tiles and nails.
6. The battens may be fixed by a carpenter or tiler.

Type of tile	Size of tile in mm	Vertical lap in mm	Horiz. lap in mm	No. of tiles reqd. per 10 m² of roof	Metres of battens reqd. per 10 m² of roof	Kg of batten nails reqd. per 10 m² of roof	Fix battens per 10 m² of roof		Fix tiling per 10 m² of roof	
							Carp. or tiler hours	Lab. hours	Tiler hours	Lab. hours
Interlocking	380 × 205	19	75	168	33	0.35	0.86	0.43	1.50	1.50
Italian or Spanish	388 × 81	Laid to 88 mm lap		120 pairs 114 flat	35 flat 45 edge	0.70	1.74	0.87	3.25	3.25
Pantiling	338 × 245	Laid to 75 mm lap		178	40	0.40	1.00	0.50	1.40	1.40
Roman	420 × 330	38	74	97	30	0.32	0.80	0.40	0.85	0.85

Table 17.15 Multipliers for tiling – various

Note: The multipliers given are for use with the hours shown for fixing both battens and tiling.

Description	Tiler, carpenter and labourer hour multipliers
Fix battens and tiles to pitch not in excess of 45°	1.00
Fix battens and tiles to pitch exceeding 45° and not exceeding 75°	1.25
Fix battens and tiles exceeding 75° to vertical	1.50
Fix battens and tiles in small areas	2.25

Table 17.16 Sundry work to tiling

Note: The hours shown are for both sides measured in the case of cutting to hips and valleys.

Type of tile	Unit	Eaves bedded in mortar		Cutting to valleys and hips		Cutting to verges		Fix hip or ridge tiles	
		Tiler hours	Lab. hours	Tiler hours	Lab. hours	Tiler hours	Lab. hours	Tiler hours	Lab. hours
Interlocking	m	0.10	0.10	0.50	0.50	0.40	0.40	0.83	0.83
Italian and Spanish	m	0.07	0.07	0.83	0.83	–	–	0.66	0.66
Pantile	m	0.10	0.10	0.66	0.66	0.50	0.50	0.82	0.82
Roman	m	0.10	0.10	0.50	0.50	0.40	0.40	0.50	0.50

Slating

Slates are generally referred to by their size and sold at so much per 1000. They may be obtained in many sizes, and since they can be laid to different laps the number of slates required per 10 m^2 of roof is variable.

For the convenience of the estimator the estimating tables show the number of slates and slate nails and also the metres of battens and batten nails required per 10 m^2 of roof for those slates most commonly used, laid to 63 mm and 75 mm lap.

Separate data have purposely been shown for estimating the cost of fixing the battens, since this may be carried out by either the slater or a carpenter.

Table 17.17 Slating to BS 680

Notes: 1. The slates are taken as being centre nailed with 45 mm nails.
2. The hours shown are per 10 m^2.
3. For slate cutting, see Table 17.20
4. For sundry work to slating, see Table 17.21
5. Allow for waste on slates 2½ per cent, and on battening and nails 5 per cent.

| | *Materials required and slater and labourer hours per 10 m^2 of roof* | | | | | | | |
| | *63 mm lap* | | | | *75 mm lap* | | | |
Name of slate and size in millimetres	*No. of slates*	*Copper nails kg*	*Slater hours*	*Labour hours*	*No. of slates*	*Copper nails kg*	*Slater hours*	*Labour hours*
Duchess, 610 × 305 mm	120	0.85	1.85	1.85	125	0.88	1.95	1.95
Marchioness, 560 × 305 mm	146	0.95	2.25	2.25	150	1.00	2.40	2.40
Countess, 510 × 255 mm	180	1.25	2.70	2.70	185	1.30	2.80	2.80
Viscountess, 460 × 230 mm	223	1.57	3.25	3.25	230	1.62	3.43	3.43
Ladies, 405 × 205 mm	288	2.07	4.32	4.32	300	2.12	4.54	4.54

Table 17.18 Battening per 10 square metres of slating

Notes: 1. Battens are taken as being 2.5 m long, nailed with 65 mm cut nails, the rafters being at 400 mm centres.
2. The data shown are for fixing battens per 10 m^2 of roof.
3. Allow 5 per cent for waste on battening and nails.

| | *Materials required and carpenter and labourer hours per 10 m^2 of roof* | | | | | | | |
| | *63 mm lap* | | | | *75 mm lap* | | | |
Name of slate and size in millimetres	*Battens metres*	*Batten nails kg*	*Carp. hours*	*Labourer hours*	*Battens metres*	*Batten nails kg*	*Carp. hours*	*Labourer hours*
Duchess, 610 × 305 mm	36	0.30	0.94	0.47	40	0.34	1.00	0.50
Marchioness, 560 × 305 mm	39	0.34	1.04	0.52	42	0.39	1.10	0.55
Countess, 510 × 255 mm	45	0.38	1.16	0.58	47	0.44	1.24	0.62
Viscountess, 460 × 230 mm	50	0.43	1.30	0.65	53	0.49	1.38	0.69
Ladies, 405 × 205 mm	60	0.50	1.90	0.95	62	0.59	1.86	0.93

Table 17.19 Multipliers for slates and battening

Description	Slater, carpenter and labourer hour multipliers
Fixing battens and slates to pitch not in excess of 45°	1.00
Fixing battens and slates to pitch exceeding 45° and not exceeding 75°	1.25
Fixing battens and slates exceeding 75° to vertical	1.50
Slating in small areas	2.25

Table 17.20 Cutting to slates

Note: The hours shown are for single cutting. In the case of hips etc., both sides of the work must be measured.

	Slater and labourer hours per linear metre							
	Fair cuttings to valleys and gutters		Cutting to eaves		Cuttings to mitred hips		Cuttings to ridges	
Sizes of slates in millimetres	Slater hours	Labourer hours	Slater hours	Labourer hours	Slater hours	Labourer hours	Slater hours	Labourer hours
305 × 150 to 330 × 180	0.80	0.40	0.54	0.27	1.32	0.66	0.66	0.33
330 × 255 to 355 × 205	0.72	0.36	0.46	0.23	1.20	0.60	0.60	0.30
355 × 255 to 405 × 305	0.66	0.33	0.40	0.20	1.06	0.53	0.54	0.27
460 × 255 to 510 × 305	0.60	0.30	0.34	0.17	0.92	0.46	0.46	0.23
560 × 305 to 610 × 305	0.54	0.27	0.26	0.13	0.80	0.40	0.40	0.20

Table 17.21 Sundry work – slating

Description	Unit	Slater hours	Labourer hours
Bedding slates or eaves per course	m	0.07	0.07
Bedding verges	m	0.13	0.13
Cement fillets to roof	m	0.20	0.20
Felting under slates	m²	0.11	0.11
Pointing and filleting verges	m²	0.20	0.20
Rendering under slates	m²	0.32	0.32
Torching under slates	m²	0.17	0.17

Zinc, lead and copper coverings

Zinc is used in connection with such works as roof covering, eaves, gutters, rainwater pipes, etc., but it is not so extensively used as formerly. It is very light, can be bent without fracturing and requires only small roof scantlings to support it. The usual stock sizes of sheets are 2.1 × 0.9 m and 2.4 × 0.9 m, and the gauge of the sheets ranges from No. 8 to No. 16. For good roof or gutter work, Nos. 15 and 16 are generally used, while Nos. 12 to 14 are suitable for flashings and aprons. For lighter roof work, Nos. 12 to 14 are used.

In laying sheet zinc to roofs by the roll and cap system an allowance of 150 mm should be allowed for lap. Other allowances are for drips to flats 150 mm, drips to gutters 150 mm, and turnups to walls 150 mm.

Table 17.22 Zinc worker

Notes: 1. Add 2½ per cent for waste.
2. Allowance should be made for laps in assessing the material required. In the case of roofing, this may be taken as 150 mm for the roll-and-cap system of fixing.

	Weight of sheet zinc and fixer and labourer hours per square metre					
	No. 14 gauge 5.73 kg per m²		No. 15 gauge 6.93 kg per m²		No. 16 gauge 7.56 kg per m²	
Description	*Fixer hours*	*Labourer hours*	*Fixer hours*	*Labourer hours*	*Fixer hours*	*Labourer hours*
Dormer cheeks	3.25	3.25	3.68	3.68	4.32	4.32
Dormer flats	4.32	4.32	4.86	4.86	5.40	5.40
Flat roofs	2.70	2.70	2.90	2.90	3.25	3.25
Sloping roofs	3.25	3.25	3.68	3.68	4.32	4.32
Gutters and cesspools	5.40	5.40	5.95	5.95	6.48	6.48
Soakers	1.08	1.08	1.20	1.20	1.30	1.30
Stepped flashings	3.78	3.78	4.32	4.32	4.86	4.86
Valleys and hip and ridge coverings	2.70	2.70	3.24	3.24	3.78	3.78
Weathering to cornices or copings	3.25	3.25	3.68	3.68	4.32	4.32

Table 17.23 Zinc worker sundry works

Description	*Unit*	*Fixer hours*	*Labourer hours*
Dressing over glass	m	2.21	2.21
Dressing into hollows	m	1.32	1.32
Dressing over mouldings, per 25 mm girth	m	0.36	0.36
Labour to secret gutters	m	1.32	1.32
Perforated zinc fixed to sashes	m²	2.70	2.70
Rolls – capped ends to	Each	1.60	1.60
Rolls – saddles to	Each	1.00	1.00
Soldered joint (0.03 kg solder)	Each	1.00	1.00
Zinc nailing	m	0.50	0.50
Zinc welts	m	1.65	1.65

Lead worker

Table 17.24 Sheet lead work

Notes: 1. The data shown are for fixing sheet lead in the work per m².
2. An allowance of 2½ per cent should be made for waste, apart from the allowance for lap.
3. The sheet lead most commonly used ranges from 14 to 42 kg in weight per square metre.

| | Weight of sheet lead and labour hours per m² | | | |
| | Exceeding 14 kg and up to 28 kg per m² | | Exceeding 28 kg per m² | |
Description. Fix sheet lead	Plumber hours	Labourer hours	Plumber hours	Labourer hours
Fix in aprons, cover and step flashings, dormer cheeks and flats, gutters, and hip and ridge coverings	3.35	3.35	3.60	3.60
Fix to cisterns	3.12	3.12	3.30	3.30
Fix to dampcourse	1.08	1.08	1.20	1.20
Fix to flat roofs	2.40	2.40	2.60	2.60
Fix to floors	0.84	0.84	0.90	0.90
Fix to sinks	3.36	3.36	3.60	3.60
Fix to tanks	2.88	2.88	3.00	3.00
Weatherings to chimneys, copings, etc.	2.88	2.88	3.00	3.00

Table 17.25 Sundry work to sheet lead

Description	Unit	Plumber hours	Labourer hours
Bed edges in white lead, 0.5 kg white lead per m	m	0.40	0.40
Bossed angles to aprons	Each	0.60	0.60
Bossed angles to flats	Each	1.33	1.33
Bossed angles to gutters	Each	1.75	1.75
Bossed angles to rolls	Each	1.50	1.50
Bossed ends to rolls	Each	1.20	1.20
Bossed intersections to rolls	Each	1.75	1.75
Dressing over corrugated sheets, etc.	m	2.00	2.00
Dressing over glass	m	1.65	1.65
Dressing through shoots and to rainwater heads	Each	0.90	0.90
Dressing over skylight bars	Each	0.60	0.60
Fix 100 mm lead rainwater pipe	m	1.00	1.00
Fix 100 mm lead bends	Each	0.50	0.50
Fix 100 mm lead 'S' pipes	Each	0.70	0.70
Fix rainwater heads	Each	0.80	0.80
Fix gratings to heads	Each	0.60	0.60
Lead plugs, 0.14 kg lead each	Each	0.50	0.50
Lead wedging to flashings	m	0.33	0.33
Soldered or wiped angles, 1.36 kg solder per m	m	2.00	2.00
Soldered or wiped seams, 1.36 kg solder per m	m	1.32	1.32
Soldered dots, 0.27 kg solder each	Each	0.50	0.50

Copper worker

Table 17.26 Copper roofing work

Notes: 1. Allow 2½ per cent for waste.
2. Allow for laps in works to roofs 80 mm for welted laps, 150 mm turn up against walls, 50 mm for cover flashings.

| Description | Gauge of copper and coppersmith and labour hours fixing per square metre | | | |
| | I.S.W.G. 0–20 | | I.S.W.G. 20–24 | |
	Coppersmith hours	Labourer hours	Coppersmith hours	Labourer hours
Dormer cheeks	4.30	4.30	6.50	6.50
Dormer flats	5.40	5.40	7.90	7.90
Flat roofs	3.80	3.80	5.40	5.40
Sloping roofs	4.32	4.32	6.30	6.30
Gutters and cesspools	6.30	6.30	8.65	8.65
Soakers	1.30	1.30	2.60	2.60
Stepped flashings	3.90	3.90	5.40	5.40
Valleys and hip and ridge coverings	2.70	2.70	3.45	3.45
Weathering to cornices and copings	3.24	3.24	4.32	4.32

Table 17.27 Sundry works to copper roofing

Description	Unit	Fixer hours	Labourer hours
Dressing over glass	m	3.00	3.00
Dressing into hollows	m	1.32	1.32
Dressing over mouldings, per 25 mm girth	m	0.36	0.36
Labour to secret gutters	m	2.16	2.16
Rolls – capped ends to	Each	1.40	1.40
Rolls – saddles to	Each	0.80	0.80
Soldered joint (0.03 kg solder)	Each	1.00	1.00
Copper nailing	m	0.50	0.50
Copper welts	m	2.00	2.00

18 Surface finishes

Granolithic paving

Granolithic paving is laid where a hard-wearing surface is required and is composed of angular granite chips and Portland cement, a proportion of clean granite dust often being added.

It may be mixed by hand or mechanically in concrete mixing plant, and the material is usually superimposed on a concrete foundation laid ready to receive it, or on prepared existing concrete such as floors etc.

In the tables below are estimating data from which to estimate the material cost of the Grano, and the labour or plant and labour cost of mixing it. The table shown for laying the Grano refers to the labour cost involved in laying only.

Table 18.1 Granolithic paving

Note: Allow for mixing by hand 6.5 labour hours per cubic metre.

Type of mixing plant	Plant and labour hours mixing only Grano per cubic metre	
	Mixing plant hours	Labour hours
5/3½	1.33	2.66
7/5	0.80	2.40
10.7	0.53	2.12

For wheeling up to a 22 m run, in barrows, allow 1.2 labour hours per cubic metre

Table 18.2 Material required per cubic metre of grano

Notes: 1. Cement weighing 1,442 kg per m³.
2. Granite chips weighing 1.47 tonnes per m³.

Mix			Material required per cubic metre of Grano		
Granite chips	Granite dust	Cement	Granite chips cubic metre	Granite dust cubic metre	Cement tonnes
2	–	1	0.77	–	1.11
3	–	1	1.10	–	0.53
1½	1	1	0.70	0.46	0.67
2	1	1	0.77	0.40	0.56

Table 18.3 Granolithic paving

| Laying only granolithic paving, single coat work. Finished thickness in millimetres | Cubic metre of material required per square metre of paving | Pavior and labourer hours laying only granolithic paving per square metre | | | |
| | | Floated or screeded | | Steel trowelled | |
		Pavior hours	Labourer hours	Pavior hours	Labourer hours
12	0.018	0.25	0.25	0.29	0.29
18	0.027	0.26	0.26	0.30	0.30
25	0.035	0.27	0.27	0.31	0.31
38	0.053	0.29	0.29	0.33	0.33
50	0.070	0.31	0.31	0.35	0.35

Table 18.4 Sundry works to granolithic paving

Description	Unit	Pavior hours	Labourer hours
Rounded angles, internal or external up to 50 mm girth.	m	0.50	0.50
Rounded angles internal or external 100 to 150 mm girth.	m	0.83	0.83
Bullnosing edges.	m	0.20	–
Applying carborundum powder and similar nonslip preparations.	m^2	0.06	0.06

Internal plastering

Table 18.5 Internal plastering to ceilings

Notes: 1. The hours shown do not allow for mixing the mortar.
2. For carrying out work at height see Table 18.8
3. For sundry work to internal plastering see Table 18.7
4. For rendering with 1 of Sirapite to 2 of sand 13 mm thick, allow 5.2 kg of Sirapite and 0.01 m³ of sand per square metre.
5. For setting coat of Sirapite 3 mm thick allow 2.6 kg of Sirapite per square metre.
6. For metal lathing allow 10 per cent for laps and waste.

Description	Unit	Thickness of coat in millimetres			Plasterer hours	Labourer hours
		First coat	Setting coat	Total thickness		
Metal lathing weighing 3–4 kg per m²	m²	–	–	–	0.27	0.27
Carlite undercoat, second coat and set with fine 3 mm coat	m²	5	5	3	0.72	0.72
Lath, plaster and set with fine two coat work	m²	12	3	15	0.84	0.84
On coat board plaster finish steel trowelled on and including 9.5 mm gypsum plasterboard, fixing with nails and fill joints with plaster and scrim cloth	m²	–	–	3 (finish)	0.65	0.65

Table 18.6 Internal plastering to walls

Notes: 1. For rendering to walls with 1 of Sirapite to 2 of sand 13 mm thick allow 5.2 kg of Sirapite and 0.01 m³ of sand per square metre.
2. For rendering to walls with 1 of cement to 1 of lime to 6 of sand.

Description	Unit	Thickness of coat in millimetres			Total thickness	Plasterer hours	Labourer hours
		First Coat	Floating Coat	Setting Coat			
Render with Sirapite and set with fine 3 mm Sirapite coat	m²	12	–	3	15	0.60	0.60
Render with cement lime mortar and set with fine 3 mm Sirapite coat	m²	10	9	3	22	1.08	1.08

Table 18.7 Sundry work to internal plastering

Description	Unit	Plasterer hours	Labourer hours
Angles rounded up to 50 mm girth.	m	0.40	0.40
Angles rounded 100 to 150 mm girth.	m	0.53	0.53
Arris.	m	0.33	0.33
Bead – double, under 75 mm girth.	m	0.33	0.33
Cornices 75 mm girth.	m	0.50	0.50
Cornices 225 mm girth	m	0.83	0.83
Mouldings and skirtings to walls 75 mm girth.	m	0.60	0.60
Mouldings and skirtings to walls 300 mm girth.	m	1.00	1.00
Make good plaster to sash frames.	m	0.33	0.33

Table 18.8 Multiples for carrying out works in plaster

Description	Plasterer and labourer hour multipliers
For carrying out work to ceiling or walls 0 to 3 m high.	1.00
For carrying out work to ceiling or walls 4.5 to 6 m high.	1.20
For carrying out work to ceiling or walls 7.5 to 9 m high.	1.40
Circular work.	2.00
Groined work.	3.50
Plastering in quantities of less than 1 square metre.	2.00
Plastering to sides and soffit of beams.	2.00

Table 18.9 Tiling to walls

Note: Allow 0.007 m^3 mortar per square metre of tiling.

Tiling	Unit	No. of tiles required per unit	Plasterer hours	Labourer hours
150 × 150 mm tiles fixed.	m^2	42	1.68	1.68
100 × 100 mm tiles fixed.	m^2	97	3.84	3.84
Angles 150 mm in length.	m	66	0.83	0.83
Skirtings 150 mm in height.	m	66	0.66	0.66
Circular cutting to glazed tiles.	m	–	2.00	2.00
Raking cutting to glazed tiles.	m	–	1.32	1.32

Table 18.10 Cement mortar – external and internal work to walls

Notes: 1. The hours shown do not allow for mixing the mortar.
2. In rough casting allow 8 kg of pea shingle per square metre.
3. In pebble dashing allow 10 kg of pebbles per square metre.

Description	Unit	Thickness of coat in mm	Plasterer hours	Labourer hours
Render on walls first coat and score for second coat.	m²	13	0.24	0.24
Ditto but 19 mm thick.	m²	19	0.36	0.36
Second coat, screeded and floated for finishing coat.	m²	6	0.36	0.36
Finishing coat trowelled or stucco.	m²	6	0.42	0.42
Render and wet rough cast.	m²	19	0.96	0.96
Render and pebble dash.	m²	19	1.08	1.08

Table 18.11 Cement mortar – external and internal work to floors

Note: The hours shown do not allow for mixing the mortar.

Description	Unit	Thickness of coat in mm	Plasterer hours	Labourer hours
Screed in one coat, screed and trowel.	m²	20	0.37	0.37
Ditto but 25 mm thick	m²	25	0.41	0.41

Table 18.12 Sundry works to cement mortar

Description	Unit	Plasterer hours	Labourer hours
Angles rounded 10 to 100mm radius.	m	0.80	0.80
Arris.	m	0.40	0.40
Cement wash floors.	m²	0.12	0.12
Cornices 150 mm girth.	m	0.55	0.55
Cornices 300 mm girth.	m	0.83	0.83
Hack face of brick walls for key.	m²	0.36	0.36
Hack existing cement rendering for key.	m²	0.48	0.48
Mouldings and skirtings to walls 75mm girth.	m	1.19	1.19
Mouldings and skirtings to walls 300mm girth.	m	2.00	2.00
Rake out joints of brick walls for key before mortar is set.	m²	0.36	0.36
Rake out joint of brick walls for key. Perished mortar.	m²	0.72	0.72
Throated drip.	m	0.20	0.20

Table 18.13 Wood block floors

Note: Allow 1.9 kg of mastic per square metre of flooring.

Description	Unit	Wood block layer hours	Labourer hours
225 × 75 mm blocks 32 mm thick laid herringbone in mastic.	m²	0.95	0.95
Lay border, single course.	m	0.40	0.40
Circular cutting.	m	0.83	0.83
Raking cutting.	m	0.66	0.66
Traversing for polishing.	m²	0.72	0.72

Table 18.14 Sundry works to floors

Description	Unit	Labour hours
Lay linoleum	m²	0.40
Cutting linoleum – straight	m	inc.
Cutting linoleum – circular	m	inc.

Painting and decorating

Table 18.15 Paintwork to brick, concrete, plaster and woodwork

Note: (1) The data shown are per coat.
    (2) For work to ceilings use a multiplier of 1:33 with the painter hours shown.

Description (Paintwork carried out by hand, per coat)	Unit	Paint required in litres	Painter hours
Priming coat on planed timber	m²	0.10	0.24
First coat of paint on priming	m²	0.15	0.29
Second coat	m²	0.14	0.26
Oil gloss finishing coat	m²	0.11	0.23
First coat of paint on previously coated timber	m²	0.10	0.23
Second coat	m²	0.14	0.27
Oil gloss finishing coat	m²	0.13	0.25
Narrow mouldings, skirtings, etc. 75 mm girth	100 m	0.37	3.30
Narrow mouldings, skirtings, etc. 150 mm girth	100 m	0.71	4.90
Narrow mouldings, skirtings, etc. 225 mm girth	100 m	3.30	6.60
Priming coat on concrete, stone or compo	m²	0.28	0.49
First coat of paint after priming	m²	0.21	0.38
Oil gloss paint as finishing coat	m²	0.14	0.34
Rafter ends per 300 mm length or less	Each	0.02	0.03
Sash frames	100 m	3.50	7.20
Sash squares	m²	0.11	0.36
Extra large sash squares	m²	0.06	0.18
Sash edges	100 m	1.50	7.70
Shelf edges	100 m	0.50	3.30

19 Glazing

Glass used for general building purposes are of six classes:

1. Transparent
2. Translucent
3. Rough cast
4. Ordinary Georgian wired
5. Polycarbonate standard sheet
6. Float glass

Glazing is the term used for the fixing and securing of various kinds of glass in prepared work, such as window sashes, door panels, partitions, etc.

In this chapter, tables are shown giving estimating data for fixing the glass, including the glazier hours taken to fix glass of various weights.

The glazier hours shown in the estimating tables are for securing glass of cut sizes, and it should be noted that if glass is cut from random sheets the glazier hours multipliers shown should be used in conjunction with the glazier hours appearing in the tables.

In the case of sheet glass fixed to steel casements or wooden sashes the glazier hours allow for front and back puttying, and in the case of wooden sashes, for sprigging. Plate glass is taken as being glazed with beads or with beads and wash-leather strips.

A separate table is shown for cutting glass of various types and thicknesses for both straight and circular cutting.

In estimating the cost of glass of irregular shape, the area should be taken as the smallest rectangle from which the irregular shape can be cut.

Table 19.1 Glazier – sheet glass

Notes: 1. Allow for waste 5 per cent on cut sheets and 10 per cent on uncut sheets.
2. The labour hours shown allow for front and back puttying and are applicable to such glass as ordinary sheet glass, cathedral, fluted, muranese, etc., and are for glazing with cut sheets.

Description Glazing with cut sheets (If glazing with glass cut from random sheets use a labour multiplier of 1.20 with the glazier hours shown)	Putty required per m² in kg	Weight of glass in kg per m² and glazier hours per m²		
		0.9 kg per m²	9–13 kg per m²	13–18 kg per m²
Glazing to steel casements				
Panes up to 0.15 m² in area	1.32	2.15	2.58	3.08
Panes 0.15–0.2 m² in area	1.04	1.72	1.94	2.48
Panes 0.2–0.3 m² in area	0.63	1.29	1.61	1.94
Panes 0.3–0.4 m² in area	0.59	1.07	1.29	1.61
Panes 0.4–0.5 m² in area	0.54	0.97	1.18	1.29
Panes 0.5–0.6 m² in area	0.49	0.86	1.07	1.18
Panes over 0.6 m² in area	0.44	0.75	0.86	1.07
Glazing to wood sashes				
Panes up to 0.15 m² in area	1.07	1.50	1.72	2.15
Panes 0.15–0.2 m² in area	0.68	1.29	1.50	1.83
Panes 0.2–0.3 m² in area	0.54	1.07	1.29	1.40
Panes 0.3–0.4 m² in area	0.49	0.86	1.07	1.18
Panes 0.4–0.5 m² in area	0.44	0.75	0.86	1.07
Panes 0.5–0.6 m² in area	0.39	0.65	0.75	0.86
Panes over 0.6 m² in area	0.34	0.54	0.65	0.75

Table 19.2 Glazier – plate glass

Notes: 1. Allow for waste 7½ per cent.
2. The labour hours shown are for plain, bevelled or embossed plate glass, and are for glazing with cut sheets with beads and wash-leather strips.

Description Glazing with cut squares (If glazing with plate, including cutting, use a labour multiplier of 1.40 with the glazier hours shown)	Beads and/or wash-leather strips reqd. m per m²	Weight of glass in kg per m² and glazier hours per m²			
		0–9 kg per m²	*9–18 kg per m²*	*18–27 kg per m²*	*27–36 kg per m²*
Plates up to 0.1 m²	14.75	3.01	3.33	3.87	4.94
Plates up to 0.2 m²	10.45	2.90	3.23	3.64	4.73
Plates up to 0.3 m²	8.70	2.80	3.12	3.43	4.52
Plates up to 0.4 m²	7.40	2.69	3.01	3.20	4.30
Plates up to 0.5 m²	6.70	2.58	2.90	2.97	4.18
Plates up to 0.6 m²	6.00	2.69	3.01	3.20	4.30
Plates up to 0.7 m²	5.70	2.80	3.12	3.43	4.52
Plates up to 0.8 m²	5.35	2.90	3.23	3.64	4.73
Plates up to 0.9 m²	5.00	3.01	3.33	3.87	4.94
Plates up to 1.0 m²	4.70	3.11	3.44	4.08	5.16
Plates up to 1.5 m²	4.00	2.97	3.66	4.40	5.38
Plates up to 2.5 m²	3.00	3.43	4.18	4.61	5.65
Plates up to 5.0 m²	2.00	4.08	4.52	5.21	6.36
Plates up to 7.5 m²	1.70	4.49	5.00	6.02	7.74
Plates up to 10 m²	1.35	5.21	5.75	6.80	8.36
Plates over 10 m²	1.00	6.28	6.40	8.16	10.46

Table 19.3 Multipliers for use with the tables for glazing

Note: In pricing the fixing of bent glass the whole area of the sheet is taken as being bent, even though part of it may be flat.

Description	Glazier hour multipliers
Cutting bent glass	2.00
Fixing bent glass	1.40
Lead glazing of irregular pattern	2.00

Table 19.4 Cutting glass – per metre

Description			Glazier hours per metre	
Type of glass	Thickness in millimetres	Weight in kg per m²	Straight cutting	Circular cutting
Plate glass	3	8.55	0.13	0.26
Plate glass	5	11.92	0.16	0.32
Plate glass	6	16.21	0.20	0.40
Plate glass	9	23.93	0.23	0.46
Plate glass	12	31.75	0.26	0.52
Plain sheet glass	2.0 to 2.3	5.52	0.10	0.20
Plain sheet glass	2.8 to 3.0	7.33	0.13	0.26
Plain sheet glass	3.0 to 3.6	8.10	0.13	0.26
Plain sheet glass	3.8 to 4.3	9.77	0.16	0.32
Wired glass	6	15.77	0.30	0.60

Table 19.5 Glazier sundry works

Description	Unit	Glazier hours
Fix lead-covered steel glazing bars	m	0.26
Fix lead glazing bars cut to length to upper surface of wooden bars	m	0.32
Hack out all kinds of broken glass excepting polish plate glass	m²	2.34
Hack out polished plate glass	m²	3.55
Remove serviceable glass such as sheet, cathedral, cast or rolled plate	m²	3.23
Remove serviceable polished plate glass	m²	4.68
Remove loose putty, paint rebate one coat ready for reglazing	m	0.16

20 *Plumbing*

Copper piping

Copper piping to BS 2871 is widely used for plumbing and services installation, having the advantage of either soldered joints or compression joints to BS 864 that require no soldering or brazing. Copper piping can readily be bent by hand to the required curvature without the necessity of filling or heating them.

In pricing the fixing of brass or copper piping the cost of fixing unions, connectors, etc. should be allowed for separately, allowing one straight coupling for, say, every 3 m.

Table 20.1 Copper piping to BS 2871 Part 1 Table Y and fixing copper fittings in trench

Note: Allow 2½ per cent for waste in long lengths and 5 per cent for short lengths.

Internal diameter of pipe in millimetres	Type of work and plumber and mate hours fixing pipes and fittings			
	Laying in trench		Fixing fittings, each	
	Plumber hours	Mate hours	Plumber hours	Mate hours
15	0.14	0.14	0.12	0.12
22	0.16	0.16	0.14	0.14
28	0.18	0.18	0.16	0.16
35	0.22	0.22	0.18	0.18
42	0.28	0.28	0.20	0.20
54	0.44	0.44	0.24	0.24

Table 20.2 Copper piping to BS 2871 Part 1 Table X and compression fittings to BS 864

Note: Allow 2½ per cent for waste in long lengths and 5 per cent for short lengths.

Internal diameter of pipe in millimetres	Type of work and plumber and mate hours fixing pipes and fittings					
	Fixing to walls per metre		Fixing compression fittings, each		Fixing capillary fittings, each	
	Plumber hours	Mate hours	Plumber hours	Mate hours	Plumber hours	Mate hours
15	0.24	0.24	0.06	0.06	0.24	0.24
22	0.25	0.25	0.07	0.07	0.28	0.28
28	0.26	0.26	0.08	0.08	0.32	0.32
35	0.28	0.28	0.09	0.09	0.36	0.36
42	0.29	0.29	0.10	0.10	0.40	0.40
54	0.31	0.31	0.12	0.12	0.48	0.48

Table 20.3 Iron and steel piping

Notes: 1. Allow 2½ per cent for waste in laying in normal lengths and 5 per cent in short lengths.
2. The data shown may be used for wrought-iron and steel tubes of gas, water and steam quality.
3. The data shown do not allow for cutting or threading, for which see Table 20.4

| Internal diameter of pipe in millimetres | Class of work and plumber and mate hours fixing pipes and fittings | | | | | |
| | Fixing to walls per metre | | Laying in trenches per metre | | Fix fittings, bends, elbows and tees, each | |
	Plumber hours	Mate hours	Plumber hours	Mate hours	Plumber hours	Mate hours
15	0.26	0.26	0.13	0.13	0.10	0.10
20	0.29	0.29	0.14	0.14	0.12	0.12
25	0.31	0.31	0.16	0.16	0.14	0.14
33	0.33	0.33	0.17	0.17	0.16	0.16
40	0.35	0.35	0.18	0.18	0.18	0.18
50	0.40	0.40	0.20	0.20	0.22	0.22

Table 20.4 Cutting and threading iron and steel piping

| Internal diameter of pipe in millimetres | Plumber and mate hours cutting pipes and threading ends, each | | | |
| | Cutting pipes, each | | Threading ends, each | |
	Plumber hours	Mate hours	Plumber hours	Mate hours
15	0.10	0.10	0.20	0.20
20	0.12	0.12	0.25	0.25
25	0.14	0.14	0.30	0.30
32	0.16	0.16	0.35	0.35
40	0.18	0.18	0.40	0.40
50	0.22	0.22	0.50	0.50

Table 20.5 Drilling and tapping cast-iron or steel mains

| Diameter of hole in millimetres | Depth of hole and Plumber hours, each | | | | | |
| | 15 mm | | 20 mm | | 25 mm | |
	Drilled	Drilled and tapped	Drilled	Drilled and tapped	Drilled	Drilled and tapped
15	0.18	0.36	0.31	0.62	0.45	0.90
20	0.24	0.48	0.42	0.84	0.60	1.20
25	0.30	0.60	0.52	1.04	0.75	1.50

Polyethylene piping

Polyethylene pipes are being increasingly used for cold-water services. These pipes have better thermal insulation than metal pipes, thus reducing the possibility of freezing up. Pipes do not corrode, in so far as they are not affected by sulphuric acid, but they are liable to attack from nitric acid, benzene, petrol, and heat. Bends are easily formed by immersing the pipe in boiling water for about five minutes; the minimum radius for such a bend is three times the outside diameter. Owing to the relatively high rate of expansion, clips are fixed closer together than for metal pipes: the interval spacing for 25 mm and larger pipes should be 12 times the outside diameter, for smaller size pipes 8 times the outside diameter, and on vertical runs about 24 times the outside diameter in various sizes. Joints in pipes (elbows, tees, etc.) or to fittings (taps, stop valves, etc.) can be either solvent-weld or compression fitting joints. The compression joint method is similar to that used in copper tubing. The size of a compression joint fitting must always be one size larger than the pipe used.

Table 20.6 Blue polyethylene piping

Note: The data shown are based on BS 6572

	Mean outside diameter, mm		Wall thickness, mm	
Nominal bore, mm	*Min.*	*Max.*	*Min.*	*Max.*
Nominal gauge				
20	20	20.3	2.3	2.6
25	25	25.3	2.3	2.6
32	32	32.3	3.0	3.4
50	50	50.4	4.6	5.2
63	63	63.4	5.8	6.5

Table 20.7 Black polyethylene

Note: The data shown are based on BS 6730

	Mean outside diameter, mm		Wall thickness, mm	
Nominal bore, mm	*Min.*	*Max.*	*Min.*	*Max.*
Nominal gauge				
20	20	20.3	2.3	2.6
25	25	25.3	2.3	2.6
32	32	32.3	3.0	3.4
50	50	50.4	4.6	5.2
63	63	63.4	5.8	6.5

Table 20.8 Fixing polyethylene piping and fittings

Internal diameter of pipe, mm	*Gauge of pipe*	Fixing to walls per metre BS 6730		Laying in trenches per metre BS 6572		Fixing elbows BS 6730		Fixing tees BS 6730	
		Plumber hours	*Mate hours*	*Plumber hours*	*Mate hours*	*Plumber hours*	*Mate hours*	*Plumber hours*	*Mate hours*
15	Normal	–	–	0.06	0.06	–	–	–	–
20	Normal	0.25	0.25	0.09	0.09	0.20	0.10	0.30	0.10
25	Normal	0.28	0.28	0.12	0.12	0.25	0.13	0.37	0.15
32	Normal	0.32	0.32	0.13	0.13	0.30	0.15	0.40	0.20
50	Normal	0.35	0.35	0.15	0.15	0.35	0.17	0.47	0.24
63	Normal	0.40	0.40	0.16	0.16	0.40	0.20	0.52	0.26

Table 20.9 Plumber sundry works

Note: The hours shown allow for making all joints to the various fittings. For fixing iron and copper piping see
Tables 20.1, 20.2 and 20.3

Description	Unit	Solder kg	Red lead g	Red lead putty kg	Plumber hours	Mate hours
13 mm bib cock screwed for iron pipe	Each	–	2.8	–	0.35	0.35
19 mm bib cock screwed for iron pipe	Each	–	4.3	–	0.40	0.40
25 mm bib cock screwed for iron pipe	Each	–	5.8	–	0.45	0.45
31 mm bib cock screwed for iron pipe	Each	–	7.1	–	0.50	0.50
38 mm bib cock screwed for iron pipe	Each	–	8.5	–	0.55	0.55
13 mm screwdown stopcocks screwed for iron pipes	Each	–	5.6	–	0.45	0.45
19 mm screwdown stopcocks screwed for iron pipes	Each	–	8.5	–	0.50	0.50
25 mm screwdown stopcocks screwed for iron pipes	Each	–	11.3	–	0.55	0.55
31 mm screwdown stopcocks screwed for iron pipes	Each	–	14.2	–	0.60	0.60
38 mm screwdown stopcocks screwed for iron pipes	Each	–	17.0	–	0.65	0.65
50 mm screwdown stopcocks screwed for iron pipes	Each	–	19.8	–	0.70	0.70
13 mm ferrules, one end screwed for iron and the other for a lead joint	Each	0.34	2.8	–	0.80	0.80
19 mm ferrules, one end screwed for iron and the other for a lead joint	Each	0.45	4.3	–	1.00	1.00
25 mm ferrules, one end screwed for iron and the other for a lead joint	Each	0.56	5.7	–	1.20	1.20
Fix bath complete with 2 19 mm pillar taps, 38 mm trap, overflow and joints	Each	2.38	–	0.45	5.50	5.50
Fix bath, shower type, complete with 2 19 mm pillar taps, 38 mm trap, overflow and joints	Each	2.38	–	0.45	7.00	7.00
Fix boiler	Each	–	14.2	–	4.00	4.00
Fix cold water cistern, including jointing-up pipes and overflow	Each	–	31.0	–	4.50	4.50
Fix cylinder, including jointing-up pipes	Each	–	28.4	–	5.00	5.00
Fix cylinder lagging	Each	–	–	–	1.00	1.00
Fix draining-boards	Each	–	–	–	1.50	1.50
Fix kitchen sinks, 750 × 500 × 150 mm or 600 × 450 × 250 mm, complete with brackets, 38 mm trap, 2 taps and joints	Each	1.59	5.6	0.45	5.00	5.00
Fix heavy factory-type sinks, complete with brackets, 38 mm trap, 2 taps and joints	Each	1.91	5.6	0.54	7.50	7.50
Fix one-piece rustless metal sinks with combined drain-board and complete with brackets, 38 mm trap, 2 taps and joints	Each	2.50	5.6	0.45	4.50	4.50
Fix mirrors	Each	–	–	–	0.60	0.60
Fix soap and sponge fittings	Each	–	–	–	0.50	0.50
Fix toilet-paper fittings	Each	–	–	–	0.50	0.50
Fix towel racks, bracket type	Each	–	–	–	0.60	0.60
Fix wash hand basin on brackets or pedestal, complete with 2 12 mm pillar taps, plug and waste trap and joints	Each	2.04	5.6	0.36	5.50	5.50
Fix W.C. pan, seat, cistern and flush pipe, complete	Each	–	–	0.45	2.70	2.70
Fix W.C. combined pan and flushing cistern type, complete	Each	–	–	0.36	1.70	1.70

Table 20.10 Composition gas piping

Notes: 1. Add 5 per cent for waste on piping.
2. For joints to floor cocks, fittings, etc. use a labour multiplier of 2.00 with the hours shown.
3. Allow for joints two clips or pipe-hooks per metre.
4. For fixing gas cookers or fires allow plumber 1.50 hours and mate 1.50 hours, plus joints.

Internal diameter of pipe in mm	Fix pipes to walls per metre			Make joints, each		
	Plumber hours	Mate hours	Petrol litres	Solder grammes	Plumber hours	Mate hours
9	0.36	0.36	0.054	4.3	0.13	0.13
13	0.40	0.40	0.068	4.8	0.15	0.15
15	0.43	0.43	0.090	5.1	0.17	0.17
19	0.46	0.46	0.109	5.9	0.19	0.19
25	0.50	0.50	0.136	6.5	0.21	0.21

Table 20.11 External gutters, rainwater pipes, etc. in plastic

Notes: 1. The data shown for gutters in standard 4mm lengths and pipes in standard 2.5 m, 3 m and 4 m lengths.
2. Half round gutters are 102 mm diameter, square section gutters are 117 mm wide × 51 mm deep, allow one gutter joint bracket and two screws for each length of gutter and one support bracket and two screws for each metre length of pipe.
3. Swan-necks (offsets) are formed with two bends.
4. Pipes to be fixed with adjustable pipe and fitting clip.

Description	Unit	Plumber hours	Mate hours
Eaves gutters, half round	m	0.25	0.25
Eaves gutters, square	m	0.30	0.30
Gutter outlets and angles half round	Each	0.16	0.16
Gutter outlets and angles square	Each	0.18	0.18
Gutter stop ends, half round	Each	0.10	0.10
Gutter stop ends, square	Each	0.12	0.12
Rainwater pipes 63 mm	m	0.25	0.25
Heads	Each	0.17	0.17
Swan-necks	Each	0.25	0.25
Single branch	Each	0.25	0.25
Shoes	Each	0.15	0.15

Table 20.12 External gutters, rainwater pipes, etc. in cast iron

Notes: 1. Allow 5 per cent for waste.
2. The data shown for gutters and pipes are based on standard 1.8 m lengths.
3. The data shown for pipes are applicable to rainwater, soil, waste, vent and flue pipes.
4. For half-round gutters allow half a gutter bolt per metre.
5. For ogee gutters allow half a gutter bolt and one 30 mm screw per metre.
6. For stop ends allow one gutter bolt each.
7. For gutter brackets allow three 25 mm screws each.
8. For gutters etc., in zinc work use a multiplier of 0.80 with the data shown.
9. The data shown for pipework and fittings are for light or medium sections, i.e. for metal 3 and 4.5 mm thick. For heavy sections, use a multiplier of 1.40 with the data shown.

Description	Unit	Mastic jointing required per unit in kg	Plumber hours	Mate hours
Eaves gutters, half round, 100 mm	m	0.10	0.33	0.33
Eaves gutters, half round, 125 mm	m	0.15	0.44	0.44
Eaves gutters, ogee, 100 mm	m	0.15	0.44	0.44
Eaves gutters, ogee, 125 mm	m	0.20	0.55	0.55
Gutter outlets and angles, 100 mm	Each	0.20	0.20	0.20
Gutter outlets and angles, 125 mm	Each	0.17	0.25	0.25
Gutter stop ends, 100 mm	Each	0.08	0.12	0.12
Gutter stop ends, 125 mm	Each	0.10	0.15	0.15
Gutter brackets – fixing	Each	–	0.12	0.12
Pipes, rainwater, etc., 75 mm	m	0.07	0.33	0.33
Pipes, rainwater, etc., 100 mm	m	0.10	0.43	0.43
Pipes, rainwater, etc., 125 mm	m	0.12	0.55	0.55
Pipes, rainwater, etc., 150 mm	m	0.15	0.72	0.72
Rainwater heads up to 300 mm	Each	0.10	0.60	0.60
Rainwater heads over 300 mm and up to 450 mm	Each	0.10	0.80	0.80
Stays or holderbats, 75 mm	Each	–	0.20	0.20
Stays or holderbats, 100 mm	Each	–	0.25	0.25
Stays or holderbats, 125 mm	Each	–	0.30	0.30
Stays or holderbats, 150 mm	Each	–	0.35	0.35
Bends and swan-necks, 75 mm	Each	0.15	0.50	0.50
Bends and swan-necks, 100 mm	Each	0.17	0.60	0.60
Bends and swan-necks, 125 mm	Each	0.19	0.70	0.70
Bends and swan-necks, 150 mm	Each	0.22	0.80	0.80
Single junctions, 75 mm	Each	0.22	0.60	0.60
Single junctions, 100 mm	Each	0.25	0.70	0.70
Single junctions, 125 mm	Each	0.30	0.80	0.80
Single junctions, 150 mm	Each	0.33	0.90	0.90
Double junctions, 75 mm	Each	0.30	0.90	0.90
Double junctions, 100 mm	Each	0.33	1.00	1.00
Double junctions, 125 mm	Each	0.40	1.10	1.10
Double junctions, 150 mm	Each	0.44	1.20	1.20
Shoes and reducers, 75 mm	Each	0.10	0.40	0.40
Shoes and reducers, 100 mm	Each	0.12	0.45	0.45
Shoes and reducers, 125 mm	Each	0.14	0.50	0.50
Shoes and reducers, 150 mm	Each	0.17	0.55	0.55
Fix wire cowls, 75 mm	Each	–	0.25	0.25
Fix wire cowls, 100 mm	Each	–	0.30	0.30
Fix wire cowls, 125 mm	Each	–	0.35	0.35
Fix wire cowls, 150 mm	Each	–	0.40	0.40

Table 20.13 External gutters, rainwater pipes, etc. in fibre cement

Notes: 1. Allow 5 per cent for waste.
2. The data shown for gutters and pipes are based on standard 1.8 m lengths.
3. The data shown for pipes are applicable to rainwater, soil, waste, vent and flue pipes.
4. The data shown includes for jointing pipes and fittings with mastic. If no jointing is carried out to these items delete the mastic shown and use a labour multiplier of 0.80 with the hours given.
5. For half-round gutters allow half a gutter bolt and one 33 mm screw per metre.
6. For gutter brackets allow three 25 mm screws each.
7. For stop ends allow one gutter bolt each.

Description	Unit	Mastic jointing required per unit in kg	Plumber hours	Mate hours
Eaves gutters, half round, 100 mm	m	0.10	0.27	0.27
Eaves gutters, half round, 125 mm	m	0.15	0.35	0.35
Eaves gutters, half round, 150 mm	m	0.20	0.44	0.44
Eaves gutters, O.G., 100 mm	m	0.15	0.35	0.35
Eaves gutters, O.G., 125 mm	m	0.22	0.44	0.44
Eaves gutters, O.G., 150 mm	m	0.30	0.53	0.53
Gutter outlets and angles, 100 mm	Each	0.34	0.16	0.16
Gutter outlets and angles, 125 mm	Each	0.40	0.20	0.20
Gutter outlets and angles, 150 mm	Each	0.45	0.24	0.24
Gutter stop ends, 100 mm	Each	0.34	0.10	0.10
Gutter stop ends, 125 mm	Each	0.40	0.12	0.12
Gutter stop ends, 150 mm	Each	0.45	0.14	0.14
Gutter brackets	Each	–	0.12	0.12
Gutters, valley up to 600 mm girth	m	0.55	0.33	0.33
Gutters, over 600mm and up to 750 mm girth	m	0.70	0.44	0.44
Gutters, wall up to 600 mm girth	m	0.55	0.66	0.66
Gutters, over 600mm and up to 750 mm girth	m	0.70	0.88	0.88
Stop ends and outlets to ditto	Each	1.40	0.25	0.25
Pipes, rainwater, etc., 75 mm	m	0.08	0.30	0.30
Pipes, rainwater, etc., 100 mm	m	0.10	0.40	0.40
Pipes, rainwater, etc., 125 mm	m	0.12	0.50	0.50
Pipes, rainwater, etc., 150 mm	m	0.15	0.65	0.65
Rainwater heads up to 300 mm	Each	0.11	0.50	0.50
Rainwater heads over 300 mm and up to 450 mm	Each	0.11	0.60	0.60
Bends and swan necks, 75 mm	Each	0.15	0.45	0.45
Bends and swan necks, 100 mm	Each	0.18	0.54	0.54
Bends and swan necks, 125 mm	Each	0.20	0.63	0.63
Bends and swan necks, 150 mm	Each	0.22	0.72	0.72
Single junctions, 75 mm	Each	0.22	0.54	0.54
Single junctions, 100 mm	Each	0.25	0.63	0.63
Single junctions, 125 mm	Each	0.30	0.72	0.72
Single junctions, 150 mm	Each	0.35	0.81	0.81
Double junctions, 75 mm	Each	0.30	0.81	0.81
Double junctions, 100 mm	Each	0.35	0.90	0.90
Double junctions, 125 mm	Each	0.38	1.00	1.00
Double junctions, 150 mm	Each	0.42	1.10	1.10
Shoes and reducers, 75 mm	Each	1.10	0.36	0.36
Shoes and reducers, 100 mm	Each	0.13	0.40	0.40
Shoes and reducers, 125 mm	Each	0.15	0.45	0.45
Shoes and reducers, 150 mm	Each	0.18	0.50	0.50

21 Quarry work, roads, paths and platelaying

Quarrying of stone

Stone is used for:

1. Pitching or hard bottoming for road foundations etc.
2. Aggregates for concrete, road metal or tarmacadam
3. Walling stone.

Example: Calculate the weight of 1 m^3 of broken stone graded from 38 to 12 mm, assuming that the weight of the solid stone is 2.563 tonnes per m^3.

By referring to Table 21.1 shown below it will be seen that the percentage voids for graded broken stone is 40.

Since the percentage voids in 1 m^3 is 40% the percentage of solid stone is $100\% - 40\% = 60\%$ of 2.563 tonnes

Weight of 1 m^3 of broken stone = 1.538 tonnes

Table 21.1 Percentage voids in broken stone of various gradings

Description	Percentage voids
Uniformly broken stone, all sizes 62 mm down to 12 mm	45%
Graded stone 62 mm down to 12 mm all gradings.	40%
100 to 150 mm pitching or hardcore.	37½%
150 to 225 mm pitching or hardcore.	35%
Spalls for filling interstices of pitching	40%
12 mm to dust.	30%

Table 21.2 Quarrying sandstone for pitching, bottoming etc. by hand labour

Description	Gelignite required per tonne of rock in kilogrammes	Detonators required per tonne of rock, No.	Fuse required per tonne of rock, metre	Labour hours per tonne
Bar out rock easily removed and break to 150 to 225 mm sizes	–	–	–	1.20
Remove localised hard pockets by blasting and break to 100 to 150 mm sizes	0.04	0.03	0.15	1.70

Table 21.3 Quarrying solid rock such as granite, limestone and sandstone

Description	Type of rock quarried	Gelignite required per tonne of rock in kg	Detonators per tonne of rock, No.	Fuse required per tonne of rock metre	Plant and labour hours per tonne				
					Compressor hours	Labour on drills hours	Shot firer hours	Blacksmith hours	Labourer hours
Quarrying the rock	Sandstone	0.08	0.03	0.007	0.18	0.36	0.18	0.18	0.72
and breaking to sizes	Limestone	0.09	0.04	0.007	0.20	0.40	0.20	0.20	0.80
suitable for pitching,	Granite	0.13	0.06	0.007	0.27	0.54	0.27	0.27	1.08
rushing, etc.									

Broken stone for concrete aggregates, road metal and tarmacadam

The production of broken stone is carried out by stone-crushing plant of the stonebreaker and granulator type.

Stonebreakers are designed to crush material down to approximately 50 mm. By fitting special jaws it is possible to produce even smaller material, but this practice is to be deprecated, as it is not economical, particularly when using hard or abrasive stone. If small material is required the usual practice is to use a machine of the granulator type which is designed for this purpose.

The percentage of each grade of product obtained when crushing

The percentage of each grade of broken stone produced by crushing depends on:

● The nature of the stone crushed.
● The type of plant used.

Representative results obtained in crushing different types of stone using stone-breakers and granulators are given in Table 21.4.

From Table 21.4 it will be seen that the actual percentage of gradings differ for the different types of machines and also for the stone crushed. The actual percentage of gradings of a plant crushing a particular type of stone is best found for the crusher in question by actual experience.

If a greater quantity of fine is required it is sometimes the practice to pass the primary breaking or a portion of it through the plant again for further reduction. This adds to the cost of the product per tonne, and should be borne in mind when estimating the cost of crushing.

Crushing plant consists of a crusher to which stone is fed by elevator or hopper, depending on the layout of the plant. If possible, hopper loading should be adopted, as the stone can then be directly loaded to the crusher. After the stone is crushed it is passed into a rotating screen where the dust is extracted and the various gradings of stone are sorted and passed into storage hoppers. These hoppers are usually elevated so that wagons or lorries may run directly under them, where they can be speedily loaded by one man controlling the outlet gate. This type of loading considerably cuts down haulage costs as the time the vehicle stands while being loaded is reduced to a minimum.

Table 21.4 Gradings obtained by crushing stone

Grading produced. Size in millimetres	Percentage of gradings obtained			
	Stonebreaker on granite	Granulator on granite	Granulator on shingle	Granulator on elvanstone
Passing 1.5	–	–	–	7
Passing 3	–	12	10	–
Passing 6	8	–	–	–
3 to 6	–	20	12	22
6 to 9	–	8	20	27
6 to 12	6	–	–	–
9 to 12	–	22	25	–
9 to 15	–	–	–	37
12 to 15	–	18	17	–
12 to 18	5	–	–	–
15 to 18	–	–	–	4
15 to 22	–	17	16	–
Over 18	–	–	–	3
Over 22	–	3	–	–
18 to 25	9	–	–	–
25 to 31	15	–	–	–
31 to 38	20	–	–	–
38 to 42	16	–	–	–
42 to 50	14	–	–	–
Over 50	7	–	–	–

Estimating the cost of crushing

In estimating the cost of crushing stone the estimator must consider:

1. The total output in tonnes per hour produced of all gradings.
2. The percentage of the various gradings produced.
3. The number of times the stone is passed through the plant in order to produce the requisite quantity of the desired grading.

The builder or contractor who crushes stone is generally only interested in particular gradings, and to achieve this it may be necessary for him to pass the primary crushing through the plant more than once. This recrushing should be taken into account.

The question of transporting, setting up and dismantling the plant should also be allowed for. This may be assessed at so much per tonne, knowing the total tonnage required.

In the data shown for stonebreakers, rotary impact breakers and granulators, it is assumed that the rock is of suitable size for primary crushing.

Table 21.5 Stone breakers

Note: The plant and labour hours shown are for crushing to 56 mm.

Size of breaker in millimetres	Plant hours per tonne	Labour hours per tonne
300 × 150	0.20	1.20
300 × 200	0.16	1.12
375 × 175	0.13	1.04
400 × 225	0.11	1.00
500 × 200	0.09	0.90
500 × 250	0.08	0.88
500 × 300	0.07	0.84
500 × 375	0.05	0.65

Table 21.6 Rotary impact breakers

Note: The size of feed varies according to type of material used. The plant hours shown are for breaking to 56 mm.

Approximate size of feed in millimetres	Plant hours per tonne	Labour hours per tonne
50	0.12	1.08
62	0.07	0.70
75	0.04	0.56
100	0.02	0.32

Table 21.7 Granulators

Note: The plant and labour hours shown are for crushing to 18 mm.

Size of granulator in millimetres	Plant hours per tonne	Labour hours per tonne
300 × 75	0.50	2.50
400 × 100	0.33	2.00
500 × 125	0.20	1.40
600 × 150	0.14	1.12
750 × 150	0.10	0.90

Table 21.8 Coating the stone per tonne

Notes: 1. The stone is taken as being crushed to the requisite size ready for loading to the plant.
2. Approximate litres of tar required per tonne of stone:

Stone 38 to 50 mm = 32 litres per tonne
Stone 13 to 38 mm = 34 litres per tonne
Stone 6 to 13 mm = 36 litres per tonne

The actual quantity to use is usually specified, as is the temperature to which the tar is to be raised before mixing.
3. For loading and transporting tarmacadam, see chapter 12.

Capacity of mixing plant per batch cubic metres	Plant and labour hours per tonne	
	Plant hours	Labour hours
0.30	0.18	0.90
0.45	0.12	0.72
0.60	0.09	0.63
0.75	0.07	0.56
0.90	0.06	0.54

Roadworks

The following data for estimating purposes is given in the Tables:

1. The weight of roller on which to base the estimate in rolling various types of foundation and surfacing materials.
2. Laying and rolling road foundation materials:
   - Clinker, sand and gravel, laid on the formation.
   - Brick, chalk and concrete hardcore foundations.
   - Stone pitching foundations.
3. Laying surface materials.
   - Asphalt.
   - Granite sett paving.
   - Tarmacadam.
4. Scarifying roads.
5. Surface dressed roads with tar, tar bitumen compounds and cold bituminous emulsion.
6. Fixing road signs, street nameplates, etc.
7. Fixing road gully pots and gully gratings.

Table 21.9 Weight of roller required

Nature of material rolled	Approximate weight of roller to use
Clinker, gravel or sand, laid on formations.	2 to 4 tonnes
Hardcore or stone pitching, laid on loamy clay, or soft formations.	6 to 8 tonnes
Hardcore or stone pitching, laid on firm foundations such as compact sand or gravel.	8 to 10 tonnes
Asphalt surfacing.	6 to 8 tonnes
Tarmacadam.	6 to 8 tonnes
Scarifying.	8 to 10 tonnes

Road foundations

The most common materials are:

- Clinker, sand or gravel.
- Brick or concrete hardcore.
- Stone pitching.
- Concrete *in situ*.

Clinker, sand or gravel are generally used as a sub-base being laid directly on the road formation to consolidated thickness ranging from 50 to 150 mm thick.

In the case of the road foundation or base the underlayer or sub-base tends to firm up the foundation and hold up the foundation material proper while it is being rolled, thus securing through consolidation without undue loss of material. It also minimises the tendency for the hardcore or stone pitching to work upwards into the interstices of the foundation material itself.

Hardcore and stone pitched foundations are used in conjunction with surfacing material such as asphalt, tarmacadam, and waterbound macadam, the consolidated thickness of foundation material laid depending on the nature of the subsoil and the weight of traffic the road has to carry. The consolidated thickness generally met with ranges from 50 to 150 mm.

Concrete *in situ* foundations are associated with surfacing of the asphalt, tarmacadam or granite sett type.

Example.   Calculate the cost of supplying, laying and rolling with a 2 tonne power driven roller 75 mm of consolidated clinker laid on the formation of a new road the following being the assumed rates:

Working cost of 2 tonne petrol roller, including driver, fuel and oil	£10.00 per h
Labour rate	£ 6.00 per h
Material cost of clinker delivered on the site	£20.00 per m^3
Material cost of clinker/m^2 = 0.099 @ £20/m^3	= 1.32 per m^2
Hire of 2 tonne roller = 0.022 h @ £10.00 per h	= 0.22 per m^2
Labour 0.088 h @ £6.00 per h	= 0.53 per m^2
Total cost of 75 mm consolidated clinker	= £2.07/m^2

Table 21.10 Lay and roll clinker per square metre

Consolidated thickness of clinker laid in millimetres	Material required and roller and labour hours per square metre			
	Cubic metres	Tonnes	Roller hours	Labour hours
25	0.033	0.028	0.012	0.048
50	0.066	0.056	0.017	0.068
100	0.132	0.112	0.024	0.096
150	0.198	0.168	0.028	0.112

Table 21.11 Lay and roll sand per square metre

Note: The data shown are based on sand weighing 1.52 tonnes per cubic metre.

Consolidated thickness of sand laid in millimetres	Material required and roller and labour hours per square metre			
	Cubic metres	Tonnes	Roller hours	Labour hours
25	0.030	0.043	0.013	0.052
50	0.060	0.086	0.018	0.072
100	0.120	0.172	0.025	0.100
150	0.180	0.258	0.029	0.116

Table 21.12 Lay and roll gravel per square metre

Note: The data shown are based on gravel weighing 1.76 tonnes per cubic metre.

Consolidated thickness of gravel laid in millimetres	Material required and roller and labour hours per square metre			
	Cubic metres	Tonnes	Roller hours	Labour hours
25	0.032	0.056	0.014	0.056
50	0.064	0.112	0.019	0.076
100	0.128	0.224	0.026	0.104
150	0.192	0.334	0.031	0.124

Table 21.13 Lay and roll brick hardcore per square metre

Note: The data shown are based on brickwork weighing 2.08 tonnes per cubic metre in the solid.

Consolidated thickness of brick hardcore laid in mm	Material required and roller and labour hours per square metre			
	Cubic metres	Tonnes	Roller hours	Labour hours
100	0.133	0.156	0.030	0.180
150	0.200	0.234	0.032	0.192
200	0.266	0.312	0.035	0.210
300	0.400	0.468	0.039	0.234

Table 21.14 Lay and roll stone pitching per square metre

Notes: 1. The volume and tonnage of material shown allow for consolidation.
2. If the pitching is hand pitched, use a multiplier of 2.50 with the labour hours shown and a multiplier of 1.05 with the amount of material required per square metre.

Consolidated thickness laid in millimetres, material required and roller and labour hours laying and consolidating stone pitching per square metre

150 mm				225 mm				300 mm			
Material required per m^2		Roller hours	Labour hours	Material required per m^2		Roller hours	Labour hours	Material required per m^2		Roller hours	Labour hours
m^3	tonnes			m^3	tonnes			m^3	tonnes		
0.200	0.266	0.036	0.216	0.330	0.440	0.039	0.234	0.400	0.533	0.043	0.258

Road surfacing

Asphalt surfacing
Data is shown for laying asphalt on prepared surfaces for the following types:

- Single-coat cold asphalt
- Two-coat cold asphalt
- Single-coat hot asphalt
- Two-coat hot asphalt

Example.   Calculate the cost of supplying, laying and rolling two-coat hot asphalt on the prepared formation of a road whose area is in excess of 400 m². The finished thickness of asphalt is 63 mm, the bottom coat being 38mm and the top coat 25 mm thick. The material is delivered to the site in mechanical tipping lorries and the following are the assumed plant, labour and material rates.

Asphalt for bottom course   = £22.00 per tonne delivered on the site
Asphalt for top course   = £22.00 per tonne delivered on the site
Working cost or hire rate of
   roller, including driver,
   fuel and oil, etc.   = £10.00 per h
Labour etc.   = £6.00 per h

Material cost of bottom coat	= 0.086 tonnes @ £22.00 per tonne	= £1.89/m²
Material cost of top coat	= 0.062 tonnes @ £22.00 per tonne	= £1.36/m²
Hire of roller	= 0.047 h at £10.00 per h	= £0.47/m²
Labour	= 0.20 h at £6.00 per h	= £1.20/m²
Cost per square metre		= £4.92

Table 21.15 Single-coat cold asphalt

Note: The data shown is based on asphalt weighing 2.242 tonnes per cubic metre. The data shown is for laying asphalt in areas in excess of 500 m².

Consolidated thickness of asphalt laid in millimetres	Lay and roll asphalt per square metre		
	Asphalt required per square metre tonnes	*Roller* hours	*Labour* hours
13	0.029	0.018	0.07
25	0.058	0.021	0.08
50	0.115	0.024	0.09
75	0.173	0.030	0.12

Table 21.16 Two-coat cold asphalt

Note: The data shown is based on asphalt weighing 2.114 tonnes per cubic metre for the bottom coat and 2.242 tonnes per cubic metre for the top coat.

	Lay and roll asphalt per square metre					
	Asphalt required per square metre					
	Bottom coat		Top coat			
Consolidated thickness of asphalt laid in millimetres	Thickness in millimetres	tonnes	Thickness in millimetres	tonnes	Roller hours	Labour hours
50	38	0.084	12	0.028	0.035	0.12
75	56	0.125	19	0.042	0.043	0.14
100	63	0.139	37	0.083	0.050	0.17

Table 21.17 Single-coat hot asphalt

Note: The data shown is based on asphalt weighing 2.242 tonnes per cubic metre.

	Lay and roll asphalt per square metre		
Consolidated thickness of asphalt laid in millimetres	Asphalt required per square metre tonnes	Roller hours	Labour hours
25	0.062	0.022	0.10
50	0.125	0.028	0.12
75	0.187	0.034	0.14

Table 21.18 Two-coat hot asphalt

Note: The data shown is based on asphalt weighing 2.114 tonnes per cubic metre for the bottom coat and 2.242 tonnes per cubic metre for the top coat.

	Lay and roll asphalt per square metre					
	Asphalt required per square metre					
	Bottom coat		Top coat			
Consolidated thickness of asphalt laid in millimetres	Thickness in millimetres	tonnes	Thickness in millimetres	tonnes	Roller hours	Labour hours
50	25	0.058	25	0.062	0.042	0.19
63	38	0.086	25	0.062	0.047	0.20
75	50	0.115	25	0.062	0.049	0.22
100	63	0.144	37	0.094	0.055	0.25

Table 21.19 Multipliers for surfacing roads with asphalt where the area is less than 500 m^2

Area of surfacing laid in square metres	Roller and labour hour multipliers
0–20	2.50
20–50	2.10
50–100	1.80
100–200	1.60
300–400	1.20
over 500	1.00

Tarmacadam surfacing

Example. Calculate the cost of supplying, laying and rolling tarred slag on the prepared foundation of a new road whose area is in excess of 500 m^2. The finished thickness of the tarmacadam is 75 mm, the bottom coat being 57 mm and the top coat 18 mm thick. The material is delivered to the site in mechanical tipping lorries and is laid by hand. The cost of the tarred slag delivered on the site and the assumed plant and labour rates are as follows:

Tarred slag for the bottom coat	= £20.00 per tonne
Tarred slag for the top coat	= £21.00 per tonne
Working cost of Bomag 100 AD roller or hire rate, including driver, fuel, oil, etc.	= £10.00 per h
Labour rate	= £6.00 per h

Material cost of bottom coat	= 0.120 tonne at £20.00 per tonne	= £2.40/m^2
Material cost of top coat	= 0.039 tonne at £21.00 per tonne	= £0.82/m^2
Hire of roller	= 0.15 h at £10.00 per h	= £1.50/m^2
Labour	= 0.15 h at £6.00 per h	= £0.90/m^2
Cost per square metre		= £5.62

Surfacing with mechanical paver plant

Plant and labour cost per hour

1 Foreman or Charge hand.
1 Mechanical paver plant and driver.
1 Screwman on the machine regulating the feed.
Power driven roller or rollers.
5 Labourers

The above unit is in general capable of handling all normal outputs, but in cases where a machine is fed to an amount approaching its capacity a sixth labourer should be allowed for in addition. Such factors as cost incurred in connection with the transport of the plant, attendant lorry if required, etc., must also be taken into account in building up the price.

Cost incurred in transporting the plant
The cost incurred in transporting the plant is best allowed for on a per tonne of material laid basis. Knowing the cost of the haulage involved and the total tonnage of material to be laid, the haulage cost per tonne of material laid is computed by dividing this cost by the total tonne of material required.

Example.   Estimate the cost of supplying, laying and rolling bituminous granite per tonne and per square metre in single coat work, the material being laid to a consolidated thickness of 19 mm, the guaranteed rate of material delivery being 20 tonnes per hour.
For purposes of this example the following are assumed:

1. The working cost of hire rate of the paver plant inclusive of driver, screwman, fuel, oil, etc. = £ 45.00 per h
2. The working cost or hire rate of a power-driven roller inclusive of driver, fuel and oil, etc. = £ 10.00 per h
3. The working cost or hire rate of a lorry inclusive of driver, fuel, oil, etc. = £ 7.00 per h
4. Foreman's rate of pay. = £ 7.00 per h
5. Labourer's rate of pay. = £ 6.00 per h
6. Cost of transporting the plant. = £180.00
7. The total tonnes of material to be laid. = 2000 tonnes
8. The cost of bituminous granite surfacing material delivered to site. = £ 36.00 per tonne
9. One lorry is chargeable.

Number of rollers required. The plant must handle the guaranteed rate of material delivery which is 20 tonnes per hour.
The covering capacity of bituminous granite laid to a consolidated thickness of 19 mm is 20.40 m^2 per tonne.

Area of surfacing laid per hour = 20 tonnes $\times$ 20.40 m^2 = 408 m^2

Since a single roller is capable of rolling up to 250 m^2 per hour the number of rollers required is given by:

$$\text{Number of rollers required} = \frac{408\, \text{m}^2}{250\, \text{m}^2}$$

$$= 1.63$$

Actual number of rollers required   = 2

The total plant and labour cost involved per hour is therefore:

1 Foreman	= £ 7.00 per h
1 Mechanical paver plant, driver and screwman	= £ 45.00 per h
2 Power-driven rollers at £10.00 per h	= £ 20.00 per h
1 Lorry at £7.00 per h	= £ 7.00 per h
5 Labourers at £6.00 per h	= £ 30.00 per h
Total plant and labour costs for 20 tonnes	= £109.00 per h

$$\text{Cost per tonne} \ \frac{109.00}{20} = £5.45 \text{ per tonne}$$

Cost of transporting the plant per tonne of material laid. The cost of transporting the plant is £180 and the total tonnage of material to be laid is 2000 tonnes.

Cost of transporting the plant per tonne = £180/2000
                                         = £0.090 per tonne

The total cost of the surfacing material supplied, laid and rolled is therefore:

Plant and labour cost	= £ 5.45 per tonne (see previous calculation)
Plant and transporting cost	= £ 0.09 per tonne (see previous calculation)
Material cost	= £36.00 per tonne (assumed)
Total cost	= £41.54

Since the area covered per tonne of material is 20.40 m², the total cost of the surfacing supplied, laid and rolled per square metre is given by:

Total cost of surfacing per square metre = £41.54 per tonne/20.40
                                          = £2.04 per m²

Table 21.20 Cost of laying and rolling surfacing

Rate of material delivery tonnes per hour A	Covering capacity of material square metres per tonne B	Area of surfacing laid per hour square metres A × B	Number of rollers required A × B 250 m²/h	Plant and Labour Cost per hour (£) Employing one roller	Plant and Labour Cost per hour (£) Employing two rollers	Estimated cost of laying and rolling surfacing per tonne (£)
10	20.40	204.0	1	47.45	–	4.745
15	20.40	306.0	2	–	53.45	3.563
20	20.40	408.0	2	–	53.45	2.673

Table 21.21 The normal capacity of surfacing plant to various consolidated thicknesses

Note: The data shown relate to a Barber-Greene English Model No. 8978 machine or a machine of comparable output, and represent a fair average output for work carried out under normal conditions.

Consolidated thickness of surfacing material laid in millimetres	Mechanical paver plant Normal capacity of plant in tonnes per hour Single coat work	Mechanical paver plant Normal capacity of plant in tonnes per hour Two coat work
12	20	–
25	24	–
50	32	24
75	–	32
100	–	40

Table 21.22 The covering capacity per tonne of surfacing material laid to different thicknesses

Consolidated thickness of surfacing material laid in millimetres	Type of material laid and square metres of consolidated surfacing material laid per tonne				
	Asphalt		Tarmacadam or bituminous macadam		
	Cold	Hot	Gravel	Granite	Limestone
13	29.70	27.39	31.75	30.45	33.68
25	14.85	13.80	15.39	15.24	16.84
38	9.90	9.14	10.30	10.31	11.25
44	8.66	8.00	9.04	9.00	9.84
50	7.43	6.86	7.52	7.71	8.84
75	4.99	4.57	5.14	5.08	5.63
100	3.86	3.60	3.86	3.81	4.20

Scarifying roads

The cost of scarifying is variable, depending on the nature of the material scarified. Successive layers of tar dressed waterbound macadam may be as hard and as costly to remove as tarmacadam. Assessment should be made as to how hard the surface is in conjunction with the thickness which has to be removed, and discretion used in selecting the data shown in the table.

The cost of tine sharpening
In hiring a roller it is the usual custom to charge an increased rate of hire for any period of time during which the roller is actually scarifying. The normal charge while rolling might be £10.00 per hour, but when scarifying an additional £4.00 per hour might be charged, making the scarifying hire rate £14.00 per hour.

Table 21.23 Scarifying

Nature of material scarified	Depth scarified in millimetres and roller and labour hours scarifying and loading into transport vehicles per square metre					
	50 mm		75 mm		100 mm	
	Roller hours	Labour hours	Roller hours	Labour hours	Roller hours	Labour hours
Granite macadam: tar dressed	0.022	0.154	0.034	0.238	0.043	0.301
Limestone macadam: tar dressed	0.018	0.126	0.029	0.203	0.037	0.259
Tarred granite	0.024	0.168	0.036	0.252	0.046	0.322
Tarred limestone	0.023	0.161	0.035	0.245	0.044	0.308

Surface dressing

Data is provided in the following Tables:

Table 21.24 giving the square metres covered by grit of various sizes per cubic metre and per tonne.

Table 21.25 giving the labour hours taken per square metre to sweep down the surface and apply chippings etc. of various sizes.

Table 21.26 showing the litres of surface dressing material required per square metre and the surface dressing plant, roller, lorry and labour hours taken per square metre to apply the dressing material only, using portable tar boilers or bitumen tanks ranging in size from 670 to 1140 litres capacity.

Table 21.27 showing the litres of surface drawing material required per square metre and the surface dressing plant, roller, lorry and labour hours taken to apply the dressing material per square metre, using large mobile pressure-spraying plant of 3500 to 4500 litres capacity.

Table 21.28 giving the cold bituminous emulsion required per square metre and the plant, roller, lorry and labour hours taken per square metre to apply the dressing material using a 180 litre hand-operated portable pressure-spraying plant.

Example.   Calculate the cost of tar dressing a road, using a medium sized portable boiler of 900 litres capacity, the rate of application specified being $0.84 \, \text{m}^2$ per litre and the chips to be used 9 mm gravel.

For purposes of this example it is assumed that a single lorry will be in constant attendance and that the boiler will be moved by roller.

The following are the assumed plant, labour and material rates:

Hire of portable tar boiler including fuel	= £ 2.00 per h
Inclusive working cost of roller with driver, fuel, oil, etc.	= £10.00 per h
Inclusive working cost of lorry with driver, fuel, oil, etc.	= £7.00 per h
Labour rate	= £6.00 per h
Tar delivered on the site in 180 litre barrels	= £0.30 per litre
Blinding gravel delivered on the site	= £18.00 per tonne

From Table 21.24 The covering capacity of 9 mm gravel is $122 \, \text{m}^2$ per tonne.

$$\text{Cost of gravel per square metre} = \frac{18.00}{122} = \text{£0.15 per m}^2$$

From Table 21.25 Sweep down and grit with 9 mm chips.

Labour hours = 0.029 h at £6.00 per h = £0.17

From Table 21.26.

Cost of tar = 1.08 litres at £0.30 per litre	= £0.32 per m^2
Hire of tar boiler = 0.007 h at £2.00 per h	= £0.01 per m^2
Hire of roller = 0.007 h at £10.00 per h	= £0.07 per m^2
Hire of lorry = 0.007 h at £7.00 per h	= £0.05 per m^2
Labour hours applying dressing = 0.021 h at £6.00 per h	= £0.13 per m^2
Total cost of dressing	= £0.58 per m^2

Table 21.24 The square metres covered per cubic metre and per tonne by various sizes of grit, chips or gravel

Size of grit in millimetres	Per cubic metre	Square metres covered			
		Per tonne			
		Sand	Granite	Limestone	Gravel
Sand	196	169	–	–	–
3	160	–	158	176	163
6	142	–	139	154	142
9	135	–	118	132	122
13	97	–	91	102	93
19	80	–	73	84	76

Table 21.25 The labour hours taken to sweep surfaces and apply grit of various sizes per square metre

Note: The labour hours shown are for sweeping and gritting only and do not allow for the labour involved in applying the dressing. See Tables 21.26, 21.27 and 21.28.

Size of grit in millimetres	Sweep down surfaces and apply grit only per square metre. Labour hours
Sand	0.022
3	0.024
6	0.026
9	0.029
13	0.031
19	0.036

Table 21.26 Surface dressing with tar, tar bitumen compounds and cold bituminous emulsion, using hand-operated tar boilers, per square metre

Notes: 1. The data shown are for plant of 680 to 1140 litres capacity. The plant may be moved by the roller or lorry.
2. For heating tar allow 0.03 kg of fuel per litre of tar.
3. The data shown are for applying the dressing and rolling only. For sweeping down and gritting, see Table 21.25.
4. The lorry hours shown refer to those of a single lorry only.

Rate of application Square metres per litre	Applying dressing and rolling only, per square metre				
	Litres of dressing required	Plant hours	Roller hours	Lorry hours	Labour hours applying dressing only
0.50	1.80	0.011	0.011	0.011	0.033
0.67	1.40	0.008	0.008	0.008	0.024
0.84	1.08	0.007	0.007	0.007	0.021
1.00	0.90	0.006	0.006	0.006	0.018
1.17	0.76	0.005	0.005	0.005	0.015

Table 21.27 Surface dressing with tar, tar bitumen compounds and cold bituminous emulsion, using pressure-spraying plant, per square metre

Notes: 1. The data shown are for dressing carried out by large pressure-spraying plant running under their own power and the dressing pressure sprayed on the surfaces.
2. The data shown are for applying the dressing and rolling only. For sweeping and gritting surfaces, see Table 21.25.
3. The lorry hours shown refer to those of a single lorry only.

Rate of application Square metres per litre	Applying dressing and rolling only, per square metre				
	Litres of dressing required	Pressure-spraying plant hours	Roller hours	Lorry hours	Labour hours in attendance on plant applying dressing only
0.50	1.80	0.034	0.034	0.034	0.068
0.67	1.40	0.025	0.025	0.025	0.050
0.84	1.08	0.022	0.022	0.022	0.044
1.00	0.90	0.017	0.017	0.017	0.034
1.17	0.76	0.014	0.014	0.014	0.028

Table 21.28 Surface dressing with cold bituminous emulsion, using 180 litre hand-operated portable pressure-spraying plant, per square metre

Notes: 1. The data shown are for applying the dressing and rolling only. For sweeping surfaces and gritting, see Table 21.25.
2. The lorry hours shown refer to those of a single lorry only.

Rate of application Square metres per litre	Applying dressing and rolling only, per square metre				
	Litres of dressing required	Plant hours	Roller hours	Lorry hours	Labour hours applying dressing only
0.50	1.80	0.012	0.012	0.012	0.036
0.67	1.40	0.009	0.009	0.009	0.027
0.84	1.08	0.007	0.007	0.007	0.021
1.00	0.90	0.006	0.006	0.006	0.018
1.17	0.76	0.005	0.005	0.005	0.016

Concrete roads

Concrete is used for the construction of roads, car parks, airport runways, depot yards, etc. and also a foundation for surfacing such as asphalt, granite setts, etc.

The data shown in this section are applicable to both concrete roads themselves and to concrete foundation work to roads.

In the construction of roads for housing development, this type of road is commonly adopted, for not only can they be laid at a fast rate of progress but they also provide excellent access for incoming material for the building operations.

Their initial cost compares favourably with the roads of the hardcore and stone-pitched type, surfaced with tarmacadam etc. and their cost of upkeep is relatively small.

Table 21.29 Concreting, using mixing plant loaded mechanically

Type of mixing plant	Bucket capacity of dragline or grab m³	Mixing and loading plant and labour hours, mixing, placing and manually tamping the concrete per cubic metre		
		Mixing plant hours	Loading plant hours	Labour hours
Revolving drum type 10/7	0.20	0.28	0.28	2.24
Revolving drum type 32/21	0.60	0.12	0.12	1.56
Revolving drum type 84/56	1.60	0.05	0.05	1.00

Table 21.30 Concreting to roads and large areas, mixing and placing only, with mixers of the drum and hopper type, loaded by hand and mechanically 10/7 Concrete mixer

Plant and labour	Unit	Method of loading, type of work, and plant and labour hours per m³							
		By hand		By Hi-lift shovel		By dragline		By grab	
		Large areas, roads, or mass concrete	Normal foundations, floors, etc.	Large areas, roads, or mass concrete	Normal foundations, floors, etc.	Large areas, roads, or mass concrete	Normal foundations, floors, etc.	Large areas, roads, or mass concrete	Normal foundations, floors, etc.
		10/7 Concrete mixer							
Concrete mixer and driver	h/m³	0.33	0.40	0.27	0.41	0.26	0.39	0.25	0.38
Hi-lift shovel and driver	h/m³	–	–	0.27	0.41	–	–	–	–
Dragline and driver	h/m³	–	–	–	–	0.26	0.39	–	–
Grab and driver	h/m³	–	–	–	–	–	–	0.25	0.38
Labourer	h/m³	2.60	3.15	2.16	2.46	2.08	2.34	2.00	2.28
		14/10 Concrete mixer							
Concrete mixer and driver	h/m³	0.26	0.30	0.22	0.29	0.21	0.27	0.20	0.26
Dragline and driver	h/m³	–	–	–	–	0.21	0.27	–	–
Labourer	h/m³	2.35	2.70	1.96	2.04	1.88	2.10	1.80	1.96
		21/14 Concrete mixer							
Concrete mixer and driver	h/m³	0.18	0.24	0.16	0.20	0.16	0.18	0.14	0.17
Dragline and driver	h/m³	–	–	–	–	0.16	0.18	–	–
Labourer	h/m³	2.00	2.58	1.56	1.50	1.52	1.44	1.44	1.40

Table 21.31 Concreting to roads and large areas, mixing and placing only, using central batching plant or continuous type mixers

Notes: 1. This table shows mix and place only, including tamping.
2. Mechanical grabs of draglines should be used to suit the mixer in order to feed it at correct rate.
3. The labour hours do not include for curing, bullnosing to edges, or screeding.
4. For reinforcement and sundry works to concrete road see Table 21.33.

Size of concrete mixer	Concrete output per hour	Mixer hours per m³	Mechanical grab or dragline hours per m³	Labourer hours per m³ (mix and place only)
10/7	7	0.140	0.140	1.70
14/10	9	0.110	0.110	1.55
21/4	14	0.070	0.070	1.40
42/28	28	0.035	0.035	0.80
84/56	56	0.020	0.020	0.50

Table 21.32 Multipliers for use with the previous table in carrying out various classes of work

Nature of work carried out	Multiplier
Roads, runways and large areas.	1.00
Mass foundations.	1.10
Mass walls.	1.30

Table 21.33 Sundry works to concrete roads

Description	Unit	Labourer hours
1. Curing with hessian.	m²	0.10
2. Curing with sand.	m²	0.24
3. Curing with waterproof paper.	m²	0.10
4. Expansion joint – bitumen or felt fixed.	m²	1.63
5. Expansion joint – run in elastic filler.	m	0.33
6. Reinforcement sheets, laid weight 1.63 kg per m².	m²	0.05
7. Reinforcement sheets, laid weight 2.18 kg per m².	m²	0.06
8. Reinforcement sheets, laid weight 2.70 kg per m².	m²	0.07
9. Round edges with bullnosed trowels.	m	0.13
10. Screed and trowel surfaces.	m²	0.48
11. Screed boards – timber, straight and afterwards remove.	m²	2.70
12. Screed boards – steel forms and afterwards remove.	m²	1.72
13. Silicate of soda applied per dressing.	m²	0.02
14. Studded roller finish.	m²	0.07
15. Tamp with steel-shod tampers.	m²	0.14
16. Tamp with leather or canvas belt finish.	m²	0.40
17. Bullnosing to edges of concrete with bullnosed trowel, per edge.	m	0.13
18. Covering the concrete with a canvas covered frame, the frame being fixed so that the canvas does not touch the concrete and afterwards remove.	m²	0.18
19. Fix expansion jointing material 150 mm deep.	m	0.17
20. Fix steel road forms and remove. Straight or to slow sweeps.	m²	1.51
21. Fix steel road forms and remove. Curved to quick sweeps.	m²	1.94
22. Form butt joint 150 mm deep, painted one coat tar, 0.27 litre tar.	m	0.33
23. Form rebated joint.	m	0.50
24. Fix dowel bar and sleeves.	Each	0.10
25. Lay waterproof paper on sub-base.	m²	0.02

Table 21.34 Road signs, nameplates etc.

Description	Unit	Material required	Fixed hours	Labourer hours
1. Fix metal sheet nameplates:				
on brickwork.	Each	Plugs and screws as required	0.50	0.50
on concrete walls.	Each	Plugs and screws as required	1.25	1.25
on timber fences.	Each	Screws as required	0.25	0.25
2. Fix road signposts in artificial paving.	Each	1 m²paving 0.067m³ concrete	1.20	1.20
3. In asphalt paving.	Each	1 m² asphalt on 150 mm concrete	2.00	2.00
4. Fix road sign already prepared to posts area:				
0.90 to 1.35 m²	Each	Screws as required	0.80	0.80
5. Paint white road lines by hand:				
on asphalt.	m	0.0018 kg paint	–	0.07
on tarmac.	m	0.0027 kg paint	–	0.09
6. Paint white road lines by hand propelled				
machine.	m	0.0036 kg paint	–	0.02
7. Cut grooves 18 mm deep for permanent white mastic lines:				
in asphalt.	m	–	–	0.75
in tarmac.	m	–	–	0.60
8. Fix permanent 18 mm white mastic lines.	m	5.5 kg white mastic	0.40	0.40
9. Fix reflector studs:				
in asphalt carriageways.	Each	2.7 kg mastic	0.60	0.60
in tarmac carriageways.	Each	2.7 kg mastic	0.50	0.50

Table 21.35 Excavate and fix gully pots

Notes: 1. The data shown allow for excavating, off loading and fixing the gully pots and also where shown concreting the foundation only or wholly surrounding the gully with 150 mm concrete.
2. Where gully pots are set on or surrounded with 150mm of concrete add only the material cost of the concrete involved.
3. The data shown are applicable to concrete or clayware gully pots.

Size of gully pot		Excavate and fix gully pot. No concrete to foundations or in surround. Labour hours each	Excavate and fix gully pot and mix and place 150 mm concrete foundation and 150 mm concrete surround to gully		Excavate and fix gully pot and mix and place 150 mm concrete in foundation to gully	
Internal diameter in millimetres	Depth in millimetres		Concrete cubic metres	Total labour hours each	Concrete cubic metres	Total labour hours each
225	375	1.20	0.045	1.50	0.135	2.10
225	600	1.90	0.045	2.20	0.180	3.10
300	600	2.50	0.060	2.90	0.240	4.10
300	750	3.00	0.060	3.40	0.270	4.80
450	900	6.60	0.090	7.20	0.420	9.40
450	1200	9.00	0.090	9.60	0.520	12.40

Table 21.36 Fix gully gratings

Note: If gratings are fixed over more than one course of brickwork, allow for the additional bricks and mortar required and use the following multipliers with the fixer and labour hours shown:

Gratings fixed on one course of bricks	1.00
Gratings fixed on two courses of bricks	1.20
Gratings fixed on three courses of bricks	1.40

Internal diameter of gully pot to which grating is fixed millimetres	Grating fixed on one course of brickwork. Per grating				Grating fixed directly on the concrete round the top of the gully. Per grating		
	Bricks required	Mortar required cubic metres	Fixer hours	Labour hours	Mortar required cubic metres	Fixer hours	Labourer hours
225	12	0.008	0.60	0.60	0.006	0.40	0.40
300	14	0.011	0.80	0.80	0.007	0.60	0.60
375	16	0.014	1.10	1.10	0.008	0.80	0.80
450	18	0.017	1.50	1.50	0.010	1.00	1.00

Kerb, channelling and path edging

Excavation for kerb channelling and path edging may be done either by hand or by mechanical plant. Where new roadworks are being constructed the excavation is usually done by the plant as the road carriageway and footways are excavated.

In excavating for kerb etc., by hand in narrow trenches, where the kerb has to be laid to existing roads the material to be excavated may be of a hard nature and if so due allowance should be made.

Concrete foundations for kerb, channelling and path edging

The concrete required for kerb and channelling laid on a separate foundation may be mixed by hand or by machine. It is the general practice to mix this concrete by a machine of suitable size and usually a 7/5 or 10/7 mixer is used.

Laying kerb, channelling and path edging

Kerb is usually laid clear of the roadway on a separate concrete foundation and in the case of roads other than concrete roads this is the normal practice. In the case of concrete roads, however, it may be laid either clear of the road slab on a separate concrete foundation or on the road slab itself on a mortar joint.

Table 21.37 Precast concrete and stone kerb

Description	Unit	Laid on separate concrete foundation		Laid on concrete road slab	
		Kerb layer hours	Labour hours	Kerb layer hours	Labour hours
Precast concrete kerb:					
Lay and joint 100 × 250 mm edge kerb	m	0.16	0.48	–	–
Lay and joint 125 × 250 mm edge kerb	m	0.17	0.51	–	–
Lay and joint 150 × 300 mm edge kerb	m	0.22	0.66	–	–
Lay and joint 250 × 100 mm flat kerb	m	0.14	0.42	0.13	0.39
Lay and joint 250 × 125 mm flat kerb	m	0.16	0.48	0.14	0.42
Lay and joint 300 × 150 mm flat kerb	m	0.21	0.63	–	–
Stone kerb (Granite Purbeck or similar):					
Lay and joint 100 × 250 mm edge kerb	m	0.17	0.51	–	–
Lay and joint 125 × 250 mm edge kerb	m	0.18	0.54	–	–
Lay and joint 150 × 300 mm edge kerb	m	0.23	0.69	–	–

For all types of kerb laid to radius use a multiplier of 2.00 with the labour hours shown.

Table 21.38 Channelling and path edging

Description	Unit	Kerb layer hours	Labour hours
Precast concrete channel:			
Lay and joint 225 × 100 mm channel	m	0.11	0.33
Lay and joint 250 × 125 mm channel	m	0.13	0.39
Lay and joint 300 × 150 mm channel	m	0.17	0.51
Stone channel:			
Lay and joint 225 × 100 mm channel	m	0.13	0.39
Lay and joint 250 × 125 mm channel	m	0.14	0.42
Lay and joint 300 × 150 mm channel	m	0.19	0.57
Granite sett channel:			
Lay 1 row 100 × 100 mm cubes	m	0.22	0.33
Lay 1 row 100 × 100 mm random length setts	m	0.18	0.27
Lay 2 rows 100 × 100 mm random length setts	m	0.33	0.50
Path edging:			
Lay and joint 50 × 125 mm concrete edging	m	0.11	0.22
Lay and joint 50 × 150 mm concrete edging	m	0.13	0.26
Lay and joint 75 × 225 mm concrete edging	m	0.19	0.38
Lay and joint one brick on edge	m	0.16	0.32
Creosoted timber edging			
Lay 100 × 25 mm	m	–	0.18
Lay 100 × 38 mm	m	–	0.26
Lay 150 × 50 mm	m	–	0.44

For edging laid to radius use a multiplier of 2.00 with the labour hours shown.

SECTION mm	125 x 250 mm EDGE KERB		150 x 300 mm EDGE KERB
		m³	m³
200 / 150	EXCAVATION	0·056	–
	CONCRETE	0·034	–
200 / 100	EXCAVATION	0·052	–
	CONCRETE	0·026	–
275 / 150	EXCAVATION	0·075	–
	CONCRETE	0·058	–
225 / 150	EXCAVATION	0·066	–
	CONCRETE	0·047	–
225 / 100	EXCAVATION	0·060	0·06
	CONCRETE	0·042	0·04
300 / 150	EXCAVATION	–	0·083
	CONCRETE	–	0·062
300 / 100	EXCAVATION	–	0·075
	CONCRETE	–	0·054
225 / 150	EXCAVATION	–	0·060
	CONCRETE	–	0·045

SECTION mm	125 x 250 mm FLAT KERB		300 x 150 mm FLAT KERB
		m³	m³
325 / 150	EXCAVATION	0·090	–
	CONCRETE	0·053	–
325 / 100	EXCAVATION	0·075	–
	CONCRETE	0·036	–
375 / 150	EXCAVATION	–	0·105
	CONCRETE	–	0·058
375 / 100	EXCAVATION	–	0·090
	CONCRETE	–	0·040

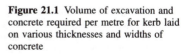

Figure 21.1 Volume of excavation and concrete required per metre for kerb laid on various thicknesses and widths of concrete

Foundation work to footways and pavings

Foundation work to footways and pavings is usually carried out with materials such as ashes, ballast, clinker, gravel, broken brick, etc. The availability of material often decides the material to be used, but clinker and brick hardcore are the most common.

After spreading and levelling the material over the area, it is rolled to the consolidated thickness required ready for the final surfacing.

In Table 21.39 for estimating purposes, the cubic metres of material required per square metre allowed for consolidation are shown.

Table 21.39 Foundation work to footways and pavings

Description and consolidated thickness laid in millimetres	Cubic metres of material reqd. per square metre	2 tonne petrol roller hours per square metre	Labour hours per square metre
Excavation: see chapter 12.			
50 mm ballast, gravel or hoggin	0.070	0.032	0.12
75 mm ballast, gravel or hoggin	0.105	0.037	0.14
100 mm ballast, gravel or hoggin	0.140	0.041	0.17
150 mm ballast, gravel or hoggin	0.210	0.048	0.22
50 mm clinker, ashes or sand	0.073	0.027	0.08
75 mm clinker, ashes or sand	0.110	0.030	0.11
100 mm clinker, ashes or sand	0.147	0.033	0.13
150 mm clinker, ashes or sand	0.221	0.041	0.18
75 mm brick hardcore	0.115	0.043	0.14
100 mm brick hardcore	0.154	0.048	0.18
150 mm brick hardcore	0.231	0.058	0.25
75 mm stone or concrete hardcore	0.112	0.048	0.16
100 mm stone or concrete hardcore	0.150	0.053	0.20
150 mm stone or concrete hardcore	0.226	0.064	0.27

Asphalt paving

Asphalt paving may be laid in either single or two-coat work. In connection with the laying cost of such surfacing it should be noted that the workability of the material has much to do with the time taken to lay it. In the case of hot asphalt the material should be laid while it is in a warm state, for if allowed to cool set takes place, whereupon it becomes hard and unworkable. In cold weather particular attention should be paid to the rate of delivery, for under such conditions it is imperative that the quantity delivered per day is such that it can be laid and rolled while it is still in an easily worked condition.

The data shown assume that the material is laid in a warm, workable state.

Notes: 1. For excavations to footways and pavings, see chapter 12.
  2. For foundation work to footways and pavings, see Table 21.39
  3. For sundry work to footways and pavings, see Table 21.45
  4. The data shown are for areas of 100 m^2 and over; for areas under 100 m^2 use the table of multipliers Table 21.42.

Table 21.40 Single coat work

Consolidated thickness of surfacing laid in millimetres	Material required tonnes per square metre	2 tonne petrol roller hours per square metre	Labour hours per square metre
18	0.050	0.034	0.136
25	0.067	0.036	0.144
37	0.101	0.042	0.168
50	0.136	0.048	0.192

Table 21.41 Two coat work

Consolidated thickness of surfacing laid in millimetres			Material required. tonnes per square metre		2 tonne petrol roller hours per square metre	Labour hours per square metre
Bottom coat	Top coat	Total thickness	Bottom coat	Top coat		
25	12	37	0.067	0.034	0.065	0.22
31	12	43	0.094	0.034	0.068	0.23
31	18	49	0.094	0.050	0.072	0.24
37	12	49	0.101	0.034	0.072	0.24
37	18	55	0.101	0.050	0.076	0.25
37	25	62	0.101	0.067	0.078	0.27

Table 21.42 Multipliers for laying single and two-coat asphalt less than 100 m^2

Area in square metres	Roller and labour hour multipliers
0–25	1.40
25–50	1.30
50–75	1.20
75–100	1.10
over 100	1.00

Table 21.43 Brick and tile paving

Note: If tile pavings are laid to narrow widths use labour multipliers as shown:

  Widths up to 150 mm – labour hour multiplier 3.00
  Widths 150–300 mm – labour hour multiplier 2.00

Nature of tiling laid	Unit	Material required per square metre		Pavior hours per unit	Labourer hours per unit
		Mortar m^3	Bricks or tiles, No.		
50 × 50 mm plain tiles	m^2	0.012	389	2.70	2.70
100 × 100 mm plain tiles	m^2	0.012	97	1.80	1.80
150 × 150 mm plain tiles	m^2	0.012	43	0.90	0.90
225 × 225 mm floor or quarry tiles	m^2	0.02	19	0.60	0.60
225 × 112 × 75 mm bricks laid on flat	m^2	0.02	38	0.84	0.84
250 × 125 × 50 mm bricks laid on flat	m^2	0.02	31	0.60	0.60
Circular cutting to tiles	m	–	–	1.32	1.32
Raking cuttings to tiles	m	–	–	0.66	0.66

Concrete *in situ* paving

The data shown for concrete *in situ* paving is for paving to such work as footpaths to houses, garage floors and washdowns, yards and concrete footways. The concrete may be mixed by hand or machine, a mixer of suitable size being used for the area concerned.

The concreting data shown are for mixing and placing the concrete only. For transporting the concrete in barrows, the labour hours per cubic metre are shown, but if the site conditions are such that transport has to be carried out by vehicles such as dumpers, lorries, etc., reference should be made to chapter 13 for suitable data.

It should be noted that the concrete when laid can have various finishes imparted to it, such as spade, trowelled or indented finish, etc. Data are shown for various finishes, and the estimator must select the relevant data for the finish specified.

It is usual to lay the concrete paving on a prepared foundation such as clinker. Foundation work is shown in Table 21.39, while preparatory excavation is dealt with under Groundworks, chapter 12.

Table 21.44 Concrete *in situ paving*

Notes: 1. The plant and labour hours shown are per cubic metre of concrete 'mixed and placed' only. For transporting the concrete by barrows up to a 22 m run use the labour hours shown, i.e. 1.07 hours per cubic metre, but if the concrete is transported by dumpers or lorries reference should be made to chapter 13 where data for this are shown.
2. For excavation, see chapter 12.

Description	Unit	Plant hours	Labour hours
Concrete mixed by hand			
Mix and place only concrete paving 75 mm thick	m³	–	7.05
Mix and place only concrete paving 100 mm thick	m³	–	6.65
Mix and place only concrete paving 125 mm thick	m³	–	6.27
Mix and place only concrete paving 150 mm thick	m³	–	5.77
Machine mixed concrete			
Mix and place only concrete paving:			
75 mm thick, using a 5/3½ mixer	m³	0.96	4.80
100 mm thick, using a 5/3½ mixer	m³	0.91	4.55
125 mm thick, using a 5/3½ mixer	m³	0.85	4.25
150 mm thick, using a 5/3½ mixer	m³	0.80	4.00
75 mm thick, using a 7/5 mixer	m³	0.67	4.69
100 mm thick, using a 7/5 mixer	m³	0.63	4.41
125 mm thick, using a 7/5 mixer	m³	0.59	4.13
150 mm thick, using a 7/5 mixer	m³	0.55	3.85
Wheeling concrete only in barrows up to a 22 m run	m³	–	1.07
Finish concrete with an indent roller finish	m²	–	0.24
Finish concrete with a screed or tamped finish	m²	–	0.18
Finish concrete with a spade finish	m²	–	0.12
Finish concrete with a trowelled finish	m²	–	0.29
Cure with hessian	m²	–	0.12
Cure with sand	m²	–	0.24
Bullnose edges of concrete	m	–	0.20
Fix 75mm straight timber side forms	m	–	0.40
Fix 100mm straight timber side forms	m	–	0.43
Fix 125mm straight timber side forms	m	–	0.50
Fix 150mm straight timber side forms	m	–	0.59

Note: For side forms fixed to radius use a labour multiplier of 2.00 with the labour hours shown.

Table 21.45 Sundry works to paving

Note: The labour hours shown are for fixing only in new work.

Description	Material required, mortar m³	Unit	Pavior hours	Labourer hours
Fix hydrant covers	0.007	Each	0.50	0.50
Fix inspection covers, 300 × 300 mm	0.006	Each	0.25	0.25
Fix inspection covers, 450 × 450 mm	0.009	Each	0.40	0.40
Fix pavement lights	0.025	m²	3.24	3.24
Fix sluice-valve covers	0.004	Each	0.40	0.40
Fix stop-cock boxes	0.003	Each	0.20	0.20
Cut and fit 50 mm paving slabs to manhole covers, etc., straight cutting	0.008	m	0.66	0.66
Cut and fit 50 mm paving slabs to manhole covers, etc., straight cutting	0.008	m	1.00	1.00
Cut hole through 50 mm paving for stop-cock boxes	–	Each	0.30	0.30

Tarmacadam paving

Tarmacadam paving may be laid in either single or two-coat work. The data shown assume that the material is laid in a warm, workable state. Care should be taken in cold weather to order the material in such quantity that it can be laid and rolled the same day, for if left overnight and allowed to harden, the laying cost is considerably increased, involving the use of heating to soften the material.

Notes: 1. For excavation to footways and paving, see chapter 12.
2. For foundation work to footways and pavings, see Table 21.39
3. For sundry work to footways and pavings, see Table 21.45
4. The data shown refers to areas over 100 m²; for areas less than this use the table of multipliers, Table 21.48.

Table 21.46 Single coat work

Consolidated thickness of surfacing laid in millimetres	Material required. Tonnes per square metre				2 tonne petrol roller. Hours per square metre	Labour hours per square metre
	Tarred slag	Tarred limestone	Tarred gravel	Tarred granite		
12	0.025	0.027	0.029	0.031	0.025	0.010
18	0.036	0.040	0.043	0.047	0.028	0.011
25	0.048	0.053	0.058	0.062	0.030	0.012
32	0.072	0.079	0.086	0.094	0.036	0.014
50	0.096	0.106	0.115	0.125	0.042	0.019

Table 21.47 Two-coat work

Consolidated thickness of surfacing laid in millimetres			Material required. Tonnes per square metre									2 tonne petrol roller. Hours per square metre	Labour hours per square metre
			Tarred slag		Tarred limestone		Tarred gravel		Tarred granite				
Bottom coat	Top coat	Total thickness	Bottom coat	Top coat	Bottom coat	Top coat	Bottom coat	Top coat	Bottom coat	Top coat			
25	12	37	0.048	0.025	0.053	0.027	0.058	0.029	0.062	0.031	0.058	0.19	
32	12	44	0.060	0.025	0.066	0.027	0.072	0.029	0.078	0.031	0.061	0.20	
32	18	50	0.060	0.036	0.066	0.040	0.072	0.043	0.078	0.047	0.065	0.22	
38	12	50	0.072	0.025	0.079	0.027	0.086	0.029	0.094	0.031	0.065	0.22	
38	18	56	0.072	0.036	0.079	0.040	0.086	0.042	0.094	0.047	0.065	0.23	
38	25	63	0.072	0.048	0.079	0.053	0.086	0.058	0.094	0.057	0.072	0.25	

Table 21.48 Multipliers for laying single and two-coat tarmacadam, less than 100 m^2

Area in square metres	Roller and labour hour multipliers
0–25	1.40
25–50	1.30
50–75	1.20
75–100	1.10
over 100	1.00

The tables for estimating the cost of surface dressing

In order to estimate the cost of surface dressing work, two tables are shown.

Table 21.49 shows the area covered by 1 tonne of sand, granite, limestone chips and crushed or pea shingle from which may be estimated the cost of the grit per square metre. It also shows the labour hours taken to sweep down and apply grit of various sizes.

Table 21.50 shows the plant and labour hours taken to apply the bituminous emulsion or tar at various rates of application per square metre and also the roller hours to roll in the grit.

By using these two tables the cost of surface dressing complete can be estimated for various combinations of application and grit sizes.

Table 21.49 Sweep down and apply grit to surfaces

Note: The weight of the various grits and sand per cubic metre is given in chapter 24.

Size of grit	Type of grit and area covered by 1 tonne in m^2				Sweep down and apply grit and sand. Labour hours per square metre
	Granite	Limestone	Shingle	Sand	
Sand	–	–	–	165	0.024
3 mm	165	188	172	–	0.026
4.5 mm	142	154	150	–	0.029
6 mm	125	137	133	–	0.031
9 mm	112	125	121	–	0.034

Table 21.50 Apply surface dressing and roll after gritting

Note: If hot tar is used allow 0.2 kg of coal per m^2.

Rate of application. Square metres per litre	2 tonne petrol roller. Hours per square metre	Apply surface dressing. Labour hours per square metre		
		Cold emulsion sprayed by 180 litres portable spraying plant	Cold emulsion applied directly from barrels and brushed in by hand	Hot tar applied by small portable tar boiler
0.55	0.022	0.108	0.102	0.114
0.72	0.019	0.096	0.091	0.101
0.92	0.017	0.084	0.079	0.089
1.00	0.014	0.072	0.068	0.076
1.29	0.012	0.060	0.058	0.064

Rail track

In dealing with the laying of rail track the civil engineer may be called upon to lay it as the permanent or temporary track.

Preparatory work in connection with laying rail track

In the data shown for estimating the cost of laying track, the ground is assumed to be level and firm. The data allows for light trimming and levelling as for temporary track only with no extensive excavation or levelling required.

For preparatory work in connection with laying rail track, involving excavation, drainage, laying clinker on the sub base, etc., see data in other sections.

Table 21.51 Lay flat bottomed rails weighing 32 kg per metre

Notes: 1. The date shown for laying rail track are for laying double rail track of 1.45 m gauge per metre centre line of track.
2. For laying bull headed rails weighing 36 to 40 kg per metre use a labour multiplier of 1.20 with the data shown for laying only.
3. For removing rail track and sleepers use a multiplier of 0.80 with the data shown for laying sleepers and track.
4. For loading rails to vehicle allow 1.20 labour hours per tonne, for concrete sleeper 0.30 labour hours each and wooden sleeper 0.25 labour hours each.
5. For preparatory work in connection with laying rail track (see above).

Description	*Unit*	*Platelayer hours*	*Labourer hours*
Unload rails etc.	Per tonne	–	1.10
Unload concrete sleepers and lay in readiness for the rails.	Each	–	0.40
Unload wooden sleepers and lay in readiness for the rails.	Each	–	0.30
Spread bottoming ballast.	m^3	–	1.60
String out and lay standard 1.45 m track with 34 kg flat bottomed rails with bolted fishplates secured to sleepers with bolts or coach screws.	Per m centre link of track	0.28	2.24
Ditto, laid to curves, necessitating the use of a Jim Crow.	Ditto	0.33	2.97
Ditto on curves where the gauge has to be opened and slotted sleepers are used.	Ditto	0.44	3.96
Ditto in turnouts measured from rail joints on either side of same.	Ditto	0.60	4.80
Pack and box up rail track.	Ditto	0.22	1.76
Fix switches.	Each	5.00	30.00
Fix buffer stops.	Each	8.00	40.00

22 Drainage

Cast-iron pipes

Cast iron pipes complying with the requirements of BS 437 are used chiefly for the conveyance of drainage below ground.

Cast iron pipes and fittings after coating shall be capable of withstanding hydrostatic pressures of 345 kPa for pipes and 170 kPa for fittings.

Pipes and fittings when subjected to these test pressures shall show no signs of leakage. The test pressure shall be applied internally and shall be not less than 15 seconds or more than 60 seconds.

Lengths of pipes

The pipes are generally supplied in standard lengths; lengths of 1.83 m, 2.74 m and 3.66 m are available. The standard lengths of the pipes of various diameters are given in tabular form below.

Table 22.1 Standard lengths of sand and spun cast iron pipes

Internal diameter of pipe	Length of pipe
50 to 225 mm inclusive	1.83, 2.74 and 3.66 mm

Joints

Various types of joint are available for jointing cast-iron pipes, but probably the most traditional is the lead joint, where either run lead or lead wool is used. It should be noted that the weights of lead, yarn, etc., shown in the estimating tables are those most commonly used. The weights to be used for various sized joints on a particular contract may, however, be specified, and if so, the specified weights should be used by the estimator in arriving at the cost of the joint.

The other types of joints used are those such as flanged joints, flexible gasket joints, Johnson couplings, etc. Data are shown in the tables for estimating the cost of all those types of joints likely to be encountered.

Laying and jointing cast-iron pipes in open trenches

The data shown in the main estimating tables are for laying and jointing cast-iron pipes to a depth not in excess of 1.5 m below the ground. A table of multipliers is shown for laying to depths in excess of this.

In order that the cost of laying and jointing the pipes complete per metre of pipe laid may be estimated, separate data are shown for laying the pipes and making the joints. As an example, in the case of a pipe 6 m long the cost of jointing per metre is the cost of one joint divided by 6 m. The cost of laying and jointing a 6 m pipe per metre is therefore the sum of:

1. The cost of laying only 1 m of the pipe.
2. The cost of making one joint divided by six.

Laying cast-iron pipe specials

Cast-iron specials refer to such pipe specials as taper pipes, bends, tees and junctions. The estimating data shown for laying these are for the extra labour involved in laying the special in the main line of pipes over and above the cost of laying a similar length of normal pipe. The method of working out the Extra Only price is shown on p. 248, where an example of the work-out of the Extra Only cost of laying a 225 mm cast-iron bend is given. The same procedure is applicable to all specials.

Haulage of pipes

Method of calculating the cost of this item is exactly the same as for clayware pipes, tables applicable to cast-iron pipes being given in the following pages.

Testing cast-iron pipes

On the completion of laying cast-iron mains, it is usual to test these to a suitable pressure. Data are shown in Table 22.19 from which may be calculated the cost of such testing per metre of pipe laid.

Pipelaying in open trenches

In laying pipes two operations have to be carried out:

1. The pipes have to be laid in position.
2. The pipes have then to be jointed.

In pricing pipelaying and jointing work, it is usual to quote a price per metre of pipe laid and jointed complete, and since the lengths of pipes are variable the data shown for estimating purposes are shown in separate tables:

1. Laying the pipes only per metre.
2. Making the joints only per joint.

By keeping the jointing data separate the estimator can calculate the cost of one joint. Having calculated the cost of the joint, this cost divided by the length of the pipe in metres gives the jointing only cost per metre laid, which, added to the laying only cost per metre, gives the cost of laying and jointing 1 m of pipe complete.

A typical example is shown, the method of procedure being applicable to all types of pipes.

Example. Calculate the cost of supplying, laying, jointing, and testing 150 mm diameter spun-iron pipes, the length of the pipes being 5.4 m each. The pipes are laid at a depth not in excess of 1.5 m. The joints are run lead and yarn joints knocked up by hand. For the purpose of this example the following rates are assumed:

Pipelayer and Jointer's rate = £ 6.50 per h
Labourer's rate   = £ 6.00 per h
Cost of pipes on the site   = £35.00 per m
Cost of lead   = £ 1.70 per kg
Cost of yarn   = £ 0.30 per kg

Laying and testing per metre: Referring to Tables 22.2 and 22.19 the cost of laying and testing per metre is given by:

	Pipelayer h	Jointer h	Labourer h
Laying per metre	0.13	–	0.39
Testing per metre	–	0.12	0.12
Total h per metre =	0.13	0.12	0.51

The total cost of off-loading, laying and testing the pipe per metre is:

Pipelayer h 0.13 at £6.50 = £0.85 per m
Jointer h   0.12 at £6.50 = £0.78 per m
Labourer h 0.51 at £6.00 = £3.06 per m

= £4.69 per m

Jointing per metre: referring to Table 22.5 the cost of one joint is:

Jointer h   1.12 at £6.50   = £7.28 per joint
Labourer h   0.56 at £6.00   = £3.36 per joint
Cost of lead   5.1 kg at £1.70   = £8.67 per joint
Cost of yarn   0.59 kg at £0.30   = £0.18 per joint
Cost of coke   say   = £0.30 per joint

Total cost = £19.79 per joint. And since each pipe is 5.4 m long:

$$\text{Cost of joint per metre of pipe} = \frac{£19.79}{5.4\,\text{m}}$$

$$= £3.67 \text{ per m}$$

The total cost of supplying, laying, jointing and testing 1 metre of pipe is:

Supply 1 metre of 150 mm diameter pipe at £35.00 per metre

= 35.00 per m

Off-load, lay and test   = 4.69 per m
Make joint   = 3.67 per m

Total   = £43.36 per m

Laying special pipe fittings in open trenches

Special pipe fittings or specials as they are usually called, comprise tapers, bends, junctions, tees, etc.

The figures given in the estimating tables for fixing these specials in the work are for the Extra Labour Hours involved in placing them in the main line of pipes as compared with the laying of a similar length of normal pipe. The reason for this is that it is the usual practice to price specials as an Extra Only item, over and above the rate tendered for laying normal pipes.

In certain cases, however, specials have to be priced at an All In figure laid complete, in which case to the Extra Only labour cost as derived from the table must be added the cost of laying a length of normal pipe equivalent to the length of the special.

In dealing with specials the estimator should bear in mind the cost of the additional joint or joints required as these also must be included.

The Extra Only estimated cost of laying and jointing a special in the main line of pipes therefore consisting of the sum of the following:

1. The cost of the special itself less the cost of a similar length of normal pipe.
2. The extra labour hours involved in laying the special.
3. The extra cost of the joint involved.

Example: Calculate the Extra Only cost over laying and jointing 225 mm spun-iron pipes for laying and jointing a 225 mm cast iron medium radius bend, with run lead joints knocked up by hand, the bend being laid in a trench not in excess of 1.5 m deep.

The prevailing labour and material rates for purposes of this example are taken as being:

Pipelayer and Jointer	= £6.50 per h
Labourer	= £6.00 per h
Cost of 150 mm Class B bend 0.90 m long	= £50.00 each
Cost of 150 mm Class B pipes	= £35.00 per m
Lead	= £1.70 per kg
Yarn	= £0.30 per kg

The Extra Labour Cost involved in laying a 225 mm bend. Referring to the Table 22.3 the Extra Labour Hours taken to lay a 225 mm cast iron bend as compared with a similar length of 225 mm ordinary pipe is:

Pipelayer	0.33 h at £6.50	=	£2.15
Labourer	1.00 h at £6.00	=	£6.00
			£8.15

The cost of the Extra Joint. Referring to Table 22.5 the cost of making a 225 mm run lead joint is:

Jointer	1.12 h at £6.50 per h	=	£ 7.28
Labourer	0.56 h at £6.00 per h	=	£ 3.36
Lead	5.10 kg at £1.70 per kg	=	£ 8.67
Yarn	0.59 kg at £0.30 per kg	=	£ 0.18
Fuel	say	=	£0.30
	Cost of Joint		£19.79

The Extra Only material cost of a 225 mm cast iron bend over a similar length of ordinary 225 mm cast iron pipe.

Material cost of a 225 mm cast iron bend	= £50.00
Deduct material cost of a 0.90 m length of ordinary 150 mm pipe	= £31.50
Extra Only material cost of bend	= £18.50
Total Extra Only cost of bend laid complete	= £ 8.15
	£19.79
	£18.50
	£46.44

If the specials are billed at an All In rate fixed complete the method of estimating the cost is as follows:

Material cost of the 225 mm cast iron medium radius bend	= £50.00
Add the Extra Only cost to lay the bend, as shown previously	= £8.15
Add the extra cost of the 225 mm run lead joint as shown previously	= £19.79

Add the cost of laying a 0.9 m length of ordinary 225 mm pipe, the data for which is shown in Table 22.2:

Pipelayer 0.14h at £6.50 per hour	= £0.92
Labourer 0.42h at £6.00 per hour	= £2.52
Total All In cost of bend laid complete	= £81.37 each

Pipelaying in tunnels

In laying pipes in tunnels the pipes may be lowered down the shafts in one of two ways:

1. Using gantries, shear legs or similar tackle operated by hand.
2. Using power-driven cranes.

Note: If the pipes concerned are light in weight it is usual to lower them by hand and rope, in which case the data shown for pipes lowering by gantries or shear legs may be used for purposes of estimating.

In using plant of the gantry or shearing type where the lowering is carried out by hand it is only practical to lower to depths of approximately 9 m below the ground. In the case of deep tunnel work a power driven crane is much more suitable and is generally used.

In considering the method by which the pipes are to be lowered it should be borne in mind that the laying of the pipes form only a small part of the tunnel work, and other operations such as removing the excavated material, concreting, refilling, etc. have to be carried out, these latter items contributing in a large measure to the cost of the work. For this reason a crane driven by power is of great use as it can be used in connection with all the operations involved.

In carrying out tunnelling work, the usual practice is to sink shafts at suitable intervals along the line of the tunnel, using manhole shafts if at all possible. Tunnels are then driven each way from the shafts, one shaft and one crane thus serving two tunnels emanating from this one shaft.

Table 22.2 Laying cast-iron pipes to BS 437 in open trenches up to 1.5 m deep

Notes: 1. For laying pipes in trenches over 1.5 m deep, see table of multipliers, Table 22.9
     2. For jointing pipes, Tables 22.5, 22.6, 22.7 and 22.8
     3. For pipes with turned and bored joints use the data shown in this table with a multiplier of 1.25.
     4. For laying specials, Table 22.3

Diameter of pipe in millimetres	Laying pipes, pipelayer and labourer hours per m	
	Pipelayer hours	Labourer hours
50	0.07	0.14
75	0.08	0.16
100	0.09	0.18
150	0.11	0.33
225	0.14	0.42

Table 22.3 Laying cast-iron pipe specials in open trenches up to 1.5 m deep

Notes: 1. The hours shown are the Extra Labour Hours involved in laying one special, but not for making the joints. For jointing, see Tables 22.5, 22.6, 22.7 and 22.8
2. For multipliers for laying cast-iron pipe specials at depths greater than 1.5 m, see Table 22.9

Diameter of pipe in mm	Taper pipes		Bends		Single junctions and tees		Double junctions and tees		Valves	
	Pipelayer hours	Labourer hours	Pipelayer hours	Labourer hours	Pipelayer hours	Labourer hours	Pipelayer hours	Labourer hours	Pipelayer hours	Labourer hours
50	0.12	0.24	0.15	0.30	0.18	0.36	0.21	0.42	0.40	0.80
75	0.14	0.28	0.17	0.34	0.21	0.42	0.25	0.50	0.50	1.00
100	0.16	0.32	0.20	0.40	0.24	0.48	0.28	0.56	0.56	1.12
150	0.20	0.60	0.25	0.75	0.30	0.90	0.35	1.05	0.70	2.10
225	0.26	0.78	0.33	1.00	0.39	1.17	0.45	1,35	0.90	2.70

Table 22.4 Fix double collars, caps and plugs in open trenches up to 1.5 m deep

Notes: 1. The hours shown do not allow for making the joints. For jointing, see Tables 22.5, 22.6, 22.7 and 22.8
2. For table of multipliers for laying the above at depths in excess of 1.5 m below ground, see Table 22.9
3. Allow 2 joints per double collar.

Diameter of pipe in mm	Double collars		Caps		Plugs	
	Pipelayer hours	Labourer hours	Pipelayer hours	Labourer hours	Pipelayer hours	Labourer hours
50	0.06	0.12	0.04	0.08	0.01	0.02
75	0.10	0.20	0.06	0.12	0.02	0.04
100	0.12	0.24	0.07	0.14	0.03	0.06
150	0.16	0.48	0.09	0.27	0.05	0.15
225	0.27	0.81	0.15	0.45	0.09	0.27

Table 22.5 Jointing cast-iron pipes and specials in open trenches up to 1.5 m deep with run lead joints

Notes: 1. The compressor hours assume two pneumatic tools being used per compressor.
2. The weight of lead and yarn per joint may vary and is usually specified.
3. For joints of the Johnson Coupling and Victaulic type, screwed gland, flexible and flange joints, see Table 22.8

Diameter of pipe in mm	Lead per joint in kg	Yarn per joint in kg	Joints knocked up by hand		Joints knocked up by pneumatic tools		
			Jointer hours per joint	Labourer hours per joint	Jointer hours per joint	Labourer hours per joint	Compressor hours per joint
50	1.18	0.08	0.28	0.14	–	–	–
75	1.81	0.13	0.40	0.20	–	–	–
100	2.15	0.17	0.52	0.26	–	–	–
150	3.06	0.34	0.76	0.38	–	–	–
225	5.10	0.59	1.12	0.56	0.45	0.15	0.22

Table 22.6 Jointing cast-iron pipes and specials in open trenches up to 1.5 m deep with lead wool joints. Single collar type – each

Note: The compressor hours shown assume that two pneumatic tools are used per compressor.

Internal diameter of pipe in mm	Material for single collar type					Joints knocked up by hand		Joints knocked up pneumatically		
	Lead wool			Yarn						
	Depth mm	Weight kg	No. of skeins	Depth mm	Weight kg	Jointer hours per joint	Labourer hours per joint	Jointer hours per joint	Labourer hours per joint	2 tool compr. hours per joint
75	32	1.19	2.5	44	0.11	0.34	0.17	–	–	–
100	32	1.70	3.5	44	0.14	0.44	0.22	–	–	–
150	35	2.41	5	56	0.23	0.66	0.33	–	–	–
225	38	3.63	7.5	66	0.45	1.00	0.50	0.60	0.30	0.30

Table 22.7 Jointing cast-iron pipes and specials in open trenches up to 1.5 mm deep with lead wool joints. Double collar type – each

Note: The compressor hours shown assume that two pneumatic tools are used per compressor.

Material for double collar joint									Joint knocked up by hand		Joints knocked up pneumatically		
Lead wool. Back of socket			Yarn for intermediate space			Lead wool. Front of socket							
Depth mm	Weight kg	No. of skeins	Depth mm	Weight kg	Depth mm	Weight kg	No. of skeins	Jointer hours per joint	Labourer hours per joint	Jointer hours per joint	Labourer hours per joint	2 tool compr. hours per joint
75	19	0.737	1.5	38	0.071	19	0.737	1.5	0.40	0.20	–	–	–
100	15	0.737	1.5	30	0.071	32	1.700	3.5	0.52	0.26	–	–	–
150	9	0.737	2.5	44	0.170	34	2.409	5	0.78	0.39	–	–	–
225	12	1.445	3	53	0.325	34	3.629	7.5	1.18	0.59	0.80	0.40	0.40

Table 22.8 Jointing cast-iron pipes in open trenches up to 1.5 m deep with mechanical coupling joints

Notes: 1. For run lead joints, see Table 22.5
2. For lead wool joints, see Table 22.6 and 22.7
3. For mechanical couplings the jointing material may consist of rubber insertion, corrugated brass or copper rings, etc.

Diameter of pipe in mm	Make joints. Jointer and labourer hours per joint					
	Flexible joints. Bolted type		Flexible joints. Loose flange type		Mechanical couplings	
	Jointer hours	Labourer hours	Jointer hours	Labourer hours	Jointer hours	Labourer hours
75	0.13	0.13	0.10	0.10	0.16	0.16
100	0.16	0.16	0.12	0.12	0.20	0.20
150	0.24	0.24	0.18	0.18	0.30	0.30
225	0.34	0.34	0.27	0.27	0.42	0.42

Table 22.9 Multipliers for laying and jointing cast-iron pipes and specials in open trenches in excess of 1.5 m deep

Depth below ground in metres	Pipelayer, jointer and labourer hours multipliers
Depth not exceeding 1.5 m	1.00
Depth exceeding 1.5m and not exceeding 3 m	1.40
Depth exceeding 3 and not exceeding 4 m	1.90
Depth exceeding 4 and not exceeding 6 m	2.50

Table 22.10 Cutting cast-iron pipes

Diameter of pipe in millimetres	Pipelayer or jointer hours per cut			
	Class A pipes	Class B pipes	Class C pipes	Class D pipes
50	0.30	0.30	0.30	0.30
75	0.45	0.45	0.45	0.50
100	0.60	0.60	0.65	0.70
150	1.00	1.00	1.10	1.20
225	1.80	1.80	2.10	2.40

Table 22.11 Sundry work to cast-iron pipes

Note: The hours shown include for making joints.

Description	Unit	Jointer hours	Labourer hours
Fix air valves, tapped and screwed	Each	2.50	2.50
Fix air valves, single-acting flanged type	Each	1.50	1.50
Fix air valves, double-acting flanged type	Each	2.20	2.20
Fix air valve covers	Each	1.25	1.25
Fix hydrants, ball type	Each	1.20	1.20
Fix hydrants, screwdown type	Each	2.50	2.50
Fix hydrant covers	Each	1.00	1.00
Fix sluice-valve covers, heavy type	Each	1.00	1.00
Fix stop-cock boxes	Each	0.50	0.50
Drill and tap cast-iron pipes, 12 mm diameter for ferrules, etc.	Each	0.30	0.30
Drill and tap cast-iron pipes, 19 mm diameter for ferrules, etc.	Each	0.35	0.35
Drill and tap cast-iron pipes, 25 mm diameter for ferrules, etc.	Each	0.40	0.40
Drill and tap cast-iron pipes, 32 mm diameter for ferrules, etc.	Each	0.45	0.45
Drill and tap cast-iron pipes, 38 mm diameter for ferrules, etc.	Each	0.50	0.50
Drill and tap cast-iron pipes, 50 mm diameter for ferrules, etc.	Each	0.60	0.60

For fixing valves, see Table 22.3 and 22.23

Clayware pipes

Clayware pipes complying with BS 65 are extensively used for soil and surface water drains or sewers and are made in the following lengths (Table 22.12):

Table 22.12 Lengths of clayware pipe:

Internal diameter of pipe	Length of pipe, millimetres
Up to and including 100 mm	1300
150 mm	1400
200 to 300 mm inclusive	1600

Laying and jointing clayware pipes in open trenches

The data shown in the main estimating tables are for laying and jointing clayware pipes to a depth not in excess of 1.5 m below the ground. A table of multipliers is shown for use with the data shown in the main estimating table for laying to depths in excess of this.

For estimating the cost of laying and jointing the pipes complete per metre, separate data are shown for laying the pipes and making the joints. As an example of the joint cost per metre of pipe laid, in the case of a 150 mm diameter pipe, which is 1.40 m long, the number of joints which have to be made per metre of pipe laid is 1 m ÷ 1.40 m = 0.72 joints.

In joining Hepseal or similar drain pipes with push-fit flexible joints the data shown in the main estimating tables for laying the pipes only should be used.

Laying clayware pipe specials

Special pipes refer to such pipes as taper pipes, bends, junctions, etc. The estimating data shown for laying these are for the extra labour involved in laying them in the main run of pipes over and above the cost of laying a similar length of normal pipe. The method of working out the Extra Only price is shown on p. 248 where an example is given showing a workout of the Extra Only cost to lay a 225 mm cast-iron bend. The method of procedure in the case of clayware pipe specials is exactly the same.

Laying clayware pipes in tunnels

The data shown for estimating the cost of laying and jointing clayware pipes in tunnels are applicable to tunnels up to 6 m deep below ground and to a length not in excess of 15 m through the tunnels. A table of multipliers is given for use with the data shown in the main estimating table for laying pipes to depths and lengths through tunnels in excess of this.

Testing clayware pipes

Data from which may be calculated the cost of testing the pipes by air, smoke or water have purposely been shown separately, as only in certain instances are the pipes so tested. It is, for insurance, the invariable rule to test all clayware pipes laid as soil sewers, but not when laid as surface water sewer. If the pipes have to be tested the estimator must allow for this by using the data shown.

Duct laying

In laying clayware pipes or ducts where no jointing is required the data shown in the main estimating tables for laying the pipes only should be used.

Hauling clayware pipes

The data shown in the main estimating tables are based on the pipes being delivered on the site by the supplier.

Table 22.13 Cement mortar joints
showing cement mortar mixes of various proportions

Notes: 1. The table is based on cement weighing 1442 kg per m^3.
   2. In the table shown for jointing clayware pipes the labour hours shown allow for mixing the mortar and making the joint complete.
   3. The data shown relating to sand refer to dry sand. If moist sand is used a multiplier of 1.25 should be used with the quantities of sand shown to allow for bulkage due to moisture content.

Mix		Material required per cubic metre of mortar	
Sand	Cement	Sand cubic metres	Cement tonnes
1	1	0.73	1.05
1½	1	0.93	0.89
2	1	1.00	0.72
2½	1	1.07	0.61
3	1	1.10	0.52
3½	1	1.15	0.44
4	1	1.20	0.39

Table 22.14 Laying and jointing clayware pipes in open trenches up to 1.5 m deep

Notes: 1. The hours shown do not include for testing.
   2. The hours shown for jointing allow for mixing up the mortar and making the joint.

Diameter of pipe in millimetres	Lay only per metre		Jointing only per joint. Socket and spigot mortar joints				Lay and joint 1 m of pipe		
	Pipelayer hours	Labourer hours	Mortar m^3 per joint	Yarn kg per joint	Jointer hours	Labourer hours	Pipelayer hours	Jointer hours	Labourer hours
100	0.06	0.06	0.0006	0.014	0.05	0.05	0.06	0.05	0.11
150	0.08	0.08	0.0008	0.027	0.07	0.07	0.08	0.07	0.15
225	0.11	0.11	0.0017	0.064	0.11	0.11	0.11	0.11	0.22

Table 22.15 Multipliers per joint

Type of joint	Jointer and labourer hour multipliers
Socket and spigot type joints	1.00
Grouted composite joints with canvas bag	1.30
'Hassall' single-lined joints	1.10
'Hassall' double-lined joints	1.20
Push-fit flexible joints	1.00
'Stanford' joints	1.00

Table 22.16 Laying clayware pipe specials in open trenches up to 1.5 m deep

Notes: 1. The hours shown are the Extra Labour hours involved in laying one special, but not for making the joints.
2. For multipliers for laying clayware pipe specials at depths greater than 1.5 m below ground. See Table 22.17.

Diameter of pipe in millimetres	Taper pipes		Bends		Single junctions		Double junctions	
	Pipelayer hours	Labourer hours	Pipelayer hours	Labourer hours	Pipelayer hours	Labourer hours	Pipelayer hours	Labourer hours
100	0.06	0.06	0.08	0.08	0.09	0.09	0.10	0.10
150	0.08	0.08	0.11	0.11	0.12	0.12	0.14	0.14
225	0.11	0.11	0.15	0.15	0.17	0.17	0.19	0.19

Table 22.17 Multipliers for laying and jointing clayware pipes in open trenches, to depths in excess of 1.5 m

Depth below ground in metres	Pipelayer, jointer and labourer hours multipliers
Depth not exceeding 1.5 m	1.00
Depth exceeding 1.5 m and not exceeding 3.0 m	1.25
Depth exceeding 3.0 m and not exceeding 4.5 m	1.50
Depth exceeding 4.5 m and not exceeding 6.0 m	1.75

Laying and jointing clayware pipes in tunnels

The cost of laying and jointing clayware pipes in tunnels is variable, depending on:

1. The depth at which the pipes are laid below the ground.
2. The length to which the pipes to be laid through the tunnel.

By referring to the table of multipliers shown the estimator can select the multiplier to use in conjunction in order to estimate the cost of laying and jointing for all combinations of shaft depths and tunnel lengths for the limits shown. The multipliers should be used with the plant, jointer and labourer hours as shown in the laying and jointing tables for tunnel work.

Table 22.18 Multipliers for laying and jointing clayware pipes in tunnels where the depths of shafts exceed 6 m and the lengths of tunnels are in excess of 15 m

Depth of tunnel below ground in metres	Length of tunnel from shaft in metres, and plant, pipelayer jointer and labourer hour multipliers			
	0–15	0–30	0–45	0–60
0–6	1.00	1.15	1.30	1.45
0–9	1.05	1.21	1.36	1.52
0–12	1.10	1.27	1.43	1.59
0–15	1.15	1.33	1.50	1.66

Table 22.19 Testing clayware pipes in open trenches with water

Notes: 1. The test is carried out with a head of water test to a suitable head.
2. The water used in testing is taken as being available on tap. If the water has to be pumped and/or hauled from a distance in tanks the cost of this should be allowed for.
3. For testing by air or smoke or testing in tunnels, use the table of multipliers shown.

Diameter of pipe in millimetres	Testing with water in open trenches per metre	
	Jointer hours	Labourer hours
100	0.03	0.03
150	0.04	0.04
225	0.07	0.07

Table 22.20 Multipliers for testing sewers in open trenches and tunnels

	Multiplier for jointer and labourer hours	
Nature of test	Testing in open trenches	Testing in tunnels
Water	1.00	1.50
Smoke	0.70	0.90
Air	0.60	0.75

Table 22.21 Clayware pipe fittings

Note: For excavating and fixing road gullies and gratings, see Tables 21.35 and 21.36

Description	Unit	Bricklayer or pipelayer hours	Labourer hours
Fix channels:			
Straight 100 mm diameter	Each	0.16	0.16
Straight 150 mm diameter	Each	0.18	0.18
Straight 225 mm diameter	Each	0.20	0.20
Fix channel bends or junctions, 100 mm	Each	0.15	0.15
Fix channel bends or junctions, 150 mm	Each	0.18	0.18
Fix channel bends or junctions, 225 mm	Each	0.20	0.20
Fix channels, tapered, 100 to 150 mm	Each	0.20	0.20
Fix channels, tapered, 150 to 225 mm	Each	0.22	0.22
Fix channels, tapered, 225 to 300 mm	Each	0.24	0.24
Fix grease traps, 100 mm diameter	Each	0.40	0.40
Fix grease traps, 150 mm diameter	Each	0.60	0.60
Fix gullies – no excavation of concrete:			
300 mm diameter × 0.6 m deep	Each	–	1.00
300 mm diameter × 0.75 m deep	Each	–	1.20
300 mm diameter × 0.9 m deep	Each	–	1.40
Fix gullies, yard type, 100 mm diameter	Each	0.40	0.40
Fix gullies, yard type, 150 mm diameter	Each	0.50	0.50
Fix interceptors, 100 mm diameter	Each	0.30	0.30
Fix interceptors, 150 mm diameter	Each	0.40	0.40
Fix interceptors, 225 mm diameter	Each	0.50	0.50
Fix rainwater shoes, 100 mm diameter	Each	0.30	0.30
Fix saddles, 100 mm diameter	Each	0.80	0.80
Fix saddles, 150 mm diameter	Each	1.10	1.10
Fix saddles, 225 mm diameter	Each	1.50	1.50
Fix siphons, 100 mm diameter	Each	0.40	0.40
Fix siphons, 150 mm diameter	Each	0.60	0.60
Fix siphons, 225 mm diameter	Each	0.90	0.90

Table 22.22 Laying and jointing flexible-jointed clayware pipes and specials

Notes: 1. The hours for jointing do not include for testing, for which see Table 22.19
2. These pipes are supplied complete with self-sealing rings and no other jointing material is required.

Diameter of pipe mm	Laying and jointing pipe, per metre					
	In trench not exceeding 1.5 m deep		In trench 1.5 to 3.0 m deep		In trench 3.0 to 4.5 m deep	
	Pipelayer hours	Labourer hours	Pipelayer hours	Labourer hours	Pipelayer hours	Labourer hours
100	0.10	0.10	0.12	0.12	0.15	0.15
150	0.15	0.15	0.18	0.18	0.22	0.22
225	0.22	0.22	0.27	0.27	0.33	0.33
Specials, each						
Bends						
100	0.05	0.05	0.06	0.06	0.07	0.07
150	0.06	0.06	0.07	0.07	0.08	0.08
225	0.08	0.08	0.10	0.10	0.12	0.12
Single junctions						
100	0.08	0.08	0.10	0.10	0.12	0.12
150	0.14	0.14	0.17	0.17	0.21	0.21
225	0.18	0.18	0.22	0.22	0.27	0.27

Flexible-jointed clayware pipes

Flexible-jointed clayware drain pipes are made in 75 mm to 300 mm diameters. These pipes are made to the same British Standard Specification as the traditional spigot and socket drains but are different in so far as each pipe is plain-ended and the joints are formed with push-fit couplings, moulded in polypropylene and incorporating rubber sealing gaskets.

Pipes can be laid all year round regardless of weather and site conditions; the flexible joint sleeve is a safeguard against possible ground movement. As a rule pipes can be laid direct on the trimmed natural trench bottom. If, owing to unusual conditions, either concrete bed and haunch or granular bed and surround are specified, see chapter-13 and Tables 22.24 and 22.25 for details of concrete bed and haunch or see Table 22.30 for granular bed and surround.

UPVC (unplasticised polyvinyl chloride) pipes to BS 4660

Pipes of 100 mm diameter are available in 1 m, 3 m and 6 m lengths and are well suited for housing work. Jointing is very simple; pipes can be supplied either with a ring seal socket at one end, or with spigots at both ends. Joints are formed by pressing the spigot into the self-sealing socket or, where more convenient, the two spigot ends are jointed with a double socket and sealed with solvent cement. Pipes can be jointed to the socket ends of cast iron or clayware pipes with a PVC adaptor.

Work in connection with manholes, haunching and backfill

Table 22.23 Sundry work in connection with brick manholes, sewage disposal, etc.

Notes: 1. The tradesmen hours shown refer to those of a bricklayer, pipelayer or jointer, as applicable.
2. In work of this class, cast-iron pipework is generally in short lengths, the cost of laying being greater than that carried out in long lengths. The data shown in this table refer to pipes laid in short lengths. For pipes laid in long lengths reference should be made to the data shown for laying and jointing various classes of pipes in open trenches shown earlier in this section.
3. The hours shown for fixing cast-iron pipes and specials are for fixing or laying them in the work only, and do not include for making the joints. For run lead, lead wool, flanged and other types of joints, see Tables 22.5, 22.6, 22.7 and 22.8
4. For excavation and timbering manholes, see chapter 12
5. For concrete foundations to manholes see chapter 13
6. For brickwork to manhole walls, see chapter 14
7. For channels, interceptors, etc. see Table 22.21
8. For precast concrete manholes, see Table 22.26 and 22.27

Description	Unit	Tradesman hours	Labourer hours
Concrete to benching hand mixed 150 mm (0.127 m^3 concrete)	m^2	0.86	0.86
Concrete to benching hand mixed 225 mm (0.191 m^3 concrete)	m^2	1.30	1.30
Brickwork:			
Cut horizontal chases, 75 mm girth	m	0.66	0.66
Cut horizontal chases, 150 mm girth	m	1.00	1.00
Cut horizontal chases, 225 mm girth	m	1.32	1.32
Cut vertical chases, 75 mm girth	m	1.00	1.00
Cut vertical chases, 150 mm girth	m	1.50	1.50
Cut vertical chases, 225 mm girth	m	2.00	2.00
Cutting circular	m^2	5.40	5.40
Cutting rough	m^2	2.70	2.70
Cutting splay	m^2	4.68	4.68
Cut holes through 112 mm walls for pipes up to 75 mm diameter and make good	Each	1.00	1.00
Cut holes through 112 mm walls for pipes 75 to 150 mm diameter and make good	Each	1.40	1.40
Cut holes through 112 mm walls for pipes 150 to 225 mm diameter and make good	Each	2.00	2.00
Cut holes through 225 mm walls for pipes up to 75 mm diameter and make good	Each	2.00	2.00
Cut holes through 225 mm walls for pipes 75 to 150 mm diameter and make good	Each	2.80	2.80
Cut holes through 225 mm walls for pipes 150 to 225 mm diameter and make good	Each	4.00	4.00
Cut holes through 350 mm walls for pipes up to 75 mm diameter and make good	Each	3.00	3.00
Cut holes through 350 mm walls for pipes 75 to 150 mm diameter and make good	Each	4.20	4.20
Cut holes through 350 mm walls for pipes 150 to 225 mm diameter and make good	Each	6.00	6.00
Hack face of brick walls for key	m^2	–	0.72
Toothing and bonding to existing work	m^2	6.30	6.30
Clinker of media placed in beds, filters, etc.	m^2	–	2.70
Concrete – Cement wash (0.36 kg cement per m^2)	m^2	–	0.12
Concrete – Hack faces of existing concrete for key	m^2	–	0.84
Concrete – Rubbing down faces	m^2	–	0.84
Fix dosing syphons	Each	3.50	3.50

Table 22.23 continued

Description	Unit	Tradesman hours	Labourer hours
Fix handstops, 150 mm wide	Each	0.70	0.70
Fix handstops, 225 mm wide	Each	0.90	0.90
Fix handstops, 300 mm wide	Each	1.10	1.10
Fix handstops, 375 mm wide	Each	1.30	1.30
Fix handstops, 450 mm wide	Each	1.50	1.50
Fix manhole covers, 50 kg (0.008 m³ mortar)	Each	0.50	1.00
Fix manhole covers, 100 kg (0.011 m³ mortar)	Each	0.60	1.20
Fix manhole covers, 150 kg (0.014 m³ mortar)	Each	0.70	1.40
Fix manhole covers, 200 kg (0.017 m³ mortar)	Each	0.80	1.60
Fix manhole covers, 250 kg (0.020 m³ mortar)	Each	0.90	1.80
Fix penstocks or disc valves, 75 mm diameter, exclusive of brackets	Each	0.50	1.00
Fix penstocks or disc valves, 100 mm diameter, exclusive of brackets	Each	0.56	1.12
Fix penstocks or disc valves, 150 mm diameter, exclusive of brackets	Each	0.70	2.10
Fix penstocks or disc valves, 225 mm diameter, exclusive of brackets	Each	0.90	2.70
Fix penstock or disc valve brackets for spindles	Each	0.80	0.80
Fix pipes, cast-iron flanged or socket and spigot in short lengths, exclusive of jointing:			
75 mm diameter	m	0.40	0.80
100 mm diameter	m	0.46	0.92
150 mm diameter	m	0.60	1.80
225 mm diameter	m	0.88	2.64
Fix flanged or socket and spigot bends or tapers, exclusive of jointing:			
75 mm diameter	Each	0.25	0.50
100 mm diameter	Each	0.30	0.60
150 mm diameter	Each	0.40	1.20
225 mm diameter	Each	0.50	1.50
Fix flanged or socket and spigot single junctions or tees, exclusive of jointing:			
75 mm diameter	Each	0.30	0.60
100 mm diameter	Each	0.40	0.80
150 mm diameter	Each	0.50	1.50
225 mm diameter	Each	0.60	1.80
Fix flanged or socket and spigot double junctions, exclusive of jointing:			
75 mm diameter	Each	0.35	0.70
100 mm diameter	Each	0.50	1.00
150 mm diameter	Each	0.60	1.80
225 mm diameter	Each	0.70	2.10
Fix ragbolts in existing brick walls	Each	0.60	–
Fix ragbolts in existing concrete walls	Each	0.80	–
Fix W.I. ladders	m	1.70	1.70
Fix W.I. screens	m²	5.44	5.44
Fix scum boards, exclusive of brackets	m²	2.70	2.70
Fix scum board brackets	Each	0.50	0.50
Fix sluice valves, exclusive of brackets:			
75 mm diameter	Each	0.50	1.00
100 mm diameter	Each	0.56	1.12
150 mm diameter	Each	0.70	2.10
225 mm diameter	Each	0.90	2.70
Fix brackets for sluice-valve spindles	Each	0.80	0.80
Fix step irons	Each	0.10	0.10

Table 22.24 Mixing and placing concrete in beds, haunches and surrounds to pipes in open trenches

Notes: 1. The data shown are for mixing, lowering and placing the concrete only per cubic metre, and do not allow
for wheeling or transporting, for which see chapter 13.
2. The data allow for lowering the concrete by means of a chute.

| | *Mixing, lowering and placing concrete, using concrete mixing plant of the sizes shown. Per cubic metre* | | | | | |
| | 7/5 | | 10/7 | | 14/10 | |
Description *Concreting, mixing, lowering and placing.* *The hours shown do not allow for wheeling* *and transporting. See note above*	*Mixing plant hours*	*Labour hours*	*Mixing plant hours*	*Labour hours*	*Mixing plant hours*	*Labour hours*
Concrete in beds, haunches and surrounds to pipes not exceeding 300 mm in diameter	0.88	5.28	0.64	5.12	0.56	5.04
Ditto, exceeding 300 mm and not exceeding 600 mm in diameter	0.85	5.10	0.63	5.04	0.55	4.95
Ditto, exceeding 600 mm and not exceeding 900 mm in diameter	0.83	4.98	0.61	4.88	0.53	4.97
Ditto, exceeding 900 mm and not exceeding 1200 mm in diameter	0.81	4.86	0.60	4.80	0.52	4.68
Exceeding 1200 mm in diameter	0.80	4.80	0.59	4.72	0.51	4.59

Table 22.25 Cubic metres of concrete per metre of pipe for various cross sections

Note: Concrete taken in all cases as being 150 mm wider than outside diameter of pipes on each side.

| *Cross section of concrete* | *Internal diameter of pipe in mm and volume of concrete in cubic metres per metre* | | | | | | | | | | | | | | | | | | |
	100	*150*	*225*	*300*	*375*	*450*	*525*	*600*	*675*	*750*	*825*	*900*	*975*	*1050*	*1125*	*1200*	*1275*	*1350*	*1500*
150 mm	0.09	0.10	0.12	0.13	0.18	0.20	0.26	0.30	0.33	0.39	0.44	0.46	0.51	0.54	0.58	0.63	0.66	0.70	0.75
150 mm	0.11	0.12	0.15	0.18	0.23	0.26	0.30	0.33	0.39	0.44	0.48	0.53	0.59	0.64	0.70	0.77	0.81	0.86	0.95
150 mm surround 150 mm	0.16	0.19	0.25	0.32	0.37	0.45	0.53	0.58	0.64	0.70	0.77	0.80	0.84	0.90	0.96	1.05	1.15	1.34	1.67
150 mm 100 mm	0.18	0.21	0.26	0.32	0.36	0.44	0.50	0.57	0.64	0.72	0.83	0.88	0.97	1.07	1.16	1.30	1.40	1.50	1.70
150 mm 150 mm	0.20	0.24	0.29	0.35	0.41	0.48	0.54	0.61	0.71	0.79	0.88	0.96	1.05	1.14	1.24	1.37	1.49	1.57	1.80
100 mm	0.05	0.06	0.07	0.08	0.09	0.10	0.11	0.12	0.13	0.14	0.15	0.15	0.16	0.17	0.18	0.20	0.21	0.21	0.23
150 mm	0.07	0.08	0.10	0.13	0.14	0.14	0.15	0.16	0.18	0.19	0.20	0.22	0.24	0.25	0.27	0.28	0.30	0.33	0.35

Precast circular concrete manholes

To estimate the cost of fixing precast concrete manholes, two tables are shown.

Table 22.26.   This shows:

1. The volume of concrete required in walls below the main manhole rings on which the first ring is set, this being given in m³ per 300 mm of depth. The width of the wall is taken as being 225 mm.
2. The volume of concrete required to surround the manhole rings and tapers with concrete 150 mm thick.

Table 22.27.   This gives data from which to estimate the cost of building the manhole.

The data shown assume that the manhole parts are delivered to the site adjacent to where they are fixed in the work and allow for offloading and fixing.

If the nature of the site is such that the manhole parts have to be off loaded and, at a later date, reloaded and distributed on the site, using suitable transport to do so, the cost of this rehandling should be allowed for in assessing the material cost of the various parts delivered on the site.

Table 22.26 Precast concrete manholes to BS 5911

Description	Unit	Internal diameter of manhole in metres and cubic metres of concrete required per 300 mm of depth				
		0.9 m	*1.02 m*	*1.20 m*	*1.35 m*	*1.50 m*
Concrete *in situ* below rings 225 m wide to form base on which to place the first ring	m³ per 300 mm of depth	0.27	0.30	0.37	0.42	0.48
Concrete surround 150 mm thick to rings:	m³ per 300 mm					
Internal diameter, 675 mm	of depth	0.14				
Internal diameter, 900 mm	"	0.17				
Internal diameter, 1050 mm	"	0.20				
Internal diameter, 1200 mm	"	0.22				
Internal diameter, 1350 mm	"	0.24				
Internal diameter, 1500 mm	"	0.28				
Concrete surround 150 mm thick to tapers, 600 mm high	Per taper					
900/675	"	0.30				
1050/675	"	0.33				
1050/675	"	0.37				
1050/675	"	0.42				
1050/675	"	0.45				

Table 22.27 Building precast circular concrete manholes

Notes: 1. For excavation and timbering of manholes, see chapter 12
   2. For fixing clayware or concrete channels, see Table 22.21
   3. For concreting manhole foundations, etc. see chapter 13

Description	Unit	Concrete required in benchings, m^3	Mortar required per joint m^3	Manhole fixer hours	Labourer hours
Fix 675 mm diameter shaft rings	m		0.006	1.00	3.00
Fix 900 mm manhole rings	m		0.009	1.65	4.95
Fix 1200 mm manhole rings	m		0.018	2.64	7.92
Fix 1500 mm manhole rings	m		0.035	3.63	10.89
Fix 900/675 mm tapers	Each		0.009	1.00	3.00
Fix 1200/675 mm tapers	Each		0.018	1.60	4.80
Fix 1500/675 mm tapers	Each		0.035	2.20	6.60
Fix 900 mm diameter precast bases	Each			0.90	2.70
Fix 1200 mm diameter precast bases	Each			1.50	4.50
Fix 1500 mm diameter precast bases	Each			2.10	6.30
Fix cover slabs to 675 mm rings	Each		0.007	0.40	1.20
Fix cover slabs to 900 mm rings	Each		0.009	0.50	1.50
Fix cover slabs to 1200 mm rings	Each		0.024	0.70	2.10
Fix cover slabs to 1500 mm rings	Each		0.042	1.10	3.30
Concrete to benchings, hand mixed:					
150 mm thick	m^2	0.127		0.86	0.86
225 mm thick	m^2	0.191		1.30	1.30
300 mm thick	m^2	0.255	–	1.62	1.62
375 mm thick	m^2	0.319	–	1.94	1.94
450 mm thick	m^2	0.382	–	2.27	2.27
Render benchings 25 mm thick	m^2	–	0.028	0.90	0.90
Fix manhole covers, 50 kg	Each	–	0.085	0.60	0.60
Fix manhole covers, 100 kg	Each	–	0.099	0.70	0.70
Fix manhole covers, 150 kg	Each	–	0.113	0.80	0.80
Fix manhole covers, 200 kg	Each	–	0.127	0.90	0.90
Fix manhole covers, 250 kg	Each	–	0.141	1.00	1.00
Fix manhole covers, 300 kg	Each	–	0.155	1.10	1.10
Fix drum centring:					
900 mm diameter per m of height	Each	–	–	0.66	1.98
1050 mm diameter per m of height	Each	–	–	0.82	2.46
1200 mm diameter per m of height	Each	–	–	1.00	3.00
1350 mm diameter per m of height	Each	–	–	1.15	3.45
1500 mm diameter per m of height	Each	–	–	1.30	3.90
Build in pipes and make good in concrete *in situ* base:					
100 mm diameter	Each	–	–	0.20	–
150 mm diameter	Each	–	–	0.25	–
225 mm diameter	Each	–	–	0.35	–

Table 22.28 Widths of trench for laying clayware pipes and concrete tubes

Notes: 1. The widths shown are average widths for the depths shown and allow for concrete beds, haunching and surrounds to pipes if required, the concrete being 150 mm wider each side than the pipes.
2. The widths shown are for normal grounds such as firm sand, loamy soil or clay, chalk, gravel, etc., timbering being allowed for at the greater depths.

Diameter of pipe in millimetres	Depth of trench in metres and average width of trench required in metres										
	0.6	0.9	1.2	1.5	1.8	2.1	2.4	2.7	3.0	3.3	3.6
100	0.43	0.46	0.53	0.61	0.68	0.76	0.76	0.76	0.76	0.76	0.76
150	0.48	0.48	0.53	0.61	0.68	0.76	0.76	0.76	0.76	0.76	0.76
225	0.58	0.61	0.61	0.68	0.76	0.76	0.76	0.76	0.76	0.76	0.76

Table 22.29 Widths of trench for laying cast iron pipes

Notes: 1. The widths shown are average widths for the depths shown and allow for concrete beds, haunching and surrounds. If such should be required the data shown for stoneware pipes and concrete tubes in the previous table may be used.
2. The widths shown are for normal grounds such as firm sand, loamy soil or clay, chalk, gravel, etc., timbering being allowed for at the greater depths.

Diameter of pipe in millimetres	Depth of trench in metres and average width of trench in metres					
	0.60	0.90.	1.20	1.50	1.80	2.10
75	0.37	0.45	0.52	0.60	0.65	0.75
100	0.37	0.45	0.52	0.60	0.65	0.75
125	0.40	0.45	0.52	0.60	0.65	0.75
150	0.42	0.45	0.52	0.60	0.65	0.75
175	0.45	0.48	0.52	0.60	0.72	0.75
200	0.48	0.50	0.52	0.60	0.72	0.75
225	0.50	0.52	0.52	0.60	0.72	0.75

Granular soil trench bed and backfill for UPVC Pipe

The bed should comprise a 100 mm thick mixture of one part free-draining coarse sand to two parts gravel, broken stone, crushed brick or broken concrete to pass a 25 mm gauge and be retained on a 12 mm gauge. Selected filling from bed to top of pipe should comprise filling material free from rubbish or stones retained on a 25 mm gauge. The next 300 mm depth of backfilling may be trench excavation material lightly rammed with a hand rammer. Remaining filling may be filled and rammed by machine or hand as for normal backfilling.

Table 22.30 Granular bed and backfill

Description	Unit	Labourer hours
Bed, surround and backfilling as described		
to 100 mm pipe	m	0.20
to 150 mm pipe	m	0.30

Times for labour, mechanical excavation, planking and strutting are shown in chapter 12. The above labour times are for grading and filling with approved material to bed and surround and for extra labour ramming by hand to a depth of 300 mm above pipe per linear metre of trench.

Polypropylene manholes

Polypropylene manholes are increasingly used on housing sites due to the advantages of major savings in time and cost.

23 Landscaping

Landscaping is an important element in major building and civil engineering projects, no doubt due to the increasing importance or emphasis given to the environment by interested clients and official bodies such as the National Trust and English Heritage.

Good landscaping associated with new development contributes to greater confidence in both commercial and residential areas and to widespread benefits. Many derelict sites have been transformed and Garden Festival Sites have been successful with consequent advantages to these areas.

The cumulative impact of small scale schemes can be seen in the encouraging initiative of large scale property owning and managing companies who have taken the opportunity to implement much new landscape to the benefit of local habitat and landscape interest in association with the long term programmes of works improvements. In the housing sector, it is encouraging to see that the Green Leaf Housing awards are being given in an ever increasing number.

Table 23.1 Soft landscaping – site works

Description	Unit	Labourer hours
Hedges – grub up and burn not exceeding 1.50 m high.	m	3
Hedges – grub up and burn 1.50 to 2.00 m high.	m	3.75
Hedges – grub up and burn 2.00 to 3.00 m high.	m	4.50
Trees – felling, cutting up and load or burn on site not exceeding 600 mm girth.	Each	16
Trees – felling, cutting up and load or burn on site 600 mm to 1500 mm girth.	Each	42
Trees – felling, cutting up and load or burn on site 1500 mm to 3000 mm girth.	Each	52

Table 23.2 Soft landscaping – seeding

Description	Materials	Exceeding 1500 m²	1001–1500 m²	501–1000 m²	N/e 500 m²
Clear site of rubbish	–	0.03	0.03	0.03	0.03
Strip site of surface vegetation	–	0.02	0.02	0.04	0.04
Cultivate 150 mm deep, remove stones and vegetable matter	–	0.02	0.04	0.04	0.07
Grade	–	0.03	0.03	0.03	0.03
Imported topsoil 150 mm deep spread and levelled	0.15 m³	0.33	0.33	0.34	0.36
Fork and rake	–	0.01	0.01	0.01	0.01
Dress with bonemeal lightly raked in	80 g per m²	0.04	0.04	0.04	0.04
Sow with seed mixture	45 g per m²	0.05	0.05	0.08	0.08
Twice roll		0.02	0.02	0.02	0.02
Scythe to top		0.007	0.007	0.007	0.007
Keep grass mown to height of 50 mm during contract		0.02	0.02	0.02	0.02
Maintenance items:					
Roll in two directions		0.01	0.01	0.01	0.01
Scythe to reduce		0.01	0.01	0.01	0.01
Twice box mow		0.03	0.03	0.03	0.03
Dress with fish manure	25 g per m²	0.04	0.04	0.04	0.04
Water as necessary		0.011	0.011	0.011	0.011

Table 23.3 Soft landscaping – turfing

Description	Materials	Exceeding 1500 m²	1001–1500 m²	501–1000 m²	N/e 500 m²
Clear site of rubbish		0.03	0.03	0.03	0.03
Strip site of surface vegetation		0.02	0.02	0.04	0.04
Lifting turf for preservation and stacking		0.35	0.35	0.40	0.40
Grade		0.03	0.03	0.03	0.03
Imported topsoil 100 mm deep spread and levelled	0.10 m²	0.22	0.23	0.24	0.25
Fork and rake		0.01	0.01	0.01	0.01
Dress with bonemeal lightly raked in	80 g per m²	0.04	0.04	0.04	0.04
25 mm turfs and laying		0.24	0.26	0.27	0.28
Twice roll		0.02	0.02	0.02	0.02
Scythe to top		0.009	0.009	0.009	0.009
Scythe to reduce		0.009	0.009	0.009	0.009
Keep grass mown to height of 50 mm during contract		0.02	0.02	0.02	0.02
Maintenance items:					
Roll in two directions		0.01	0.01	0.01	0.01
Scythe to reduce		0.01	0.01	0.01	0.01
Twice box mow		0.03	0.03	0.03	0.03
Top dress with fine sifted soil brushed into joints	0.007 m³	0.013	0.013	0.013	0.013
Dress with fish manure	25 g per m²	0.04	0.04	0.04	0.04
Water as necessary		0.011	0.011	0.011	0.011

Table 23.4 Soft landscaping – work to sports fields etc.

Description	Unit	Plant	Plant hours	Labourer hours
Roll grass one way	hectare	2 tonne petrol	1.25 (0.001 m²)	–
Roll topsoil 150 mm thick one way with tractor drawn sheepsfoot roller	hectare	tractor drawn sheepsfoot roller	0.75 (0.001 m²)	–
Roll hoggin 150 mm thick one way with tractor drawn sheepsfoot roller	hectare	tractor drawn sheepsfoot roller	0.80 (0.001 m²)	–
Cultivate	hectare	tractor drawn cultivator	1.00 (0.001 m²	–
Cut land drain trench	metre	trenching plough	0.006	0.006
Dress fertiliser 1 tonne per hectare	hectare	driver and drill	3.00	–
Grass seeding 40 to 50 kg per hectare	hectare	driver and seeder	1.50	3.00
Harrow once over	hectare	tractor drawn disc harrow	0.75	–
Lime over area	hectare	driver and drill	5.50	
Plough once over	hectare	tractor drawn disc plough	4.50	
Watering	hectare	water cart and spraying bar	2.50	5.00
Weed spraying	hectare	water cart and spraying bar	3.00	6.00

Table 23.5 Soft landscaping – trees

Description	Unit	Material	Labourer hours
Clear rubbish, debris and vegetation, excavate pit, dispose of excavated material.			
Pit size 0.90 × 0.90 × 0.60 m deep	Each	–	2.90
Pit size 1.20 × 1.20 × 1.00 m deep	Each	–	6.30
Pit size 2.00 × 2.00 × 1.25 m deep	Each	–	12.00
Cultivate 150 mm deep to bottom of pit.			
Pit size 0.90 × 0.90 m	Each	–	0.05
Pit size 1.20 × 1.20 m	Each	–	0.10
Pit size 2.00 × 2.00 m	Each	–	0.27
Manure 100 mm deep to bottom of pit.			
Pit size 0.90 × 0.90 m	Each	0.081 m³	0.10
Pit size 1.20 × 1.20 m	Each	0.144 m³	0.15
Pit size 2.00 × 2.00 m	Each	0.400 m³	0.31
Imported topsoil to:			
Pit size 0.90 × 0.90 × 0.60 m deep	Each	0.486 m³	1.07
Pit size 1.20 × 1.20 × 1.00 m deep	Each	1.44 m³	3.20
Pit size 2.00 × 2.00 × 1.25 m deep	Each	5.00 m³	10.50
Finish soil to even levels to:			
Pit size 0.90 × 0.90 m	Each		0.03
Pit size 1.20 × 1.20 m	Each		0.04
Pit size 2.00 × 2.00 m	Each		0.12
Clear weed growth and rubbish before planting to:			
Pit size 0.90 × 0.90 m	Each		0.03
Pit size 1.20 × 1.20 m	Each		0.09
Pit size 2.00 × 2.00 m	Each		0.42

Table 23.5 continued

Description	Unit	Material	Labourer hours
Peat 50 mm deep forked with upper layer of topsoil to:			
Pit size 0.90 × 0.90 m	Each	0.04 m³	0.13
Pit size 1.20 × 1.20 m	Each	0.07 m³	0.23
Pit size 2.00 × 2.00 m	Each	0.20 m³	0.63
Plant only the following:			
Transplants	Each		0.08
Whips	Each		0.10
Feathered tree	Each		0.13
Half standard tree	Each		0.15
Light standard tree	Each		0.17
Standard tree	Each		0.17
Selected standard tree	Each		0.20
Heavy standard tree	Each		0.22
Extra heavy standard tree	Each		0.25
Semi-mature tree	Each		2.25
Water at time of planting:			
Pit size 0.90 × 0.90 m	Each		0.03
Pit size 1.20 × 1.20 m	Each		0.07
Pit size 2.00 × 2.00 m	Each		0.07
Fertiliser John Innes based to:			
Pit size 0.90 × 0.90 m	Each	56 g	0.01
Pit size 1.20 × 1.20 m	Each	100 g	0.02
Pit size 2.00 × 2.00 m	Each	280 g	0.03
Lightly cultivate after planting to:			
Pit size 0.90 × 0.90 m	Each		0.02
Pit size 1.20 × 1.20 m	Each		0.04
Pit size 2.00 × 2.00 m	Each		0.11
Turf around tree position after planting:			
0.90 × 0.90 to form opening		2 No.	
0.60 × 0.60 m	Each	turfs	0.23
1.20 × 1.20 to form opening		4 No.	
0.60 × 0.60 m	Each	turfs	0.44
2.00 × 2.00 to form opening		14 No.	
0.60 × 0.60 m	Each	turfs	1.18
Stake – material 100 mm dia. × 3 m in length, hazel or chestnut.	Each		0.29
Ties – material – 3 No. rubber covered canvas strip 25 mm wide, 12 No. 40 mm galvanised nails.	Each		0.10
Guard – material – 1 No. galvanised metal mesh size 1.79 × 0.943 mm, 6 No. galvanised metal clips.	Each		1.53
Remove weed growth and rubbish, light cultivation to tree position 0.60 × 0.60 m	Each visit		0.08
Watering to tree position 0.60 × 0.60 m	Each visit		0.04
Mulch 80 mm deep to tree positions 0.60 × 0.60 m. Material spent compost or pulverised bark 0.065 m³			0.16
Prune tree	Each visit		0.15
Spats to protect against rabbit damage	Each		0.02

Table 23.6 Hard landscaping – precast concrete paving

Description	Unit	Labourer hours
Take up, wheel and load onto lorry 50 mm pcc paving slab	m²	0.30
" " " 63 mm "	m²	0.36
" " " 75 mm "	m²	0.42
Take up, wheel and stack for re-use 50 mm pcc paving slab	m²	0.80
" " " 63 mm "	m²	0.85
" " " 75 mm "	m²	0.90
Take from stack and relay 50 mm pcc on new mortar bed	m²	0.70
" " 63 mm "	m²	0.85
" " 75 mm "	m²	1.00
Cutting out fitting 50 mm pcc paving slab around 1.20 × 1.20 tree pit	Each	1.20
Ditto. 63 mm	Each	1.45
Ditto. 75 mm	Each	1.70
Take up, wheel away and load onto lorry Grasscrete concrete precast slabs size 366 × 274 × 100 mm	m²	4.60

Table 23.7 Hard landscaping – stone paving and gravel surfacing

Description	Unit	Material required			Paving	Plant 2 tonne roller	Labour	
		Mortar m³ per m²	Stone tonne per m²	Gravel m³ per m²			Pavior	Labourer
50 mm artificial stone paving	m²	0.03	–	–	1.00	–	0.36	0.36
63 mm artificial stone paving	m²	0.03	–	–	1.00	–	0.48	0.48
75 mm artificial stone paving	m²	0.03	–	–	1.00	–	0.60	0.60
50 mm crazy paving	m²	0.04	0.12	–	–	0.031	1.20	1.20
50 mm gravel surfacing	m²	–	–	0.084	–	0.036	–	0.13
75 mm gravel surfacing	m²	–	–	0.126	–	0.046	–	0.15
50 mm stone paving	m²	0.03	–	–	1.00	–	0.48	0.48
75 mm stone paving	m²	0.03	–	–	1.00	–	0.79	0.79

Table 23.8 Hard landscaping – paving

Description	Unit	Materials	Labour
Grasscrete concrete precast concrete slabs 366 × 274 × 100 mm thick with 3 No. longitudinal and 4 No. transverse ribs enclosing 6 No. grass cells.	m²	10 No. slabs 0.03 m³ fine soil filling 0.02 m sand bedding 45 gm seed.	1.15
Speed regulator ('sleeping policeman') precast concrete paving unit.	each	1 No. 500 × 500 × 130 mm high overall 0.003 m³ mortar	0.35

Table 23.9 Hard landscaping – granite setts

	Unit	Material required	Pavior hours	Labourer hours
Remove old setts	m²	–	–	0.33
Clean old setts for reuse	m²	–	–	1.05
25 mm bed of cement and sand	m²	25 mm of semi-dry cement and sand bed	0.21	0.21
38 mm bed of cement and sand	m²	38 mm of semi-dry cement and sand bed	0.29	0.29
Relay old setts and grout in.	m²	0.034 m³ cement mortar	0.67	0.67
Lay new setts				
100 × 100 × 100 mm long and grout in	m²	0.012 m³ cement mortar	0.21	0.21
Lay new setts				
100 × 125 × 200 mm long and grout in	m²	0.028 m³ cement mortar	0.50	0.50
Raking cutting	m		0.27	0.07
Curved cutting	m		0.11	0.11

Table 23.10 Multipliers for use with the table for granite sett paving for areas less than 50 square metres

Area m²	Pavior and labourer hours multipliers
0–2	1.84
2–5	1.67
5–10	1.50
10–25	1.34
25–50	1.17
Over 50	

Table 23.11 Cobble or pebble paving – materials

Sizes	Nominal mm	Laid flat touching square metre/tonne	Laid vertically square metre/tonne
Small	38–50	10.00	6.66
Medium	51–75	8.33	5.50
Large	76–100	6.66	4.33

Table 23.12 Lay cobble or pebble paving

Description	Unit	Material required	Pavior hours	Labourer hours
Small	m²	0.09 m³ cement mortar	0.42	0.42
Medium	m²	0.24 m³ cement mortar	0.51	0.51
Large	m²	0.39 m³ cement mortar	0.61	0.61

Table 23.13 Precast concrete block paving

Description	Unit	Washed sharp sand	Blocks	Plate vibrator hours	Half bond Pavior	Half bond Labourer	Herringbone Pavior	Herringbone Labourer
65 mm chamfered rectangular 'Charcon' Europa interlocking concrete block paving 200 mm long × 100 mm wide	m²	0.001	50	0.09	1.10	0.55	1.20	0.60
80 mm chamfered rectangular 'Charcon' Europa interlocking concrete block paving 200 mm long × 100 mm wide	m²	0.001	50	0.09	1.25	0.63	1.35	0.67
Raking cutting 65 mm	m	–	–	–	0.10	0.05	0.13	0.07
Curved cutting 65 mm	m	–	–	–	0.13	0.07	0.16	0.08
Straight cutting	m	–	–	–	–	–	0.13	0.13

	300 mm girth Pavior	300 mm girth Labourer	1000 mm girth Pavior	1000 mm girth Labourer
Cutting and fitting paving around manhole covers, bollards, seats, etc.	0.05	0.02	0.07	0.03

Table 23.14 Charcon 'Safeticurb' precast concrete surface water drainage system

Type	Size	Unit	Pavior hours	Labourer hours
DBA	250 × 250 × 914 mm long block with 125 mm bore.	m	0.80	0.40
DBG/CI	250 × 250 × 914 mm long block with cast iron grids and 125 mm bore.	m	0.80	0.40
DBM	250 × 250 × 914 mm long block with 125 mm bore and cast iron inset along the full length of the drainage slot.	m	0.80	0.40

Table 23.15 Hard landscaping – site furniture

Description	Unit	Materials	Labourer hours
Seats:			
Unload, wheel, assemble and place in position 1200 mm long seat	Each		0.33
● Ground fixing:			
Excavate for foundations 0.30 × 0.30 × 0.60 m deep and dispose	Each		1.20
Concrete foundation, wheel and place	Each	0.108 m³ concrete	0.96
● Pedestal fixing:			
Excavate for base 0.60 × 0.375 × 0.150 m deep and dispose	Each		
Concrete foundation, wheel and place	Each	0.668 m³ concrete	
Fix bolts	Each	4 No. G18 galvanised raw bolts	0.55
● Wall mounting:			
Drill mortice in brickwork	Each		0.60
Fix bolts	Each	4 No. G18 galvanised raw bolts	0.55
Litter bins:			
Unload, wheel and place in position steel/hardwood litter bin 465 mm diameter × 830 mm high	Each		0.75
● Ground fixing:			
Excavate for foundations 0.30 × 0.30 × 0.30 m deep and dispose	Each		0.55
Concrete foundations, wheel and place	Each	0.027 m³ concrete	0.24
● Pedestal fixing:			
Excavation as above	Each		0.55
Concrete as above	Each	0.027 m³ concrete	0.24
Fix bolts	Each	4 No. D20 galvanised	0.55
Bollards:			
Unload, wheel and place in position oak bollard 0.12 × 0.12 × 1.50 m long set 0.40 m in ground	Each		0.60
Excavate 0.50 × 0.50 × 0.40m deep and dispose	Each		2.04
Concrete foundations	Each	0.10m³	0.88
Planters:			
Unload, wheel and place in position free standing precast concrete planter 0.60 m diameter × 0.47 m high	Each		1.50
Cycle blocks:			
Unload, wheel and place in position precast concrete block 0.60 × 0.30 × 0.095 m deep overall	Each		0.18

Fencing

This section deals with those fences most commonly erected, where the posts are of concrete, wood, angle or tee iron.

Boundary walls constructed of brickwork, concrete or stone also come under the category of fences, and are really fences in a very substantial form. They are not, however, included under this section, since data for estimating their cost can be found under the appropriate trade sections:

Brick walls: see chapter 14
Concrete walls: see chapter 13

In fences of the post or standard types, not only may the posts be fixed at various centres, but they may have fixed to them various attachments for the fence proper, such as:

Barbed wire   Galvanised wire
Cleft chestnut pales   Wire netting
Chain link   Wood rails etc.
Close boarding

Estimating the cost of erecting fencing

In estimating the cost of erecting fencing, which is generally billed per metre complete, two points have to be considered:

1. The distance apart at which the posts have to be fixed.
2. The attachments which have to be fixed to the posts.

In (1), the posts may have to be fixed at distances apart of other than one metre, and in most cases they are actually fixed at centres considerably in excess of this, 2 to 2.5 m apart being quite common.

In (2), having fixed the posts at the requisite distance apart, there may be attached to these posts various types of fencing, such as galvanised wire, barbed wire, chain link mesh, etc.

Many combinations of fencing may, therefore, present themselves, and the estimator must have at his disposal data from which to estimate the cost of erection for all possible combinations. The tables shown under this section are drawn up in such a way that:

● The cost of excavating and fixing one post can readily be calculated.
● The cost of fixing wire, chain link, wooden rails, etc. can be estimated per metre of fence.

The method of calculating the cost of erecting fencing per metre complete with all posts and attachments then simply becomes:

$$\text{Cost of erecting fence complete per metre} = \frac{\text{Cost of excavating and fixing one post}}{\text{Metres between posts}} + \frac{\text{Cost of fixing attachments per metre of fence}}{}$$

Note: 1. The attachments may be wire, chain link or other fencing material.
2. The labour hours shown in the tables refer to those of a tradesman, or a labourer, as is appropriate, taking into consideration the nature of the work.

Table 23.16 Fences with angle or tee iron standards and stays

Description	Unit	Labour hours
Excavate holes for standards 0.6 m deep	Each	0.50
Excavate holes for standards 0.9 m deep	Each	0.90
Fix standards 1.5 m overall length	Each	0.27
Fix standards 1.8 m overall length	Each	0.40
Fix standards 2.1 m overall length	Each	0.80
Fix standards 2.7 m overall length	Each	1.50
Drive standards 0.45 m into ground	Each	0.20
Concrete to feet of standards, mix, wheel up to 22 m and place	m^3	9.80
Fix stays to 1.6 m standards	Each	0.45
Fix stays to 1.8 m standards	Each	0.63
Fix stays to 2.1 m standards	Each	1.20
Fix stays to 2.7 m standards	Each	2.10
Fix single strand No.6 gauge wire	m	0.07
Fix single strand No.10 gauge wire	m	0.04
Fix single strand No.14 gauge wire	m	0.03
Fix single line of barbed wire	m	0.11
Fix round bars up to 15 mm diameter through standards	m	0.17
Fix square bars up to 19 mm square through standards	m	0.22
Fix chain link fencing 50 mm mesh, No.12 gauge, 0.9 m high	m	0.17
Fix chain link fencing 50 mm mesh, No.12 gauge, 1.2 m high	m	0.28
Fix chain link fencing 50 mm mesh, No.12 gauge, 1.5 m high	m	0.39
Fix chain link fencing 50 mm mesh, No.12 gauge, 1.8 m high	m	0.50
Fix wire netting 0.6 m high	m	0.10
Fix wire netting 0.9 m high	m	0.12
Fix wire netting 1.2 m high	m	0.15
Fix wire netting 1.5 m high	m	0.22
Fix wire netting 1.8 m high	m	0.31

Table 23.17 Fences with concrete posts

Description	Unit	Labour hours
Excavate post holes 0.6 m deep	Each	0.50
Excavate post holes 0.9 m deep	Each	0.90
Fix concrete posts 1.5 m overall length	Each	0.38
Fix concrete posts 1.8 m overall length	Each	0.54
Fix concrete posts 2.1 m overall length	Each	1.10
Fix concrete posts 2.7 m overall length	Each	2.00
Concrete to foot of posts, mix, wheel up to 22 m and place	m^3	8.78
Fix struts to 1.5 m posts	Each	0.60
Fix struts to 1.8 m posts	Each	0.80
Fix struts to 2.1 m posts	Each	1.60
Fix struts to 2.7 m posts	Each	2.90
Fix eyebolts	Each	0.20
Fix single strand No.6 gauge wire	m	0.07
Fix single strand No.10 gauge wire	m	0.04
Fix single strand No.14 gauge wire	m	0.03
Fix single line of barbed wire	m	0.11
Fix chain link fencing 50 mm mesh, No.12 gauge, 0.9 m high	m	0.22
Fix chain link fencing 50 mm mesh, No.12 gauge, 1.2 m high	m	0.33
Fix chain link fencing 50 mm mesh, No.12 gauge, 1.5 m high	m	0.44
Fix chain link fencing 50 mm mesh, No.12 gauge, 1.8 m high	m	0.55
Fix wire netting 0.6 m high	m	0.10
Fix wire netting 0.9 m high	m	0.12
Fix wire netting 1.2 m high	m	0.15
Fix wire netting 1.5 m high	m	0.22
Fix wire netting 1.8 m high	m	0.31
Fix 25 mm diameter galvanised tubing	m	0.33
Fix 50 mm diameter galvanised tubing	m	0.66

Table 23.18 Cleft chestnut fencing

Note: The data shown are for fixing the fencing complete with posts per metre the posts being driven in.

Description	Unit	Posts at 1.8 m centres Labour hours	Posts at 2.7 m centres Labour hours
Fix fencing 0.9 m high	m	0.39	0.29
Fix fencing 0.9 m high	m	0.50	0.40
Fix fencing 0.9 m high	m	0.60	0.50

Table 23.19 Fences with wooden posts

Description	Unit	Labour hours
Excavate post holes 0.6 m deep	Each	0.50
Excavate post holes 0.9 m deep	Each	0.90
Fix wooden posts 1.5 m long overall	Each	0.30
Fix wooden posts 1.8 m long overall	Each	0.45
Fix wooden posts 2.1 m long overall	Each	0.90
Fix wooden posts 2.7 m long overall	Each	1.20
Drive posts 0.45 m into ground	Each	0.25
Tar foot of wooden posts, 0.22 litres tar per post	Each	0.12
Concrete to foot of posts, mix, wheel up to 22 m and place	m^3	8.78
Fix struts to 1.5 m posts	Each	0.50
Fix struts to 1.8 m posts	Each	0.70
Fix struts to 2.1 m posts	Each	1.30
Fix struts to 2.4 m posts	Each	1.80
Fix arris rails out of 100 × 75 mm	m	0.20
Fix timber rails out of 100 × 50 mm	m	0.17
Fix 25 mm diameter galvanised tubing	m	0.33
Fix 50 mm diameter galvanised tubing	m	0.66
Fix shiplap boarding 0.9 m high	m	0.33
Fix shiplap boarding 1.2 m high	m	0.40
Fix shiplap boarding 1.5 m high	m	0.46
Fix shiplap boarding 1.8 m high	m	0.53
Fix gravel boarding	m	0.26
Fix single strand No.6 gauge wire	m	0.07
Fix single strand No.10 gauge wire	m	0.05
Fix single strand No.14 gauge wire	m	0.04
Fix single line of barbed wire	m	0.11

Table 23.20 Sundry fencing works

Description	Unit	Labour hours
Fix barrel bolts	Each	0.50
Fix gates – field type; hang only	Each	0.80
Fix gates – double hung; hang only	Each	1.40
Fix hinges, heavy type	Per pair	2.00
Hooks for gate hinges	Per pair	0.60
Fix latches	Each	0.50
Creosote timber fences, close boarded, 0.43 litres per square metre	m^2	0.30
Tarring timber fences, close boarded, 0.54 litres tar per square metre	m^2	0.40

Sports facilities

Sports and leisure facilities are increasingly incorporated into major building and civil engineering projects – In many instances as planning gain for large supermarket or retail park schemes. No doubt this is due to the increasing importance or emphasis given to Sports and Leisure by interested clients, including local authorities and the activities of the Sports Council with funding from the National Lottery.

The Sports Council was incorporated by Royal Charter in 1972 and its main objectives are to increase participation in sport and physical recreation, to increase the quantity and quality of sports facilities, to raise standards of performance and to provide information about sport. The Technical Unit for Sport produces a number of very useful publications.

The National Playing Field Association Cost Guide for budget costing for Sports Facilities, Buildings and Childrens Play Facilities is also a very useful guide.

Table 23.21 Sports facilities

Description	Time/ganger hours
Fix shot stop board with bolts	1
Fix hammer/shot circle cast into concrete	2
Discus circle	2
Discus cage to comply with IAAF (eight posts)	8
Long jump/triple jump foundation trough	1
Long jump/triple jump blanking board	0.5
Long jump/triple jump take off board	0.5
Long jump/triple jump no-jump indicator	0.2
Set of finishing posts with sockets	1

24 *Weight of materials*

Table 24.1 General

Material	Tonnes per cubic metre	Cubic metres per tonne	Litres per tonne
Ashes	0.85	1.18	
Ballast, all in	1.80	0.56	
Bitumen	1.37	0.73	
Bituminous emulsion	–	–	1000
Blockwork concrete	1.44	0.69	
Brickwork, solid – pressed bricks	2.12	0.47	
Brickwork, solid – ordinary	1.92	0.53	
Cement – aluminous	1.40	0.71	
Cement – Portland	1.44	0.70	
Cement – rapid hardening	1.28	0.79	
Chalk	2.24	0.44	
Clay	1.92	0.52	
Coke	0.57	1.70	
Concrete – ballast	2.24	0.45	
Earth top soil	1.60	0.62	
Earth vegetable	1.23	0.82	
Flint	2.59	0.38	
Gravel	1.50	0.67	
Gravel, coarse with sand	1.76	0.57	
Lime, ground – quicklime	0.96	1.04	
Lime, slaked	0.48	2.08	
Loam	1.60	0.66	
Marl	1.76	0.57	
Media – filter	0.88	1.13	
Pitch	1.16	0.87	
Sand, fine, clean pit	1.44	0.70	
Sand, medium pit	1.53	0.65	
Sand, Thames or washed river	1.69	0.59	
Shale	2.60	0.38	
Slag	1.51	0.67	
Slate	2.89	0.34	
Snow – freshly fallen	0.12	8.33	
Snow – old lying and compacted	0.52	1.92	
Stone – solid, basalt	2.77	0.36	
Stone – solid, Bath	2.00	0.50	
Stone – solid, granite	2.67	0.38	
Stone – solid, Kentish rag	2.64	0.38	
Stone – solid, limestone	2.41	0.41	
Stone – solid, Portland	2.44	0.41	
Stone – solid, Purbeck	2.60	0.39	
Stone – solid, sandstone	2.33	0.44	
Stone – solid, traprock	2.73	0.37	
Stone – solid, whinstone	2.77	0.36	
Tar	–	–	873

Table 24.2 Weight of crushed materials

Note: The data shown are based on 45 per cent voids.

Nature of material	Weight in the solid. Kilograms per cubic metre	Weight of crushed material	
		Tonnes per cubic metre	Cubic metres per tonne
Basalt	2809	1.53	0.65
Brick	2123	1.16	0.86
Clinker	–	0.80	1.24
Concrete	2286	1.24	0.81
Granite	2711	1.48	0.67
Kentish rag	2694	1.47	0.68
Limestone	2449	1.33	0.69
Sandstone	2367	1.29	0.75
Shingle	–	1.51	0.66
Slag, cold blast	2580	1.40	0.72
Slag, hot blast	2531	1.37	0.73
Whinstone	2809	1.53	0.65

Table 24.3 Weight of road foundation materials

Note: The data shown are based on 45 per cent voids

Nature of material	Weight in the solid. Kilograms per cubic metre	Weight in the loose as delivered on the site	
		Tonnes per cubic metre	Cubic metres per tonne
Ashes	–	0.96	1.04
Clinker	–	0.80	1.24
Gravel	–	1.52	0.66
Hardcore, brick	2123	1.16	0.86
Hardcore, chalk	2286	1.24	0.81
Hardcore, concrete	2286	1.24	0.81
Pitching, granite	2711	1.48	0.68
Pitching, limestone	2367	1.29	0.78
Pitching, sandstone	2449	1.33	0.75
Pitching, whinstone	2804	1.33	0.65

Table 24.4 Weight of road surfacing materials

Nature of material	Weight in the solid. Kilograms per cubic metre	Weight in the loose as delivered on the site	
		Tonnes per cubic metre	Cubic metres per tonne
Asphalt, bottom course	–	2.12	0.51
Asphalt, top course	–	2.24	0.45
Asphalt. mastic	2711	2.33	0.43
Tarred granite	–	1.53	0.62
Tarred gravel	2449	1.53	0.65
Tarred limestone	2531	1.57	0.64
Tarred slag	2804	1.40	0.70
Tarred whinstone	–	1.56	0.61

Table 24.5 Approximate weight of bricks

Nature of brick	Approximate weight of one brick in kg	Approximate weight of 1000 bricks in tonnes	Approximate number of bricks per tonne
Flettons	2.54	2.50	400
Facing bricks	2.72	2.68	373
Stocks	3.06	3.00	332
Firebricks	3.17	3.12	320
Wirecuts	3.26	3.21	311
Pressed bricks	3.62	3.57	280
Blue Staffs	3.97	3.90	256
Heavy engineering bricks	4.20	4.13	242

Table 24.6 Approximate weight of brickwork

Type of brick	Approximate weight of brickwork in kg per m^3
Facing bricks	2025
Flettons	1602
Staffordshire blue wirecuts	2123
Staffordshire blue pressed facing bricks	2188
Stocks	1945
Wirecuts	2091
Heavy engineering bricks	2286

Table 24.7 Approximate weight of standard (4N) concrete block

Thickness of slab in millimetres	Weight per square metre in kilograms
75	56
100	75
150	113
200	151

Table 24.8 Approximate weight of wet mortar

Type of mortar	Approximate weight of wet mortar Tonnes per m^3	m^3 per tonne
Grey stone lime mortar	1.87	0.54
Blue lias lime mortar	1.93	0.53
Cement lime mortar	1.96	0.51
Lime ash mortar	1.73	0.59
Portland cement mortar	2.00	0.50

Table 24.9 Weight of floor tiles

Size of tile in millimetres	Weight per 1000 tiles in tonnes
100 × 100 × 18	0.45
150 × 75 × 22	0.60
150 × 150 × 18	0.80
150 × 150 × 22	1.15
225 × 112 × 31	1.80
225 × 225 × 31	3.60
300 × 300 × 50	10.00

Table 24.10 Weight of kerb per metre

Size of kerb in millimetres	Type of kerb and weight in tonnes per metre	
	Concrete	Granite
125 × 50	0.011	0.013
175 × 50	0.022	0.025
250 × 100	0.066	0.077
250 × 125	0.077	0.088
250 × 150	0.088	0.099
250 × 200	0.110	0.132
300 × 150	0.099	0.121
300 × 200	0.132	0.165

Table 24.11 Average weight of building stone

	Cubic metres per tonne
Bath	0.44
Portland	0.41
Sandstone	0.41
York	0.40
Limestone	0.40
Purbeck	0.39
Granite	0.38
Marble	0.37

Table 24.12 Weight of artificial stone paving

Thickness of stone	Tonnes per square metre
50 mm	0.12
63 mm	0.15

Table 24.13 Weight of York and Purbeck stone paving

Thickness in millimetres	Square metres per tonne	
	York	Purbeck
50	7.8	7.5
63	6.4	6.0
75	5.3	5.0
100	3.9	3.7

Table 24.14 Weight of precast concrete manhole parts

Diameter of manhole in mm	Precast concrete bases 600 mm high. Weight per base in kg	Tapers		Cover slabs 150 mm thick	
		Height of taper in mm	Weight per taper in kg	Diameter of cover slab in mm	Weight per cover slab in kg
900	600	600	265	1050	225
1050	900	600	350	1200	300
1200	1300	600	500	1350	425
1350	1750	600	615	1500	625
1500	2000	900	775	1800	915
1800	2500	900	975	2000	1100

Table 24.15 Weight of precast concrete manhole rings

Diameter of ring in millimetres	Weight per metre in kg
675	281
900	467
1050	624
1200	760
1350	869
1500	1036
1800	1449

Table 24.16 Weight of agricultural drain pipes

Diameter of pipe in millimetres	Kilograms per 1000
50	900
62	1250
75	1750
100	2300
1500	5000

Table 24.17 Cast iron socket and spigot pipes weights of standard lengths

Pipes 3.6 m long

Test pressure Nominal bore mm	Class B 120 m head Weight kg	Class C 180 m head Weight kg	Class D 240 m head Weight kg
100	116	119	132
125	147	159	184
150	186	206	235
175	219	257	285
200	268	309	352
225	332	369	420
250	360	426	490
300	494	572	651
350	577	717	821
375	630	794	904
400	698	872	989
450	832	1043	1190
500	973	1158	1326
525	1072	1275	1476
600	1277	1598	1795
675	1629	2005	2296

Pipes 3.6 m long

Test pressure Nominal bore mm	Class B 120 m head Weight kg	Class C 180 m head Weight kg	Class D 240 m head Weight kg
75	60	60	61
100	80	83	91
125	102	110	123
150	128	140	161
175	154	177	200
200	185	216	242
225	214	254	288
250	249	293	336
300	318	394	446
350	399	495	563
375	435	547	620
400	482	600	678
450	574	718	816

Pipes 3 m long

Test pressure Nominal bore mm	Class B 120 m head Weight kg	Class C 180 m head Weight kg	Class D 240 m head Weight kg
50	32	32	32

Pipes 4 m long

Test pressure Nominal bore mm	Class B 120 m head Weight kg	Class C 180 m head Weight kg	Class D 240 m head Weight kg
75	64	64	66

Table 24.18 Cast iron socket and spigot vertical pipes lengths, thicknesses and weights of pipes for various test pressures.

Nominal internal diameter of pipe	Length of pipe	Class A Test pressure 60 m head		Class B Test pressure 120 m head		Class C Test pressure 180 m head		Class D Test pressure 240 m head	
		Thickness	Weight	Thickness	Weight	Thickness	Weight	Thickness	Weight
75	2.7	9.65	59	9.65	59	9.65	59	10.16	61
100	2.7	9.90	78	9.90	78	10.16	79	11.68	88
100	3.6	9.90	100	9.90	100	10.16	103	11.68	115
125	2.7	11.43	101	11.43	101	11.43	109	13.20	122
125	3.6	10.41	130	10.41	130	11.43	141	13.20	158
150	2.7	10.92	126	10.92	125	12.44	139	14.78	158
150	3.6	10.92	162	10.92	162	12.44	181	14.78	205
175	2.7	11.43	151	11.43	151	13.46	174	15.49	195
175	3.6	11.43	196	11.43	196	13.46	226	15.49	255
200	2.7	11.94	170	11.94	179	14.78	215	16.51	239
200	3.6	11.94	236	11.94	236	14.78	279	16.51	312
225	2.7	12.45	212	12.45	212	15.24	252	17.53	283
225	3.6	12.45	274	12.45	274	15.24	327	17.53	369
250	2.7	13.21	259	13.21	259	16.00	291	18.54	330
250	3.6	13.21	321	13.21	321	16.00	380	18.54	431
300	2.7	13.97	307	14.48	316	17.58	393	20.32	446
300	3.6	13.97	397	14.48	410	17.53	510	20.32	585
350	3.6	14.48	483	15.49	513	19.05	649	21.84	728
375	3.6	14.98	535	16.00	565	19.59	709	23.60	804
400	3.6	15.24	559	16.51	622	20.32	783	23.37	889
450	3.6	16.00	683	17.53	740	21.58	923	24.89	1056
500	3.6	16.51	781	18.54	865	22.61	1080	25.47	1228
525	3.6	17.20	835	19.05	932	23.36	1170	26.16	1325
550	3.6	17.27	905	19.56	1009	23.87	1268	27.43	1420
600	3.6	18.03	1028	20.32	1179	24.89	1373	28.70	1563
650	3.6	18.80	1157	21.08	1280	25.91	1554	29.97	1832
675	3.6	19.05	1217	21.59	1348	26.06	1645	30.48	1926
700	3.6	19.30	1278	21.84	1423	26.92	1739	30.98	2029
750	3.6	20.06	1366	22.61	1578	27.68	1971	32.00	2240
800	3.6	20.83	1515	23.64	1738	28.70	2178	33.27	2480
825	3.6	21.08	1739	23.88	1830	29.21	2274	33.78	2596
900	3.6	22.10	1870	24.89	2078	30.48	2592	35.05	2923
950	3.6	22.86	2038	25.65	2257	31.24	2804	36.07	3181
975	3.6	23.11	2113	25.91	2338	31.75	2946	36.58	3310
1000	3.6	23.37	2191	26.16	2421	32.00	3018	37.08	3439
1100	3.6	24.89	2457	27.43	2778	33.87	3498	39.37	3979
1150	3.6	25.40	2731	28.19	2997	34.54	3740	39.87	4248
1200	3.6	26.16	2932	28.70	3189	35.05	3960	40.64	4515

Table 24.19 Weight of concrete pipes

Note: The data show average weight, as there is slight variation in weight between the pipes supplied by various manufacturers. The data shown are for unreinforced pipes. If pipes are reinforced add 2½ per cent to the weights shown, and reduce the metres of pipes per tonne accordingly.

Internal diameter of pipe in millimetres	Ogee joints	
	Metres per tonne	*Tonnes per metre*
300	11.89	0.083
375	8.74	0.115
450	6.93	0.144
525	5.07	0.197
600	4.15	0.240
675	3.52	0.283
750	2.82	0.355
825	2.48	0.402
900	2.12	0.473
975	1.89	0.529
1050	1.64	0.610
1125	1.36	0.732
1200	1.26	0.796
1350	0.99	1.100
1500	0.82	1.222

Table 24.20 Weight of steel pipes BS 534

Pipe		Class A		Class B		Class C		Class D	
Inside diameter in mm	*Outside diameter in mm*	*Weight kg per m*	*Gauge and thickness in mm*	*Weight kg per m*	*Gauge and thickness in mm*	*Weight kg per m*	*Gauge and thickness in mm*	*Weight kg per m*	*Gauge and thickness in mm*
50	59	3.81	12g	4.20	11g	4.60	10g	5.14	9g
75	87	6.29	11g	7.84	9g	9.37	7g	10.17	6g
100	112	8.99	10g	10.05	9g	12.19	7g	13.26	6g
125	137	12.36	9g	13.69	8g	15.00	7g	16.33	6g
150	162	14.67	9g	17.83	7g	19.41	6g	21.36	5g
175	187	20.65	7g	22.50	6g	24.75	5g	29.04	6.4
200	212	23.47	7g	25.56	6g	28.15	5g	33.04	6.4
225	237	28.63	6g	31.05	5g	37.05	6.3	41.50	7.1
300	312	37.86	6g	41.75	5g	49.05	6.3	54.06	7.1
375	387	46.13	6g	53.56	5.5	61.08	6.3	68.58	7.1
450	462	56.32	6g	64.06	5.5	73.11	6.3	91.06	7.9
525	537	74.59	5.5	85.12	6.4	95.65	7.1	105.65	7.9
600	612	97.14	6.4	121.11	7.9	133.05	8.7	144.95	9.5
675	693	123.78	7.1	135.68	7.9	156.00	8.7	165.00	9.5
750	768	153.00	7.9	168.00	8.7	198.00	9.5	195.00	10.3
825	843	168.00	7.9	201.00	9.5	217.00	10.3	224.00	11.1
900	918	183.00	7.9	219.00	9.5	237.00	10.3	255.00	11.1
1050	1075	256.00	9.5	277.00	10.3	299.00	11.1	340.00	11.1
1200	1225	292.00	9.5	340.00	11.1	389.00	12.7	436.00	14.3
1350	1375	382.00	11.1	436.00	12.7	490.00	14.3	544.00	15.9
1500	1525	485.00	12.7	544.00	14.3	605.00	15.9	665.00	17.5

Table 24.21 Weight of clayware pipes

Internal dia. of pipe in millimetres	Metres per tonne	Tonnes per metre
100	82	0.012
150	49	0.020
225	27.7	0.036
300	16.5	0.060
375	11.2	0.091
450	7.3	0.137
525	5.4	0.183
600	4.1	0.244
675	3.2	0.310
750	2.7	0.366
900	1.8	0.515

Table 24.22 Clayware gully pots

Size of gully pot in mm		Number of gully pots per tonne
Diameter	Depth	
300	750	17
300	900	16
375	750	12
375	900	10
375	1050	9 ½
450	750	9
450	900	8
450	1050	6 ½
450	1200	6

Table 24.23 Weight of steel tubes to BS 1387

Note: The weight of screwed and socketed tube is based on: 4.5 m length for 3.2 mm nominal bore; 5.7 m length for 6.4 m to 150 mm nominal bore inclusive

Nominal bore in millimetres	Approx. outside diameter in mm Class A	Weight per metre in kilograms					
		Plain ends			Screwed and socketed		
		Class A	Class B	Class C	Class A	Class B	Class C
3.2	10.3	0.356	0.409	0.492	0.370	0.414	0.496
6.4	13.5	0.522	0.586	0.723	0.526	0.589	0.726
9.5	17.5	0.682	0.859	1.039	0.689	0.866	1.044
13.0	21.4	0.964	1.237	1.473	0.975	1.246	1.480
19.0	27.0	1.411	1.729	2.119	1.426	1.773	2.131
25.0	34.0	2.023	2.521	3.013	2.046	2.497	3.030
32.0	43.0	2.594	3.435	4.231	2.620	3.561	4.254
38.0	48.0	3.301	4.482	5.280	3.349	4.522	5.316
50.0	60.0	4.168	5.692	6.732	4.241	5.754	6.786
64.0	76.0	5.839	7.900	9.456	5.968	8.062	9.555
76.0	89.0	6.865	9.369	11.233	7.050	9.521	11.310
89.0	102.0	8.817	10.768	12.538	9.009	10.932	12.990
100.0	114.0	9.970	12.175	14.453	10.246	12.429	14.765
125.0	138.0	–	15.000	17.956	–	15.393	18.308
150.0	165.0	–	17.837	21.361	–	18.452	21.937

Table 24.24 Half hard light gauge copper tubes for water, gas and sanitation to BS 2871: Table X

Note: These tubes are suitable for the following working pressures:

$$1 \text{ bar} = 10^5 \text{ N/m}^2 = 100 \text{ kPa}$$

Nominal size in millimetres	Maximum outside diameter millimetres	Thickness millimetres	Maximum working pressures bar	Weight in kg per metre
6	6.045	0.6	133	0.09
8	8.045	0.6	97	0.12
10	10.045	0.6	77	0.16
12	12.045	0.6	63	0.19
15	15.045	0.7	58	0.28
18	18.045	0.8	56	0.39
22	22.055	0.9	51	0.53
28	28.055	0.9	40	0.68
35	35.07	1.2	42	1.13
42	42.07	1.2	35	1.37
54	54.07	1.2	27	1.77
76.1	76.30	1.5	24	3.13
108	108.25	1.5	17	4.47

Table 24.25 Standard cement corrugated sheets (Eternit profile 3)

Standard lengths 1225 to 3050 mm, rising by 150 mm increments.
Nominal width 750 mm. Net covering width when laid, 648 mm.
Number of corrugations per sheet, $10\frac{1}{2}$.
Pitch of corrugations 75 mm, depth 25 mm.
End lap, 150 mm. Side lap, 102 mm.
Weight of 10 m^2 of roofing as laid, 145 kg approximately.
Weight of 1 m^2 of sheeting, 12 kg approximately.
Approximately 12.5 m^2 of sheeting cover 10 m^2.

Area and weight table

Length of sheet metres	Net area in square metres	Approximate No. of sheets per tonne
3.05	1.879	36
2.75	1.69	40
2.60	1.59	43
2.45	1.49	45
2.13	1.28	52
1.83	1.09	61
1.53	0.89	73
1.38	0.80	81
1.23	0.70	91

Table 24.26 Cement 'Bigsix' corrugated sheets (Eternit deep profile)

Standard lengths 1525 to 3050 mm, rising by 150 mm increments.
Standard width 1.05 m. Net covering width when laid, 1016 mm.
Number of corrugations per sheet, 3.
Pitch of corrugations 339 mm, depth 86 mm.
End lap, 150 mm. Side lap, 124 mm.
Weight of 10 m^2 of roofing as laid, 170 kg approximately.
Weight of 1 m^2 of sheeting, 13 kg approximately.
Approximately 11.4 m^2 of sheeting cover 10 m^2.

Area and weight table

Length of sheet metres	Net area in square metres	Approximate No. of sheets per tonne
3.05	2.95	22
2.90	2.79	23
2.43	2.32	28
2.25	2.13	30

Table 24.27 Zinc coated corrugated sheets – approximate weight per sheet in kilograms

Thickness	0.40 mm		0.80 mm		1.00 mm		2.00 mm	
Corrugations mm	8/76.2	10/76.2	8/76.2	10/76.2	8/76.2	10/76.2	8/76.2	10/76.2
Width overall mm	743	908	743	908	743	908	743	908
Width as fixed mm	610	762	610	762	610	762	610	762
Length metres	kg	kg	kg	kg	kg	kg	kg	kg
1.5	3.75	4.58	7.05	8.61	8.81	10.76	17.63	21.53
1.80	4.50	5.49	8.46	10.33	10.57	12.91	21.15	25.83
2.10	5.25	6.41	9.87	12.05	12.33	15.06	24.68	30.14
2.40	6.00	7.32	11.28	13.78	14.09	17.21	28.20	34.44
2.70	6.75	8.24	12.69	15.50	15.85	19.36	31.73	38.75
3.00	7.50	9.15	14.10	17.22	17.61	21.51	35.25	43.05

Table 24.28 Galvanised corrugated sheets

	Approximate number of sheets per tonne							
	0.40 mm		0.80 mm		1.00 mm		2.00 mm	
Length of sheets in metres	8/76.2 mm	10/76.2 mm	8/76.2 mm	10/76.2 mm	8/76.2 mm	10/76.2 mm	8/76.2 mm	10/76.2 mm
1.50	266	218	142	116	114	93	57	46
1.80	222	182	118	97	95	77	47	39
2.10	190	156	101	83	81	66	41	33
2.40	167	137	89	73	67	58	35	29
2.70	148	121	79	65	63	52	32	26
3.00	133	109	71	58	58	46	28	23

Table 24.29 Weight of sheet iron

Gauge	Thickness in millimetres	Gauge	Weight in kg per m²
3	54.17	16	12.20
4	48.80	17	10.84
5	40.36	18	9.66
6	38.69	19	8.57
7	34.47	21	6.84
8	30.66	22	6.12
9	27.31	23	5.39
10	24.40	24	4.81
11	21.73	25	4.30
12	19.32	26	3.81

Table 24.30 Weight of sheet copper

S.W.G.	Thickness in millimetres	Weight in kg per m²
15	1.83	16.4
16	1.63	14.4
17	1.42	12.8
18	1.22	11.0
19	1.02	9.7
20	0.91	8.2
21	0.81	7.2
22	0.71	6.4
23	0.61	5.5
24	0.56	5.1
25	0.51	4.6
26	0.46	4.1
27	0.41	3.8
28	0.38	3.4
29	0.33	3.1
30	0.30	2.8
31	0.28	2.6
32	0.26	2.5
33	0.24	2.3
34	0.22	2.2
35	0.20	2.0

Table 24.31 Weight of sheet lead – thickness in millimetres of various weights of sheets and the nearest S.W.G.

Weight in kg per m²	Thickness in millimetres	Nearest standard wire gauge
20	1.58	16
25	2.38	14
30	2.78	12
35	3.17	11
40	3.57	10
45	3.97	9
50	4.38	7
55	4.76	6
60	5.20	5

Table 24.32 Weight of sheet zinc

Gauge	Nearest wire gauge	Thickness in millimetres	Weight in kg per m²
3	36	0.18	1.13
4	33	0.20	1.44
5	31	0.25	1.75
6	30	0.28	2.14
7	29	0.33	2.37
8	28	0.38	2.70
9	27	0.43	3.12
10	25	0.48	3.51
11	24	0.53	4.04
12	23	0.64	4.65
13	22	0.71	5.20
14	21	0.79	5.73
15	20	0.91	6.93
16	19	1.04	7.56
17	18	1.17	8.47
18	17	1.30	9.38

Table 24.33 Weight of cast iron gutters, rainwater, waste soil pipes per metre

Note: The weights shown are per metre for 1.8 m standard lengths.

Description	Ordinary type	4.8 mm	6.3 mm
Half round gutters, 75 mm	2.5	4.00	6.00
Half round gutters, 88 mm	3.0	4.25	6.50
Half round gutters, 100 mm	3.25	4.75	7.25
Half round gutters, 112 mm	3.75	6.50	8.25
Half round gutters, 125 mm	4.25	6.75	9.00
Half round gutters, 150 mm	5.75	7.25	9.75
O.G. gutters, 100 mm	3.75	6.50	7.50
O.G. gutters, 112 mm	4.25	7.00	8.00
O.G. gutters, 125 mm	5.00	7.25	8.25
O.G. gutters, 150 mm	6.25	–	–
Rainwater and waste pipes, 50 mm	3.88	6.00	8.00
Rainwater and waste pipes, 63 mm	4.75	7.50	9.50
Rainwater and waste pipes, 75 mm	5.75	8.75	11.75
Rainwater and waste pipes, 88 mm	7.00	10.25	13.75
Rainwater and waste pipes, 100 mm	8.25	11.50	15.00
Rainwater and waste pipes, 112 mm	9.75	13.00	16.50
Rainwater and waste pipes, 125 mm	11.50	14.75	19.50
Rainwater and waste pipes, 150 mm	15.75	17.75	23.00
Soil pipes, 75 mm	–	8.75	11.75
Soil pipes, 88 mm	–	10.25	13.75
Soil pipes, 100 mm	–	11.50	15.00

Table 24.34 Weight of roof tiles

Type of roof tile	Weight per 1000 tiles in tonnes	Weight per 1000 pairs in tonnes
Interlocking tiles, 380 × 205 mm	1.50	–
Italian tiles (per 1000 pairs)	–	5.90
Pantiles, 335 × 260 mm	2.00	–
Plain tiles, 270 × 165 mm, hand made	1.15	–
Plain tiles, 270 × 165 mm, machine made	1.10	–
Roman tiles, 420 × 330 mm	3.60	–
Spanish tiles (per 1000 pairs)	–	5.70

Table 24.35 Weight of hip and ridge tiles

Type of tile	Weight per 100 tiles in tonnes
350 × 175 mm	0.50
450 × 150 mm	0.70

Table 24.36 Weight of angle and cove tiles

Type of tile	Weight per 100 tiles in tonnes
100 × 100 × 18 mm	0.04
150 × 150 × 22 mm	0.11
150 × 175 × 22 mm	0.14
200 × 200 × 29 mm	0.26
225 × 225 × 33 mm	0.36

Table 24.37 Weight of slates

Type of slate		Weight per 100 slates in tonnes
Countesses,	510 × 255 mm	2.00
Doubles,	325 × 150 mm	0.75
Duchesses,	610 × 305 mm	3.00
Ladies,	405 × 200 mm	1.25

Table 24.38 Weight of round or square bars

	Round bars			Square bars	
Diameter in millimetres	Wrought iron kg per m	Steel kg per m	Size in millimetres	Wrought iron kg per m	Steel kg per m
6	0.216	0.222	6.4	0.312	0.319
8	0.387	0.395	7.9	0.489	0.498
10	0.603	0.616	9.5	0.703	0.717
12	0.870	0.888	12.7	1.249	1.274
16	1.547	1.579	15.9	1.953	1.992
20	2.415	2.466	20.6	3.301	3.367
25	3.775	3.854	25.4	5.000	5.100
32	6.184	6.313	33.3	7.812	7.968
40	9.662	9.864	41.3	13.203	13.467
50	15.097	15.413	50.8	20.000	20.400

Table 24.39 Weight of flat bar iron

Width in millimetres	Thickness in mm and weight in kilograms per metre											
	1.6	3.2	4.8	6.4	8.0	9.6	11.2	12.7	15.9	19.0	22.2	25.0
25	0.31	0.63	0.95	1.32	1.56	1.87	2.19	2.50	3.12	3.75	4.35	5.00
28	0.36	0.71	1.06	1.41	1.75	2.16	2.46	2.82	3.51	4.21	4.92	5.62
32	0.39	0.80	1.19	1.56	1.95	2.34	2.73	3.12	3.90	4.78	5.47	6.25
35	0.43	0.87	1.31	1.72	2.15	2.58	3.01	3.43	4.29	5.16	6.01	6.87
38	0.48	0.94	1.43	1.87	2.34	2.82	3.39	3.75	4.69	5.62	6.57	7.50
41	0.51	1.02	1.55	2.02	2.53	3.05	3.55	4.06	5.08	6.08	7.11	8.28
44	0.55	1.11	1.57	2.19	2.73	3.29	3.82	4.38	5.47	6.57	7.65	8.75
50	0.63	1.26	1.89	2.50	3.12	3.75	4.38	5.00	6.25	7.50	8.75	9.25

Table 24.40 Weight of metals

Type of metal	Thickness in mm and weight in kilograms per square metre								
	1.6	3.2	6.4	9.5	12.7	15.9	19.0	22.2	25.0
Brass	13.9	27.7	55.4	83.2	108.9	139.0	168.0	190.4	222.0
Cast iron	11.5	22.1	45.7	68.7	91.8	114.4	137.0	161.0	191.2
Copper	14.1	28.2	56.4	84.6	112.8	141.0	169.2	197.4	225.6
Lead, cast	18.1	36.2	72.4	108.6	144.8	181.0	217.2	253.4	289.6
Steel	12.5	25.0	50.0	75.0	100.0	125.0	150.0	175.0	200.0
Wrought iron	12.2	24.4	48.8	73.2	97.4	121.9	146.4	170.8	193.8

Table 24.41 Weight of metals

Type of metal	Weight in kg per cubic metre
Cast brass	8.474
Cast copper	8.769
Cast iron	7.348
Milled lead	11.325
Sheet copper	8.981
Steel	7.894
Tin	7.304
Wrought iron	7.668
Zinc	7.048
Aluminium	2.585

Table 24.42 Equal angles

Size in millimetres	Thickness in millimetres and weight in kilograms per metre								
	3	4	5	6	8	10	12	15	24
25 × 25	1.11	1.45	1.77	–	–	–	–	–	–
30 × 30	1.36	1.78	2.18	–	–	–	–	–	–
40 × 40	–	2.42	2.97	3.52	–	–	–	–	–
50 × 50	–	–	3.77	4.47	5.82	–	–	–	–
60 × 60	–	–	4.57	5.42	7.09	8.69	–	–	–
70 × 70	–	–	–	6.38	8.36	10.30	–	–	–
80 × 80	–	–	–	7.34	9.63	11.90	–	–	–
90 × 90	–	–	–	8.30	10.90	13.40	15.90	–	–
100 × 100	–	–	–	–	12.20	–	17.80	21.90	–
200 × 200	–	–	–	–	–	–	–	–	71.10

Table 24.43 Channels

		Dimensions		
Depth in millimetres	Width in millimetres	Web Thickness T1 in millimetres	Flange Thickness T2 in millimetres	Weight in kg per metre
76	38	5.1	6.8	6.70
102	76	7.1	11.0	17.69
127	64	6.4	9.2	14.90
152	76	6.4	9.0	17.88
152	89	7.1	11.6	23.84
178	76	6.6	10.3	20.84
178	79	7.6	12.3	26.81
203	76	7.1	11.2	23.82
203	89	8.1	12.9	29.78
305	89	10.2	13.7	41.69
305	102	10.2	14.8	46.18

Table 24.44 Rolled steel joists

		Dimensions		
Depth in millimetres	Breadth in millimetres	Thickness T1 in millimetres	Thickness T2 in millimetres	Weight in kg per metre
75	32	5.54	6.35	6.00
75	75	7.37	9.65	15.00
88	38	7.52	7.62	9.00
88	75	8.89	8.89	15.75
100	44	8.38	9.14	12.00
100	75	7.62	10.92	18.00
115	75	10.16	10.92	21.00
125	75	10.16	11.18	22.50
125	105	11.18	11.43	28.50
125	112	8.69	14.48	33.00
125	125	9.42	14.22	36.00
150	75	9.91	11.43	24.00
150	112	11.02	10.16	30.00
150	125	10.74	13.12	37.50
175	94	6.35	9.53	24.00
200	100	10.41	14.22	37.50
200	125	10.16	15.49	45.00
200	150	11.18	15.49	52.50
225	95	7.62	11.43	30.00
225	175	19.56	20.57	87.00
250	125	12.19	15.24	52.50
250	150	12.45	18.80	67.50
300	125	11.18	16.51	58.50
300	150	10.41	18.29	66.00
300	150	12.95	22.09	81.00

Table 24.45 Weight of black bolts and nuts

Notes: 1. The data shown are for hexagon-headed bolts and nuts.
2. The length of the bolt is measured from under the head.

Length of bolt in millimetres	Diameter of bolt and weight of bolt and nut in kilograms								
	6	*9*	*13*	*16*	*19*	*22*	*25*	*31*	*38*
25	0.014	0.042	0.091	0.167	0.278	–	–	–	–
31	0.016	0.045	0.097	0.177	0.292	0.448	–	–	–
38	0.017	0.049	0.103	0.187	0.307	0.469	0.675	–	–
44	0.019	0.053	0.110	0.197	0.322	0.488	0.701	–	–
50	0.021	0.056	0.116	0.208	0.336	0.508	0.727	1.338	–
56	0.022	0.060	0.122	0.218	0.350	0.528	0.753	1.379	2.282
63	0.024	0.064	0.130	0.229	0.365	0.548	0.779	1.420	2.341
69	0.025	0.067	0.136	0.239	0.380	0.568	0.805	1.461	2.399
75	0.027	0.071	0.142	0.249	0.395	0.588	0.833	1.501	2.459
81	0.029	0.074	0.149	0.259	0.409	0.608	0.858	1.541	2.517
88	0.030	0.079	0.156	0.269	0.424	0.628	0.875	1.582	2.575
94	0.032	0.082	0.162	0.279	0.439	0.648	0.909	1.623	2.634
100	0.034	0.085	0.168	0.289	0.453	0.669	0.935	1.663	2.693
106	–	0.089	0.175	0.299	0.468	0.687	0.960	1.703	2.750
113	–	0.093	0.181	0.309	0.483	0.707	0.987	1.744	2.809
119	–	0.097	0.188	0.320	0.497	0.728	1.013	1.785	2.867
125	–	0.100	0.195	0.329	0.512	0.748	1.039	1.825	2.926
131	–	–	0.201	0.340	0.527	0.767	1.064	1.865	2.984
138	–	–	0.207	0.350	0.541	0.787	1.091	1.907	3.043
144	–	–	0.214	0.361	0.557	0.807	1.111	1.947	3.101
150	–	–	0.220	0.371	0.571	0.828	1.143	1.987	3.161
163	–	–	0.234	0.391	0.600	0.868	1.195	2.069	3.254
175	–	–	0.247	0.411	0.630	0.911	1.247	2.150	3.394
188	–	–	0.262	0.432	0.659	0.947	1.298	2.231	3.509
200	–	–	0.272	0.452	0.688	0.986	1.351	2.312	3.631
213	–	–	0.286	0.472	0.717	1.026	1.399	2.394	3.745
225	–	–	0.299	0.493	0.747	1.066	1.455	2.474	3.863

Table 24.46 Number of nails per kilogram

Oval wire nails

Length	18 mm	25 mm	31 mm	38 mm	44 mm	50 mm	63 mm	75 mm	88 mm	100 mm
No. of nails per kg	4500	2425	1365	880	620	375	240	110	100	70

Cut clasp nails

Length	25 mm	38 mm	50 mm	63 mm	75 mm	88 mm	100 mm	112 mm	125 mm	150 mm
No. of nails per kg	4500	2425	1365	880	620	375	240	110	100	70

Round wire nails

Length in millimetres	Gauge	No. of nails per kg
25	14	1322
31	13	1145
38	12	620
50	10	265
63	10	200
75	8	130
88	8	110
100	8	100
112	8	90
125	6	65
175	5	34

Floor brads

Length in millimetres	50	63	75
No. of nails per kg	285	155	110

Galvanised slate nails, 10 Gauge

Length in millimetres	38	43	50
No. of nails per kg	310	265	220

Copper slate nails

Length in millimetres	38	43	50
No. of nails per kg	330	265	200

Table 24.46 continued

Zinc slate nails

Length in millimetres	38
No. of nails per kilogram	420

Wire clout nails

Length in millimetres	19	25	32
No. of nails per kg	970	770	530

Galvanised Roofing Screws, No. 12

Length in millimetres	50	63	75
No. of screws per kg	80	67	55

Galvanised roofing washers

$8 \times 22 \times 3\,\text{mm}$	155 per kg

Galvanised roofing nails, No. 4

Length in millimetres	50	63	75
No. of nails per kg	80	67	52

25 *Useful tables*

Table 25.1 Mensuration formulae

Circumference of circle:	πd
Areas:	
Area of triangle:	$\dfrac{b \times h}{2}$
Area of circle:	πr^2
Area of sector:	$\pi r^2 \dfrac{\text{Angle}}{360}$ (in degrees)
Surface area of cylinder:	$\pi d \times h$
Area of segment of circle:	Area of sector – Area of triangle
Surface area of sphere:	πd^2
Surface Area of cone:	$\pi d \times s$ (sloping height)
Volumes:	
Sphere:	$\dfrac{\pi d^3}{6}$
Cylinder:	$\pi r^2 \times h$
Pyramid:	$\dfrac{\text{Area of base} \times h}{3}$
Segment of sphere:	$\dfrac{\pi h}{6}(3r^2 + h^2)$
Cone:	$\dfrac{\pi r^2 \times h}{3}$

Table 25.2 Area subtended by kerbs of various radii

Radius in m	Area of shaded area in square metres	Length of circumference (Quarter circle) in metres
1	0.22	1.57
2	0.86	3.19
3	1.93	4.71
4	3.44	6.28
5	5.36	7.85
10	21.46	15.71
15	48.28	26.56
20	85.85	31.42
25	134.13	39.27
30	203.25	47.12

Table 25.3 Flow of water in streams

Note: The formula shown is applicable only to streams and open channels.

Q = Quantity discharged per second
A = Cross-sectional area of waterway in stream
F = Fall in a given length L
P = Wetted perimeter or girth of bottom and sides of open channel
V = Velocity per second

Then:

$$V = \sqrt[29]{\frac{A \times F}{P \times L}} \text{ m per second.}$$

$Q = A \times V \text{ m}^3$ per second.

Flow velocity of water in concrete lined channels is approximately double that in natural streams.

Table 25.4 Head and pressure

Head of water in metres and equivalent in kilograms per square centimetre

Metres head	kg per cm²	Metres head	kg per cm²	Metres head	kg per cm²
0.3	0.030	16.5	1.674	57.0	5.835
0.6	0.061	18.0	1.827	60.0	6.088
0.9	0.091	19.5	1.978	67.5	6.854
1.2	0.121	21.0	2.150	75.0	7.593
1.5	0.152	22.5	2.284	82.5	8.366
1.8	0.182	24.0	2.439	90.0	9.139
2.1	0.213	25.5	2.587	97.5	9.913
2.4	0.239	27.0	2.741	105.0	10.686
2.7	0.274	28.5	2.890	112.5	11.389
3.0	0.304	30.0	3.044	120.0	12.163
4.5	0.456	33.0	3.348	150.0	15.297
6.0	0.608	36.0	3.655	180.0	18.289
7.5	0.759	39.0	3.958	210.0	21.303
9.0	0.907	42.0	4.260	240.0	24.396
10.5	1.064	45.0	4.569	270.0	27.419
12.0	1.217	48.0	4.872	300.0	30.444
13.5	1.370	51.0	5.174		
15.0	1.519	54.0	5.483		

Table 25.5 Pressure and head

Pressure in kilograms per square centimetre and equivalent head of water in metres

kg per cm²	Metres head	kg per cm²	Metres head	kg per cm²	Metres head
1	10.0	45	444.11	140	1384.6
2	19.9	50	494.4	150	1483.5
3	29.9	55	543.1	160	1582.4
4	39.7	60	593.4	170	1680.9
5	49.4	65	642.8	180	1779.5
6	59.2	70	692.3	190	1878.6
7	68.9	75	741.7	200	1977.6
8	79.0	80	791.2	225	2227.4
9	88.9	85	840.3	250	2477.2
10	98.9	90	889.5	275	2721.8
15	148.2	95	939.2	300	2966.4
20	197.8	100	988.8	325	3218.6
25	247.2	110	1008.7	350	3460.8
30	296.7	120	1186.8	375	3708.0
35	346.1	125	1236.2	400	3955.2
40	395.6	130	1285.6	500	4954.4

Table 25.6 Crushing strength of brickwork

	Crushing strength in tonnes per square metre	
Description	Stocks	Blue Staffs
Bricks	919	7667
Brickwork in lime mortar	157	811
Brickwork in cement mortar	163	946

Table 25.7 Quantity of brickwork (laid in mortar) and number of bricks per metre of sewer or culvert

	Half a brick thick		One brick thick	
Diameter in metres	No. of bricks	Cubic metres of brickwork	No. of bricks	Cubic metres of brickwork
0.3	72	0.15	181	0.39
0.6	127	0.26	303	0.61
0.9	177	0.37	415	0.83
1.2	240	0.48	524	1.07
1.5	279	0.59	633	1.26
1.8	353	0.70	746	1.46
2.1	406	0.81	858	1.70
2.4	462	0.92	963	1.93
2.7	514	1.03	1075	2.15
3.0	574	1.14	1188	2.37

Table 25.8 Square metres per cubic metre and cubic metres per 100 m² for timber of various thicknesses

Thickness of timber in mm	Square metres per cubic metre	Cubic metres per 100 m²	Thickness of timber in mm	Square metres per cubic metre	Cubic metres per 100 m²
12	83.3	120	50	20.0	500
19	52.6	190	57	17.6	570
25	40.0	250	63	15.9	630
32	31.2	320	75	13.5	750
40	25.0	400	89	11.3	890
45	22.2	450	100	10.0	1000

Table 25.9 Covering capacity of cement mortar

Mix	25 mm thick	
Neat	34.09 m²	
1 cement:1 sand	55.74 m²	1 cubic
1 cement:2 sand	81.94 m²	metre
1 cement:3 sand	111.48 m²	loose

Table 25.10 One cubic metre of cement will make approximate quantity of concrete

Mix	Quantity
4:2:1 concrete	4.10 m³
5:2½:1 concrete	5.00 m³
6:3:1 concrete	5.80 m³
8:4:1 concrete	7.50 m³

Table 25.11 Circular sewers: discharge through 225 mm diameter pipe in litres per minute

Gradient	Depth of flow in proportion to diameter of pipe			
	One-eighth	One-quarter	One-half	Seven-eighths
1 in 20	264	1012	3432	6978
1 in 30	218	887	2796	5660
1 in 40	182	718	2409	4932
1 in 50	168	650	2160	4432
1 in 80	136	509	1714	3491
1 in 100	118	455	1500	3100

Table 25.12 Circular sewers: discharge through 300 mm diameter pipe in litres per minute

Gradient	Depth of flow in proportion to diameter of pipe			
	One-eighth	One-quarter	One-half	Seven-eighths
1 in 30	446	1727	5796	11370
1 in 40	391	1500	5000	10164
1 in 50	345	1327	4455	9092
1 in 80	273	1068	3296	7182
1 in 100	241	963	3136	6418
1 in 200	173	659	2227	4546

Table 25.13 Circular sewers: discharge through 375 mm diameter pipe in litres per minute

Gradient	Depth of flow in proportion to diameter of pipe			
	One-eighth	One-quarter	One-half	Seven-eighths
1 in 40	682	2682	8637	17729
1 in 50	614	2387	7728	15819
1 in 80	478	1887	6090	12500
1 in 100	432	1690	5478	11183
1 in 200	305	1205	4036	7887
1 in 440	205	796	2591	5295
1 in 660	164	660	2114	4523
1 in 880	145	568	1818	3751

Table 25.14 Circular sewers: discharge through 450 mm diameter pipe in litres per minute

Gradient	Depth of flow in proportion to diameter of pipe			
	One-eighth	One-quarter	One-half	Seven-eighths
1 in 50	955	3773	12201	25003
1 in 66	827	3114	10820	21707
1 in 80	745	2841	9637	19710
1 in 100	668	2614	8651	17761
1 in 200	478	1882	6100	12480
1 in 440	318	1236	4123	8388
1 in 880	227	873	2909	5919

Table 25.15 Angles of repose

Nature of material	Angle of repose
Vegetable soil	28°
Compact soil	50°
Loamy soil	40°
Compact loamy soil	50°
Gravel and sand mixed	40°
Gravel or shingle (loose)	39°
Soft dry clay	45°
Dry sand	38°

Table 25.16 Angle of slopes

Angle	Slope
18° 25′	3 to 1
26° 35′	2 to 1
33° 42′	1½ to 1
45° 0′	1 to 1
53° 0′	¾ to 1
56° 20′	⅔ to 1
63° 30′	½ to 1

Table 25.17 Covering capacity of paints, etc., on various surfaces (applied by hand)

Description	Square metres covered by 1 litre
Special aluminium priming coat on planed timber	20
Lead priming coat on planed timber	12
First coat of oil paint after priming	16
Second coat	18
Third coat	22
First coat of oil paint on previously painted timber	18
Second coat	20
Oil gloss paint applied as a finishing coat	24
Undercoat for enamel	20
Enamel on prepared surfaces – first coat	14
Enamel on prepared surfaces – second coat	18
Flat finish for varnish	22
Varnish on prepared surfaces – first coat	16
Varnish on prepared surfaces – second coat	20
Stain applied to planed timber	16
Stain applied to unwrought timber	10
Priming coat on brick or stucco	8
Priming coat on concrete, compo or stone	10
Priming coat on plaster	14
First coat oil paint after priming	12
Oil gloss paint applied as a finishing coat	18
Priming coat on smooth iron	20
Priming coat on slightly rusted iron	14
Priming coat on rust-pitted iron	10
First coat after priming	20
Second coat after priming	24

Table 25.18 Widths of trenches for laying pipes at different depths below the ground: clayware pipes and concrete tubes

Notes: 1. The widths shown are average widths for the depths shown, and allow for concrete beds, haunches or surrounds to pipes if required, the concrete being 150 mm wider a side than the pipes.
2. The widths shown are for normal grounds such as firm sand, loamy soil or clay, chalk, gravel, etc. and allow for timbering at the greater depths.

Diameter of pipe millimetres	Depth of trench in metres and average width of trench required in metres											
	0.3 m	0.6 m	0.9 m	1.2 m	1.5 m	1.8 m	2.1 m	2.4 m	2.7 m	3.0 m	3.3 m	3.6 m
100	0.425	0.425	0.450	0.525	0.600	0.675	0.750	0.750	0.750	0.750	0.750	0.750
150	0.475	0.475	0.475	0.525	0.600	0.675	0.750	0.750	0.750	0.750	0.750	0.750
225	0.575	0.600	0.600	0.600	0.675	0.750	0.750	0.750	0.750	0.750	0.750	0.750
300	0.650	0.650	0.675	0.675	0.750	0.750	0.750	0.750	0.750	0.750	0.750	0.750
375	0.750	0.750	0.750	0.750	0.825	0.825	0.825	0.825	0.825	0.825	0.825	0.825
450	0.825	0.825	0.825	0.825	0.900	0.900	0.900	0.900	0.900	0.900	0.900	0.900
525	0.900	0.900	0.900	0.900	0.975	0.750	0.975	0.975	0.975	0.975	0.975	0.975
600	0.975	0.975	0.975	0.975	1.050	1.050	1.050	1.050	1.050	1.050	1.050	1.050
675	1.050	1.050	1.050	1.050	1.125	1.125	1.125	1.125	1.125	1.125	1.125	1.125
750	1.125	1.125	1.125	1.125	1.200	1.200	1.200	1.200	1.200	1.200	1.200	1.200
825	1.200	1.200	1.200	1.200	1.275	1.275	1.275	1.275	1.275	1.275	1.275	1.275
900	1.275	1.275	1.275	1.275	1.350	1.350	1.350	1.350	1.350	1.350	1.350	1.350
975	1.350	1.350	1.350	1.350	1.425	1.425	1.425	1.425	1.425	1.425	1.425	1.425
1050	1.425	1.425	1.425	1.425	1.500	1.500	1.500	1.500	1.500	1.500	1.500	1.500
1125	1.500	1.500	1.500	1.500	1.575	1.575	1.575	1.575	1.575	1.575	1.575	1.575
1200	1.575	1.575	1.575	1.575	1.650	1.650	1.650	1.650	1.650	1.650	1.650	1.650

Table 25.19 Widths of trenches for laying pipes at different depths below the ground: uPVC cast iron and steel pipes, water mains, etc.

Notes: 1. The widths are average widths, and do not allow for joint holes or concrete beds, haunches or surrounds. If such should be required the appropriate data shown for clayware pipes and concrete tubes should be used.
2. The widths shown are for normal grounds such as firm sand, loamy soil or clay, chalk, gravel, etc., timbering being allowed for at the greater depths.

Diameter of pipe in millimetres	Depth of trench in metres and average width of trench in metres							
	0.3 m	0.6 m	0.9 m	1.2 m	1.5 m	1.8 m	2.1 m	2.4 m
75	0.300	0.375	0.450	0.525	0.600	0.675	0.750	0.750
100	0.325	0.375	0.450	0.525	0.600	0.675	0.750	0.750
125	0.350	0.400	0.450	0.525	0.600	0.675	0.750	0.750
150	0.375	0.425	0.450	0.525	0.600	0.675	0.750	0.750
175	0.400	0.450	0.475	0.525	0.600	0.700	0.750	0.750
200	0.425	0.475	0.500	0.525	0.600	0.750	0.750	0.750
225	0.450	0.500	0.525	0.525	0.600	0.750	0.750	0.750
300	0.525	0.575	0.600	0.600	0.675	0.750	0.750	0.750
375	0.600	0.650	0.675	0.675	0.750	0.825	0.825	0.825
450	0.675	0.725	0.750	0.750	0.825	0.900	0.900	0.900
525	0.750	0.800	0.825	0.825	0.900	0.975	0.975	0.975
600	0.825	0.875	0.900	0.900	0.975	1.050	1.050	1.050
675	0.900	0.950	0.975	0.975	1.050	1.125	1.125	1.125
750	0.975	1.025	1.050	1.050	1.125	1.200	1.200	1.200
825	1.050	1.100	1.125	1.125	1.200	1.275	1.275	1.275
900	1.125	1.175	1.200	1.200	1.275	1.350	1.350	1.350
975	1.200	1.250	1.275	1.275	1.350	1.425	1.425	1.425
1050	1.275	1.325	1.350	1.350	1.425	1.500	1.500	1.500
1125	1.350	1.400	1.425	1.425	1.500	1.575	1.575	1.575
1200	1.425	1.475	1.500	1.500	1.575	1.650	1.650	1.650

Table 25.20 Bearing value of soils

Description of soil	Approximate bearing value in kN/m²
Bog and peat	Nil
Loam, marl, clay	75
Made ground	Nil
Rock	2000
Coarse gravel	600
Fine gravel	450
Sand	250

Table 25.21 Voids in materials

Nature of material	Cubic metres of voids per cubic metre
Broken bricks	0.48
Broken stone	0.45
Gravel	0.39
Shingle – clean	0.37
Sand – pit	0.24
Sand – river	0.29
Sand – sea	0.37

Table 25.22 Shrinkage of materials

Note: Materials increase in bulk when excavated and, on being deposited, shrink. Wet soils shrink more than dry. Rock increases in bulk when broken up and does not settle to less than its original volume. The increase in bulk is dependent on the size of the broken pieces, and varies between 40 and 60 per cent.

Nature of material	Percentage shrinkage
Clay	10
Gravel	8
Gravel and sand	9
Loam and light sandy soils	12
Loose vegetable soils	15

Table 25.23 Cubic metres of solid material hauled per load by various types of transporting plant

Notes: 1. In transporting by dumpers or wagons the bulkage of the material should be taken into account. Such vehicles are rated on a volume basis, their Struck Measured Capacity being stated. The vehicles, however, have a Heaped Capacity, the load being heaped in the vehicle. In the table the cubic metres hauled per cubic metre of struck measured capacity allow for normal heaping of the loads and the bulkage of the various materials, thus a 2 cubic metre dumper hauls 2 × 0.80 = 1.60 cubic metre of stiff clay (solid) per load.
2. In hauling excavated materials by lorry, the weight of load carried should not exceed that which the vehicle is designed to carry.

| Nature of material hauled | Weight of the material in the solid. Cubic metres per tonne | Haulage by lorries. Cubic metres of material (solid) hauled per load | | | | | | The cubic metres of solid material hauled by dumpers or in wagons per m³ of struck measured capacity |
		1 tonne lorry	2 tonne lorry	3 tonne lorry	4 tonne lorry	5 tonne lorry	6 tonne lorry	
Chalk	0.44	0.44	0.88	1.32	1.76	2.20	2.64	0.70
Soft or sandy clay	0.57	0.57	1.14	1.71	2.28	2.85	3.42	0.85
Stiff clay	0.53	0.53	1.06	1.59	2.12	2.65	3.18	0.80
Gravel	0.57	0.57	1.14	1.71	2.28	2.85	3.42	1.14
Loam	0.67	0.67	1.34	2.01	2.68	3.35	4.02	0.92
Marl	0.57	0.57	1.14	1.71	2.28	2.85	3.42	0.98
Sand	0.67	0.67	1.34	2.01	2.68	3.35	4.02	1.09
Soil	0.63	0.63	1.26	1.89	2.52	3.15	3.78	0.90

Table 25.24 Masonry walling

| Type of wall | Material required per cubic metre of walling | | Square metres of wall per tonne |
	Stone m³	Mortar m³	
Uncoursed rubble	0.90	0.25	–
Coursed rubble	1.02	0.20	–
Flint	0.85	0.26	–
Flint facing – whole flints	–	–	0.93
Flint facing – split flints	–	–	1.42

Table 25.25 Standard size for concrete flags

0.9 m × 0.6 m × 50 mm
0.9 m × 0.6 m × 63 mm
0.75 m × 0.6 m × 50 mm
0.75 m × 0.6 m × 63 mm
0.6 m × 0.6 m × 50 mm
0.6 m × 0.6 m × 63 mm
0.45 m × 0.6 m × 50 mm
0.45 m × 0.6 m × 63 mm

Tolerance permitted   1.6 mm

Hydraulic pressure not less than 70 kg per cm²

Slabs 50 mm thick: 1 tonne covers 8.35 m²
Slabs 63 mm thick: 1 tonne covers 6.65 m²

Table 25.26 Square metres covered by 1 cubic metre of material

Nature of foundation material	*Consolidated thickness laid in millimetres and square metres covered by 1 cubic metre of material*								
	50 mm	*75 mm*	*100 mm*	*125 mm*	*150 mm*	*175 mm*	*200 mm*	*225 mm*	*300 mm*
Clinker	14.88	9.90	7.42	5.90	4.90	–	–	–	–
Gravel	15.80	10.50	7.92	6.34	5.28	–	–	–	–
Sand	16.50	12.00	8.20	6.60	5.50	–	–	–	–
Brick and concrete Hardcore Granite Limestone Sandstone and Whinstone rubble Pitching	–	–	7.42	5.90	4.95	4.23	3.71	3.30	4.47

Table 25.27 Stone pitching square metres covered by 1 tonne of material

Weight of stone in the solid. Tonnes per cubic metre	*Consolidated thickness laid in millimetres and square metres covered by 1 tonne of material*				
	150 mm	*175 mm*	*200 mm*	*225 mm*	*300 mm*
2.082	4.30	3.69	3.22	2.86	2.15
2.162	4.18	3.59	3.14	2.78	2.08
2.242	4.06	3.49	3.06	2.71	2.03
2.322	3.95	3.39	2.97	2.63	1.98
2.402	3.84	3.29	2.89	2.56	1.92
2.482	3.73	3.20	2.81	2.48	1.87
2.562	3.61	3.10	2.72	2.41	1.81
2.643	3.50	3.01	2.65	2.33	1.76
2.724	3.39	2.91	2.56	2.26	1.70
2.803	3.28	2.82	2.47	2.18	1.65
2.883	3.16	2.72	2.39	2.11	1.59

Table 25.28 Surface dressing roads: Square metres covered by various sizes of chippings gravel or sand per cubic metre and per tonne

Size in millimetres	Per cubic metre	*Square metres covered*			
		Per tonne			
		Sand	*Granite chips*	*Gravel*	*Limestone chips*
Sand	242	168	–	–	–
3	198	–	148	152	165
6	176	–	130	133	144
9	154	–	111	114	123
13	121	–	85	87	95
19	99	–	68	71	78

Index

Access to works, 40
Accommodation on sites, 44
Administrative staff, 39
Agent, 39
Aggregates, concrete, 118–122
Agricultural land drain weight, 282
Air valve, 252
Angle dozer, 67, 70
Angles:
    brick, 164
    lead, 198
Angles of Repose, 303
Approximate Quantities contract, 8
Apron:
    brick, 164
    lead, 198
Arches, brick, 168
Architect, 3
Architrave, 176–177
Artificial stone paving, 270
Asphalt:
    surfacing, 223
    paving, 238
Ashes, 238
Auger, 53

Backacter, 67, 70
Backfilling, 67, 100
Basement excavations, 93–100
Basic labour rate, 17
Basic output bonus, 21–22
Battens, 191, 195
Beams, 138, 139
Beam formwork, 147
Bearing value of soil, 306
Beds, haunches and surrounds, 261
Bills of Quantities, 1, 2, 3–4, 10–11
Birdcage scaffold, 35
Blockwork, 167
Bolts, 296
Bonus, 16–25
Braces, 138
Bricks:
    sizes, 152
    output, 152
    mortar, 153
    waste, 153
    flettons, 154
    facings, 155–157
    pointing, 157–159
    number, 163
    centering, 168–170
Brickwork and Blockwork, 152–168
Builder's plant, 29–35
Builder's Work, 164, 259
Bulking of excavated materials, 84

Bulldozer, 82–83
Buyer, 26

Capital cost of plant, 32
Carpenter, 171–183
Cast in situ piles, 53
Cast iron drainage, 245–252
Caterpillar tractors, 80–81
Cavity form, 166
Ceilings, 202
Cement mortar, 153, 159–160
Centering, 168–170
Chain link fencing, 274–275
Channels clayware, 257
Chimney pots, 166
Chippings, 308
Clayware pipes, 253–258
Close boarded fencing, 276
Code estimating practice, 36
Cold asphalt, 223–224
Cold water:
    services, 210
    cistern, 212
Columns:
    concrete, 138
    formwork, 148
Concrete work, 118–151
Concrete:
    beds, haunches and surrounds, 259–261
    ready mixed, 139
Condam (CDM) regulations, 45
Contingencies, 13
Contract Forms, 4–10
Copper piping, 209
Corrugated roof sheets, 187–189
Cost sheets:
    labour, 21
    plant, 21
Cover, manholes, 260
Covering capacity:
    paint, 304
    surfacing materials, 228
Craftsmen:
    typical cost, 17
Cranes, 98, 105
Crazy paving, 270
Creosote fences, 276
Crushing stone, 217–219
Cutting and Priming, 162

Dado rails, 177
Damp proof course, 166
Dayworks, 12
Dead shores, 172
Decking, 172
Delivery of materials, 12

Depreciation of plant:
  straight-line method, 31
  outlay and reserve fund method, 32
  written down value method, 33
Diesel hammers, 54
Director's fees, 46
Displacement piles, 53
Disposal surplus excavated material, 76–84
Doors and door frames, 179, 180
Double acting hammer, 54
Draglines, 70, 74, 75, 78
Drainage, 245–265
Dressing, bituminous emulsion, 229–231
Drilling rig, 53
Drop hammers, 54
Dumpers, 72, 79

Earthwork support, 84–85, 108–110
  shafts, basements, 94–100
  tunnel work (timbering), 101–107
Eaves gutters, 213–215
Elements of cost, 10
English bond, 163, 165
English garden wall bond, 163
Establishment charges, 46–47
Estimating and tendering, difference between, 10
Excavation and earthworks, 66–118
Excavation:
  surface, 66–84
  below ground, 84–117
  dry ground, 84–108
  wet ground, 108–117
Expansion joints, 144, 233
External gutters, 213–215
External doors, 180
Extra over cost for brickwork, 155–157

Face shovel, 70
Facings, 155–157
Fascia, 176
Fencing, 274–276
Finishes surface, 200–205
Firm Price Tender, 49
Flags concrete, 307
Flashings, 197–199
Flemish bond, 163, 165
Flemish garden wall bond, 163
Fletton brickwork, 154–155
Floors and flooring, 173–174
Flow of water in streams, 300
Flying shores, 172
Footways, 238, 242, 270
Formwork, 145–150
Foundations:
  trench, 135
  road, 221–222
Fuel, 29

Galvanised corrugated iron roofing, 221–222
Gantries, 172
Gas piping, 213

Gates, 180
General Conditions of Contract, 3–9
Girder casings, 138
Glazing, 206–207
Granite kerb, 236
Granolithic paving, 200–201
Grass:
  seeding, 267
  turfing, 267
Gravel surfacing, 240
Grounds timber, 176–177
Gully:
  gratings, 235
  pots, 234
Gutter boards, 176
Gutters, 213–215

Handrail, 180
Hardwood, 181
Haulage of material, 307
Haunches, bed and concrete surrounds, 261
Head of water, 300
Head Office charges, 46
Headings, 106–107
Health and Safety (CDM), 45
Herringbone strutting, 173
Hire rate of plant, 29–31
Hoardings, 44
Hoisting concrete, 136–137
Holes through brick walls, 164
Hollow floor slabs, 144
Hot asphalt, 224
Humidity of the works, 45
Hydrant boxes, 245

Inclement weather, 16
Increase in bulk, 84
Independent scaffolding, 34
Individual elements of cost, 10
In situ concrete, 118–139
Insurances, 14, 17, 37, 46
Interceptors, 257
Interest plant, 32–33
Interlocking tiles, 192
Internal Plastering, 202
Internal walls block, 167
Intumescent coatings, 186
Iron pipes drainage, 246–252
Ironmongery, 182–183
Italian tiles, 192

JCT form of contract, 3–9
Joinery, 171, 176–183
Joints, pipe:
  cast iron, 245, 251–252
  clayware, 253–256
  upvc, 258
Joists:
  steel, 185
  timber, 174

Labour, 16–25
Ladders, 186
Landings concrete, 136
Landscaping:
    soft, 266–269
    hard, 270–273
Larssen sheet piles, 60–65
Latches, 182
Latham Report, 9
Lath plaster, 202
Lead sheet, 198
Lighting and power, 43
Lime sand mortar, 153
Limestone tarred, 228
Linoleum, 205
Lintels, 140
Lintels formwork, 150–151
Loading materials, 77–78
Lorries, 33, 76
Lump sum basis, 14

Macadam surfacing, 225
Maintenance landscaping, 267
Maintenance of plant, 31
Manholes, 259–263
Materials:
    generally, 26–28
    weight, 278–298
Measurement Standard Method, 2
Mechanical excavation, 69–71
    angledozer, 70, 82
    backacter, 70, 89
    bulldozer, 70, 82
    dragline, 70, 74
    shovel, 70, 74
    tractor and scraper, 70
    caterpillar, 80
Mensuration formulae, 299
Mesh reinforcement, 233
Metal formwork, 150
Metals, weight of, 293–294
Mixing concrete, volume, 118
Mortars, 153, 159–160
Moulds for precast concrete, 151
Mouldings, 176–177

Nails, weight of, 297–298
Narrow trenches, 135
National insurance, 17
Net profit, 14
Newel post, 180
Nominated sub-contractor, 13
Nominated supplier, 13
Notices and fees, 45
Nuts, weight of, 296

Oak, 181
Offices, temporary, 44

One brick wall, 155
Open trenches, pipe laying in, 245–246
Output, bricklayer, 152
Overheads, 46–47

Painting, 205
Panes, 206–207
Partitions:
    timber stud, 173
    blockwork, 167
Path edging, 235–236
Pavings (see Roads)
Percentage addition, 14
Percentage rail track, 244
Piecework, 22–24
Pilaster, 164
Pitching stone, 222
Plain tiling, 190–192
Plant, 29–35
Planting trees, 269
Plastering, 202–204
Plate glass, 207
Plates wall, 173
Plinths brick, 164
Plugging to walls, 176
Plumbing, 209–215
Pointing brickwork, 165
Polythene piping, 210–211
Posts fencing, 274–276
Precast concrete kerb, 236–237
Precast concrete manhole, 262–263
Precast floor slabs, 144
Preliminaries, 36–45
Prime Cost, 13
Profit, 48
Protection of works, 45
Provisional Sum, 12–13
Pumping, 43
Purlins, 174
Putty, 206

Quantities, Bill of, 2
Quarry works, 216–220
Quoins, 164

Rafters, 174
Ragbolt, 144
Rail track, 244
Rainwater goods, 213–215
Raking out joints, 165
Raking shores, 172
Rates labour, 17
Ready mixed concrete, 139
Recovery value formwork, 145–146
Reinforcement, 142–143
Rendering in cement mortar, 204
Repose, angles of, 303
Reveals, 162, 164
Ridge board, 174

Ridge tiles, 192
Roads, 220–235
Rolled steel joints, 185
Rolling turf, 267
Roofing, 187–199
Rough cast glass, 206
Round bars weight, 293
Rubbed facings, 165
Rubbish from site, 45

Saddles, 257
Sand bed, 222
Sanitary fittings, 212
Scaffolding, 34–35
Schedule of Rates, Form of Tender, 3–4, 7–9
Scrapers, 80–82
Screed, 204
Screed boards, 233
Setts granite channel, 236
Setting out, 44
Sewers, 253
Shafts, 93–99
Sheet metal roofing:
    zinc, 197
    copper, 198
    lead, 199
Shores, 172
Shrinkage of materials, 306
Single coat asphalt, 238–239
Single coat tarmacadam, 242–243
Site:
    charges, 36–45
    access, 38
    management, 38
    accommodation, 44
Skimmers mechanical, 74–75
Skirtings, 177
Skylights, 179
Slating, 195–196
Slope angles of, 304
Sluice valves, 260
Small plant, 29, 39
Softwoods, 171
Soils bearing value, 306
Spanish tiles, 192–193
Specification, 4
Squints brickwork, 162
Staff costs, 38, 46–47
Stairs wood, 180
Stanchions formwork, 148
Standard Method of measurement, 2, 48
Steel bar reinforcement, 142
    weight, 293
Steel formwork, 150
Steel joists, 185
Stone:
    artificial, paving, 270
    broken, 217–219
    paving, 270
    percentage voids, 216
    pitching, 216–222

weight, 278–279
Straight line method of depreciation allowance, 31–32
Street gully pots and gratings, 234
Strip site, 267
Struck joint, 157
Structural steelwork, 184–186
Strutting:
    herringbone, 173
    solid, 173
Supply of water, 37, 41–43
Surface dressings:
    paths, 243
    roads, 229–231
Surface excavation, 66–84
Surface treatment concrete, 143
Surrounds, beds and haunches, concrete, 261
Systematic estimating, 14–15

Tamping concrete, 123, 130, 134
Tapered brick walls, 161
Tar surface dressing, 230
Tarmacadam paving:
    paths, 242–243
    roads, 225–227
Tarring fences, 276
Telephones, temporary, 37, 44
Temporary fences, hoardings, 37, 44
Temporary roofs, 37
Temporary screens, 37, 44
Tendering, 48–49
Tendering and estimating, difference between, 10
Testing pipework, clayware pipes, 256
Tiles, roof, 190–194
Tiles, wall, 203
Timber:
    grounds, 177
    joists, 173–174
    bridges, 172
    path edging, 236
    shores, 172
Tines sharpen, 185
Tools, small plant, 39
Total estimated cost, 10–15
Tractors and scrapers, 80–82
Transport to tip, 69
Traps:
    grease, 257
    interceptors, 257
Travel, time to tip, 79
Travelling time, labour, 39
Trees, felling,
Trenches, 86–93, 108–112, 116–117
Tunnels, headings:
    earthworks, 101–108
    pipelaying, 249
Turf, laying, 267

Universal column fire protection, 186
Unloading, 26–27
UPVC pipes, 258
    bed and backfill, 264

Useful tables:
  mensuration formulae, 299
  kerbs area sub tended, 299
  flow of water, 300
  head and pressure, 300–301
  brickwork crushing strength, 301
  brickwork culvert, 301
  timber cubic metres, 302
  mortar covering capacity, 302
  concrete, cement content, 302
  flow in sewers, 302–303
  angles of repose, 303
  angles of slopes, 304
  paint covering capacity, 304
  width of trenches for pipes, 305
  bearing value of soil, 306
  voids in materials, 306
  shrinkage of materials, 306
  materials haulage, cubic metres, 307
  masonry walling, 307
  concrete flags, sizes, 307
  materials covering capacity, 308
  materials surface dressing, 308

Valley gutter:
  zinc, 197
  lead, 198
  copper, 199
Valley, rafter, 174
Valleys, tiles, 192
Varnish, 304
Valves, sluice, 260
Verges:
  slates, 196
  tiles, 192
Vibrating equipment, 130–131, 134
Voids, in materials, 306
Volume:
  batch concrete, 118–119
  excavation, 66, 116–117

Walls:
  block, 167
  brick, 154–157, 161, 163
  concrete, 137
  external render, 204
  formwork, 147, 150
  internal plastering, 202
  plugging, 178
Wall tiling, 203
Waste, timber, 171
Watching and lighting, 40
Water bearing ground, excavation in, 108–115
Water/cement ratios, 119
Water for the works, 37, 41–43
Water installation:
  pipes, 210–211

  cistern, 212
  boiler, 212
  cylinder, 212
Weather conditions, effect on output, 16
Weight concrete, 119
Weight of materials, 278–298
  General, 278
  crushed materials, 279
  road foundation materials, 279
  road surfacing materials, 279
  bricks, 280
  brickwork, 280
  concrete block, 280
  wet mortar, 280
  floor tiles, 281
  kerby per metre, 281
  building stone, 281
  artificial stone paving, 281
  stone paving, 282
  concrete manhold rings, 282
  agricultural drainpipes, 282
  cast iron pipes, 283, 284
  concrete pipes, 285
  steel pipes, 285
  clayware pipes, 286
  clayware gully pots, 286
  steel tubes, 287
  copper tubes, 287
  corrugated sheets, 288–289
  sheet iron, 289
  sheet copper, 290
  sheet lead, 290
  sheet zinc, 291
  cast iron gutters and pipes, 291
  roof tiles, 292
  slates, 292
  reinforcement bars, 293
  metals, 293–294
  steel angles, 294
  channels, 295
  joists, 295
  bolts and nuts, 296
  nails, 297–298
Welded joints, 213
Welfare, 37, 40
Width of trench, 116–117
Windows, 179
Window linings, 179
Wood block floors, 205
Woodwork, 171–183
Woodwork painting, 205
Working cost of plant, 29–35
Work, estimating total cost, 10–15

Yield concrete, 118
  tables, 132–134

Zincworker, roofing, 197